VAPORISERS

Selection, Design & Operation

DESIGNING FOR HEAT TRANSFER

Series editor: David Browning

Titles under preparation:

Heat exchangers — Construction & thermal design (single-phase)

Condensation and condenser design

VAPORISERS

SELECTION, DESIGN & OPERATION

R. A. Smith, MA, MIMechE

Copublished in the United States with
John Wiley & Sons, Inc., New York

Longman Scientific & Technical
Longman Group UK Limited
Longman House, Burnt Mill, Harlow
Essex CM20 2JE, England
and Associated Companies throughout the world.

Copublished in the United States with
John Wiley & Sons, Inc., 605 Third Avenue, New York, NY 10158

First published 1986

British Library Cataloguing in Publication Data

Smith, R.A.
   Vaporisers: selection, design & operation.
   ——(Designing for heat transfer)
   1. Vaporisers (Machinery)
   I. Title II. Series
   621.402'5   TP159.V/

   ISBN 0-582-49490-7

Library of Congress Cataloging-in-Publication Data

Smith, R. A., 1918–
   Vaporisers: selection, design & operation.

   "Copublished in the United States with John Wiley
& Sons, Inc., New York."
   Bibliography: p.
   Includes index.
   1. Atomizers. I. Title.
TP159.A85S65   1986   660.2'8426   86-18561
ISBN 0-470-20709-4 (Wiley, USA only)

Set in 10/12pt Monophoto Times Roman
Printed and Bound in Great Britain
at the Bath Press, Avon

Contents

Figures

Tables

Preface

This book is written for the engineer in the process industries who may wish to design a vaporiser himself, or to check a design offered by a manufacturer or a contractor. Changes in process requirements may lead to alterations in the duty of an existing vaporiser, so its suitability will need to be reassessed. Attention must be paid to safety and to the maintenance of vaporisers and their ancillary equipment. For these activities, it is desirable to have a sound practical knowledge of the types of vaporiser that are available and of the calculations in heat transfer and fluid flow that are needed in design. This book gives advice on the choice of the best type of vaporiser for many of the duties encountered in process plant. The middle chapters deal with calculations. The special problems relating to mechanical design are discussed, but the reader is referred to other documents for mechanical design. No special consideration is given to the boilers used in generating electricity or to the vaporisers used in the air-conditioning or cryogenic industries.

Heat exchangers may be designed on the basis of past experience, using achieved values of the overall heat transfer coefficient to predict how a new exchanger will perform under similar circumstances. This procedure is limited. A fundamental knowledge of heat transfer and fluid flow is needed to produce a procedure for designing heat exchangers for a wide range of conditions. It is over 50 years since William H. McAdams wrote his book *Heat Transmission*, which he described as 'designed to serve both as a text for students and as a reference for practising engineers'. This was followed in 1950 by Kern's *Process Heat Transfer*, which became the main reference on heat transfer for many process engineers. These books gave rational methods for predicting heat transfer in single-phase flow and in condensation, but gave very little on boiling.

More recently, many results have become available on heat transfer in boiling, especially on flow-boiling inside tubes. These results have been described in textbooks, and correlations have been published for predicting heat transfer coefficients in pool boiling and in-tube boiling. Although many aspects of boiling heat transfer have been dealt with very thoroughly, usually to meet the needs of the designers of boilers, there are many gaps in our knowledge, mainly on shellside boiling and the problems that arise in boiling mixtures of organic liquids. An attempt is made in Chapter 7 of this book to present the reader with a procedure for estimating the heat transfer coefficient in varporisers of the many different types, boiling various liquids, in the process industries. This is preceded by a chapter on the estimation of the heat transfer coefficient for the heating fluid, and is followed by chapters on two-phase fluid flow and the application to the design of a vaporiser, with forced or natural circulation.

The main recent advance in design methods for heat exchangers has resulted from the advent of the computer, which is especially useful when it is required to: (1) estimate the amount of fluid flowing across the tubes on the shellside of a shell-and-tube heat exchanger; (2) estimate the amount of recirculation in a natural-circulation vaporiser; and (3) carry out the integration necessary when estimating the required surface area and the pressure drop in a two-phase heat exchanger. Moreover the facility for quick calculations permits the preparation of several alternative designs, so that the most economical and reliable may be chosen. The reader may build up his own computer programs, based on correlations given in this book or elsewhere, to meet his own special needs. Or he may use a proprietary program (the main suppliers are listed).

The book is one of a series on Designing for Heat Transfer. Other books are being prepared in this series to give a similar treatment to the design of other types of heat exchanger. At the moment the following titles are in preparation:

Heat exchangers – Construction and thermal design (single phase)

Condensation and condenser design

R. A. Smith,
Middlesbrough
August 1985

Acknowledgements

My thanks are given to Dr O. J. Dunmore and Dr P. J. Nicholson for writing the last three chapters of this book.

Especial thanks for reading the complete text and making helpful suggestions are given to Mr J. G. Collier of the CEGB, Mr D. R. Browning, Dr P. J. Nicholson of ICI and my wife Louise.

Advice on specific sections has been gratefully received from Dr G. F. Hewitt, Dr R. G. Owen, Mr J. M. Robertson, Mr E. A. D. Saunders, Dr K. Cornwell and Dr D. H. Lee.

R. A. S.

CHAPTER 1

Introduction

The technical manager or engineer in the process industry is often concerned with the equipment that is used to vaporise a liquid. The manufacturers offer him a choice from several, often very different, types when he wishes to purchase a new vaporiser, and many problems arise in the operation of existing equipment.

The vaporisers in use in the process industries cover a wide range of sizes and of duties. However, the same principles of design apply to all. This book deals with vaporisers used for the following duties.

1. A *fired boiler* to generate steam, which may be needed for process heating, power generation or as a process fluid.
2. A *waste heat boiler* to utilise surplus heat in a process fluid (after a high-temperature reaction) to generate steam to supplement that generated in fired boilers.
3. A *reboiler* to provide the source of heat needed to vaporise some of the liquid collected at the bottom of a distillation column, so that it may be returned to the column as vapour.
4. A *vapour generator* to *revaporise* a gas that has been liquefied to facilitate transport and storage, or to vaporise a *heat transfer fluid*.
5. A *vapour generator* to obtain a pure vapour from an impure liquid, as in a desalination plant.
6. An *evaporator* to *concentrate* a liquid.
7. An *evaporator* to produce *crystals* from a saturated liquid.
8. A *chiller* in a refrigeration system to cool a process fluid to a temperature below ambient.

Items (1) and (2) are called 'steam generators'.

Heat is supplied by the combustion of fuel for duty (1) and occasionally for duties (3) to (6). Heat for duties (3) to (6) is usually supplied by condensing steam, but waste heat in a process fluid may be used.

A glossary at the end of the book defines the special terms that are used throughout. There is also a comprehensive index.

The reader may find the book useful when he has to:

1. Choose between several competitive tenders and check that the one chosen will be satisfactory in all respects.
2. Design a vaporiser.
3. Commission the installed plant.
4. Rectify any faults that arise in operation.
5. Make minor modifications necessitated by changes in process requirements.
6. Maintain the vaporiser and its ancillary equipment.
7. Ensure that the equipment is always safe.

The physical process of vaporisation is described in Chapter 2, with particular reference to a liquid flowing through a heat exchanger past a heated surface. The different modes of vaporisation are described and a qualitative account is given of the factors that determine which mode prevails and when the transition to another mode occurs. Finally there is a comment on the importance of maintaining an adequate rate of fluid flow through an industrial vaporiser.

Chapter 3 lists the types of vaporiser that are available for the duties listed, grouped according to their arrangement of heating surface. Direct contact heating with submerged combustion (section 3.7) is the cheapest. Consideration should be given to the possible use of plate heat exchangers (section 3.6) because they are cheaper than the more generally used tubular heat exchangers (sections 3.1 to 3.5). In some cases of the duties numbered (5), (6) and (7), it is necessary to use 'flash evaporation', which is described in section 2.9.

Chapter 4 describes the ancillary equipment that is needed for some types of vaporiser. The choice of the best type of vaporiser for a specific duty is discussed in Chapter 5. In each case, it is necessary to consider what will be the best arrangement from the point of view of safety, reliability and ease of maintenance. The failure of just one item of equipment in a complex process may lead to the plant being shut down for many days, with a loss of profit much greater than the cost of the item; unfortunately, it is often a vaporiser that fails, either by not performing its duty adequately or by developing a leak in the heated surface. Despite the importance of safe and reliable operation, the designer must also try to achieve an economic design, minimising the sum of the capital charges and the operating costs.

Chapters 6 to 10 inclusive describe the process engineering calculations in heat transfer and fluid flow that are necessary to ensure that the available surface area will be adequate for the specified heat transfer rate. It is also essential to check the adequacy of a design in other respects, taking special care to ensure that there is always a sufficient supply of liquid to the heated surface in all places, to avoid overheating of the surface.

Chapter 11 introduces the reader to special problems involved in the mechanical design of vaporisers. Further information is given by Saunders (1987). It is important to be able to calculate the maximum temperature that the heating surface can possibly reach, when this is liable to cause a serious loss in strength. For stress calculations it may be necessary to calculate the maximum differential thermal expansion between the tubes and the shell of the heater (the estimation of surface temperatures is described in section 6.6).

Section 11.3 deals with the troublesome topic of tube vibrations induced by the flow of a fluid across a bank of tubes.

Chapter 12 deals with problems relating to the materials of construction, including their selection and the methods of making adequate allowance for corrosion under the worst possible conditions of operation. The method of attaching tubes to tubesheets is very important, the failure of such joints having often been the cause of plant shutdown. Weld procedures and the requirements for stress relief must be specified. Weld materials must be compatible and the danger of electrolytic corrosion at the points where dissimilar materials are in contact must be guarded against.

Chapters 13 and 14 briefly introduce the many important problems that may arise in the operation and maintenance of vaporising equipment, including the treatment of water for steam generators.

The opinions and the recommended methods of carrying out the process engineering calculations given in this book are based on the author's personal experiences, backed by help from manufacturers and from textbooks, the most important being McAdams (1954), Coulson and Richardson (1977), Kern (1950), Bergles *et al.* (1981), Wallis (1969), Perry and Chilton (1973), TEMA (1978), Schlünder (1983) and Chisholm (1983).

Saunders (1987) deals with the mechanical design and construction of heat exchangers, and with the thermal and hydraulic design of single-phase equipment. Further books are being prepared in this series.

It is outside the scope of this book to describe the necessary calculations for the mechanical design of a vaporiser. This is dealt with in the appropriate design code. The special problems in the design of tubular heat exchangers are dealt with by TEMA (1978); it is assumed that the reader has access to a copy of this. Saunders (1987) gives details of shell-and-tube heat exchangers.

The chemical engineering drawing symbols used throughout this book are those derived from British Standards, as given by Austin (1979).

All the equations in the text are for use with the SI system of units. Conversion factors from other units are given in Appendix C. Derived SI units are used in the descriptive text where it is convenient to avoid the use of very large or very small numbers. The derived units are normally related to the pure units by powers of 1000. Thus the diameter of tubes is given in mm, rather than m. The exception is the use for pressure of bars, which are 10^5 pascals (N/m^2), as these have become widely used.

CHAPTER 2

Vaporisation

The conversion of liquid to vapour is referred to by the general term 'vaporisation', which may occur in one of two ways: when bubbles form at a heated submerged surface it is referred to as 'boiling'; when vaporisation takes place at an interface between a liquid and its vapour it is referred to as 'evaporation'. In some situations both forms of vaporisation take place simultaneously. For boiling to occur, the submerged surface must be heated to a temperature above the boiling temperature of the liquid, which is defined as the temperature at which the vapour pressure of the liquid equals the external pressure. For evaporation to occur, the liquid must be slightly superheated, i.e. above its boiling temperature, or it must be in contact with a mixture of its vapour and an incondensable gas.

With a mixture of two or more liquids, boiling usually takes place over a range of temperatures. The temperature at which bubbles first appear in a heated liquid is called the 'bubble point'. The upper temperature of the boiling range is called the 'dewpoint', which is defined as the temperature at which condensation begins when a vapour or a mixture of vapours and incondensable gases is cooled. Exceptions to the above are termed 'azeotropic mixtures', defined as mixtures for which boiling takes place at only one temperature. The subject is dealt with in standard textbooks in physical chemistry.

With an aqueous solution, the boiling temperature is greater than that of pure water at the same pressure (see section 10.4.1).

The first section of this chapter deals with studies of 'pool boiling'; the subsequent sections discuss vaporisation and heat transfer in industrial vaporisers.

A summary of the definitions in this book is contained in the glossary at the end of the book.

2.1 Pool boiling

When a heated surface is submerged in a liquid that is stationary in a vessel, this leads to the situation known as 'pool boiling'. An example of this is the domestic electric kettle, where there is no movement past the heating element other than that due to natural convection.

Many experiments have been carried out on the boiling of a liquid in a container, the heat being supplied by an electrically heated wire or disc submerged in the liquid. As the heat flux (heat transfer rate per unit area) is increased, bubbles begin to form at very small crevices or discontinuities in the surface. The bubbles grow at the nucleation sites and break away, carrying some superheated liquid with them. This is called 'nucleate boiling'. The heated surface must be superheated by a few degrees before any bubbles form on it, the amount of superheat being greater with smooth than with rough surfaces. Fresh liquid must be able to flow towards the heated surface to replace the vapour that has been generated and the liquid carried away with the bubbles. As the heat flux is increased, a point is reached when the supply of fresh liquid fails and a film of vapour forms on the heated surface. Immediately there is a large increase in the temperature of the heated surface, because the benefit of the stirring action of the departing bubbles has been lost and because the thermal conductivity of a vapour is always less than that of its liquid; this condition is called 'film boiling', or 'vapour blanketing'.

Figure 2.1 shows a plot of heat flux (measured in W/m^2 and denoted by $\dot{q}$) plotted against the temperature of the heated surface, for water boiling at atmospheric pressure. It was taken from Fig. 14.1 of McAdams (1954) and relates to experiments by Nukiyama (1934) using a wire of platinum 0.14 mm in diameter. At low heat fluxes, between points A and B, heat is transferred by natural convection only. Nucleate boiling starts when the heated surface reaches the temperature at B (where it is superheated by 5 °C). Thereafter there is a large increase in heat flux with increasing temperature, until the region marked C is reached, where the boiling process changes from nucleate to film boiling; this is due to the coalescence of the bubbles to form a stable film of vapour, which eventually covers the whole surface. The vapour acts as a thermal insulant and causes a large jump in the temperature of the heated surface, in this case from 124 °C to 1140 °C at D. Any further increase in heat input leads to a further increase in surface temperature, as from D to E, until the heated wire melts. The heat flux at the point C is called the 'critical heat flux', the 'point of departure from nucleate boiling (DNB)' or the 'maximum heat flux'. The first of these terms is used in this book. Figure 2.1 shows that the critical heat flux for water boiling at atmospheric pressure, heated by a fine wire, is 1170 kW/m^2. The subject is discussed further in section 8.1.

As the heat flux is reduced, film boiling continues below the critical heat flux until the point F, known as the 'Leidenfrost temperature', is reached.

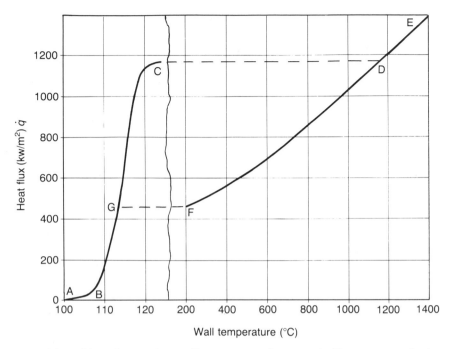

2.1 Plot of heat flux against wall temperature for water boiling at atmospheric pressure

At lower heat fluxes, the surface becomes wetted again, the temperature falls to the first curve and nucleate boiling occurs again, between points G and B.

If the heating surface were maintained at a constant temperature between the temperatures at the points C and F, boiling would be unstable, oscillating between nucleate and film boiling. Above the Leidenfrost temperature at F a drop of the liquid will dance on the surface, always separated from it by a cushion of vapour. For the case of Fig. 2.1, the Leidenfrost temperature was 194 °C.

In industrial vaporisers, the boiling process is complicated by the fact that the liquid to be vaporised flows at an appreciable velocity past the heated surface, as explained in the following sections of this chapter. This is called 'forced convective boiling'. The driving force may be provided by a pump, or by a combination of the pump and natural flow.

2.2 Single-phase liquid heat transfer

The rest of this chapter is concerned with heat transfer in industrial vaporisers. The liquid to be vaporised may be inside or outside tubes or flowing between plates. It may be below, at or above its boiling temperature.

When the liquid being heated is below its boiling temperature and the heated surface is below the temperature at which bubbles form, the transfer of heat is by single-phase forced convection. This type of heating may occur near the entrance to the heated surface, especially if boiling is being delayed by heating the liquid under pressure before passing it into a flash vessel, as described in section 2.9. Section 7.1 deals with the calculation of heat transfer in single-phase liquid convection. This region of sensible heat transfer is terminated by one of two possible events. If the temperature of the wall is superheated to the amount needed for bubbles to form on it, subcooled nucleate boiling, described in the next section, will begin. If the bulk temperature of the liquid reaches its boiling temperature, two-phase convective boiling is likely to begin, as described in section 2.4.

2.3   Subcooled nucleate boiling

Bubbles do not form on a heated surface until the degree of superheat is sufficient for the pressure inside a nucleation site to overcome the force due to surface tension tending to prevent the growth of the bubble. Thereafter bubbles will form on the heated surface, become detached and pass into the bulk of the liquid (where they may condense if the bulk of the liquid is below its boiling temperature). As the amount of superheat is increased, the number of nucleation sites where bubbles grow increases, and consequently there is a rapid increase in the rate of heat transfer. Subcooled nucleate boiling can be observed in the pool boiling inside a domestic kettle before the water has reached its boiling temperature; the heating surface produces bubbles which collapse when they pass into the bulk of the water, giving rise to the characteristic 'singing' sound. Nucleate boiling facilitates the transfer of heat because the bubbles of vapour and the liquid around them convey heat from the surface much more rapidly than when there is no boiling.

The physics of the formation of bubbles at nucleation cavities is complex, but has been dealt with in several textbooks, such as Bergles *et al.* (1981), pp. 194–200 and Collier (1981) section 4.3. Subcooled nucleate boiling ends when the bulk of the liquid reaches its boiling temperature, and the bubbles begin to form a two-phase boiling mixture.

2.4   Two-phase convective and nucleate boiling

If nucleate boiling has already begun in flow over any form of heated surface, two-phase boiling begins as soon as the bulk of the liquid reaches its boiling temperature. If, however, the bulk of the liquid reaches its boiling temperature before the onset of nucleate boiling, the liquid becomes superheated until bubbles are formed on nuclei in the liquid or until the onset of nucleate boiling at the wall. In industrial applications there is seldom a significant amount of superheating, because liquids normally contain sufficient suspended particles or dissolved gases to provide nuclei for boiling, but with a very clean liquid, significant superheating may occur.

The formation of vapour increases the velocity of flow past the heated surface, thus facilitiating the transfer of heat. The complicated problems of calculating heat transfer in two-phase convection, in nucleate boiling and in a combination of the two, are dealt with in sections 7.2, 7.3 and 7.4.

A high rate of shear at the wall tends to remove bubbles before they are fully grown and so reduce the benefits of nucleate boiling, as discussed further in section 7.2.1.

When the temperature difference is only a few degrees, as in most cryogenic applications, nucleate boiling may be completely absent at all qualities (ratio of mass flow rate of vapour to total mass flow rate). Vaporisation is then by evaporation at the interface between the slightly superheated liquid and the bubbles of vapour.

When the volumetric flow rate of vapour considerably exceeds that of the liquid, flow inside tubes is usually in an annular pattern: most of the liquid forms an annulus on the wall and a core of vapour in the centre of the tube flows faster than the liquid, some of the liquid being sheared off the wall to form mist, which may later be redeposited on the annulus of liquid flowing on the wall. Thus there is a continuous interchange of liquid between liquid annulus and gaseous core. In annular flow, there may be nucleate boiling in the annulus, but it is more likely that vaporisation will be by evaporation at the interface between the annulus of liquid and the core of vapour. A method of estimating the range of qualities for in-tube annular flow is given in sections 9.2.1 and 9.2.2. With shellside flow at similar qualities, spray flow occurs, the liquid being transported as spray, which forms a film around each tube. The subject of flow patterns in shellside flow is discussed in section 9.2.3.

It is always hoped that the heat transfer surface will be wetted until the heat required has all been transferred at the outlet from the heater. However, there are two events that can lead to dry-wall vaporisation, with a consequent reduction in the heat flux. Both are more likely to occur at a high heat flux and at a high quality. If the critical heat flux is exceeded, a vapour blanket forms between the heating surface and the bulk of the fluid, as described in section 2.5. If the volumetric flow rate of liquid is very small compared with that of the vapour, the wall becomes dry, as described in section 2.6.

2.5 Vapour blanketing of a heated wall

This can occur at a very high heat flux even before the liquid has reached its boiling temperature, but the critical heat flux decreases with increasing quality. Furthermore, an increase in the velocity of flow leads to an increase in the critical heat flux. The consequences of approaching the critical heat flux differ appreciably according to the method of heating. With upward flow inside vertical tubes, the critical heat flux is high, but if it is reached, a stable film of vapour is formed, leading to a marked increase in the temperature of the wall. This produces the serious problem of 'burnout' in fired water-tube boilers, as described in section 8.2. With boiling inside horizontal tubes the critical heat flux is lower than with vertical tubes, but the consequences of reaching it may be less serious. The temperature of the top of the tube begins to fluctuate, due to incomplete wetting of the surface, while nucleate boiling is still occurring along the sides and bottom of the tube. This is called 'dryout'. Although less dramatic than burnout, dryout can lead to internal corrosion of the tube along horizontal lines at the interface between liquid and vapour. This is dealt with further in sections 8.2.2 and 12.2.2.

When boiling liquid is flowing across a horizontal bundle of heated tubes, pockets of vapour may form along the top of each tube, leading to a loss in performance and possibly to corrosion. The estimation of the critical heat flux for a bundle of tubes is dealt with in section 8.3.

2.6   Evaporation of drops in dry-wall mist flow

When the volumetric flow rate of vapour is very much greater than that of the liquid, the vapour will become the continuous phase and the liquid will be present as droplets (see section 2.4). As the quality increases, a stage is reached when the rate of deposition of mist on the heated surface is less than the potential rate of evaporation; so the wall becomes dry, as shown in Fig. 2.2 for in-tube vaporisation. Consequently the benefits of nucleate boiling are lost and heat is transferred by single-phase convection in the vapour, which is less than in liquid at the same mass flow rate. The vapour becomes superheated and the remaining mist evaporates as a result of the transfer of heat from vapour to droplets. The estimation of heat transfer in dry-wall convection is dealt with in section 7.5.

The main difficulty in designing a vaporiser in which there is a region of dry-wall mist flow near the outlet is to determine the point at which the wall becomes dry. Since the change from two-phase convective and nucleate boiling to mist flow is accompanied by a serious reduction in the heat flux, it is important not only to be able to estimate the heat fluxes in the two regions but also to be able to determine where the change takes place. Advice on this is given in section 7.5.

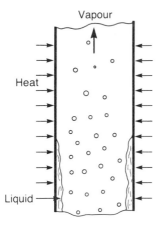

2.2   Transition from annular to mist flow (in-tube vaporisation)

2.7   Fluid flow in relation to boiling

Increasing the velocity of flow of a fluid past any type of heated surface increases the rate of heat transfer due to convection, both with single-phase and with two-phase flow. Also any tendency to fouling or scaling is reduced. It is therefore important in industrial vaporisers to maintain the required flow rate of the fluid that is being heated. This is usually achieved by recirculating a large amount of liquid from the outlet to the inlet of the vaporiser, after separating the vapour generated from the liquid to be recirculated. Thus the quality in the vaporiser is zero at inlet and increases to something in the region of 5% to 20% at outlet. The recirculation may be achieved by *natural circulation*, *assisted circulation* or *forced circulation*. With natural circulation flow is produced by the difference between the density of the recirculated liquid in the return line and the density of the two-phase mixture that is being heated in the vaporiser. With forced circulation, a pump is used to recirculate the liquid. Assisted circulation is produced by a combination of the gravity forces of natural circulation and the externally imposed forces of forced circulation. Vaporisers in which there is no recirculation are said to operate as *once-through* vaporisers.

The fluid flow calculations described in Chapter 9 are important in the design of vaporisers. It is necessary to be able to calculate the frictional pressure drop in two-phase and single-phase flow and to be able to predict the fraction of the volume occupied by the vapour in a two-phase mixture; these are needed for the estimation of the recirculation rate in natural circulation and for sizing the pump in forced circulation. It is also important that the flow should be steady, i.e. free from fluctuations in pressure, flow rate and quality.

Whenever bubbles rise to the surface of a liquid and escape into a vapour space, they carry with them droplets of the liquid. The amount of entrained liquid increases rapidly with increasing rate of boiling. In some applications a small amount of entrained liquid can be tolerated, but it is important to check that the presence of the liquid will not lead to excessive pressure drop in the vapour line or to fouling of a heat exchanger through which the vapour subsequently passes.

Chapter 9 also includes the hydraulic problems in the design of distribution manifolds, liquid/vapour separators and spray eliminators.

2.8   Evaporation from the surface of a falling film of liquid

In a falling-film evaporator, described in section 3.5, the liquid is introduced into the top of vertical tubes and flows as a falling film down the inside of the tubes, which are contained in a shell, through which the heating fluid passes. The liquid flows down the inside surfaces of the tubes as a thin film, perhaps only 0.5 mm thick. Heat is transferred by conduction across the film to the liquid/vapour interface, where evaporation takes place. The liquid is superheated. Because the film is so thin, the temperature drop across it is usually sufficiently small to avoid nucleate boiling at the wall. Figure 2.3 shows part of a tube with an evaporating film and the vapour generated flowing downwards.

The subject of filmwise condensation on the inner surface of a vertical tube has been studied extensively in the past. It is shown in section 7.6 that the equations derived for filmwise condensation inside a vertical tube may also be used for evaporation from a falling film.

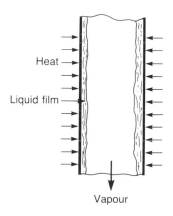

2.3   Evaporation from a falling film

2.9 Flash evaporation

When the pressure of a liquid at its boiling temperature is reduced, the liquid is at first in a superheated state. Bubbles form on nuclei in the liquid and vaporisation occurs throughout the bulk of the liquid until the latent heat of vaporisation has cooled the bulk of the liquid to the lower boiling temperature corresponding to the new (lower) pressure. This is known as 'flash evaporation'. In the absence of sufficient nuclei, this may occur in an explosive manner.

When it is desired to heat a liquid without any boiling taking place, it is necessary for the pressure in the heater to be high enough for the bulk of the liquid to be always below its boiling temperature and for the temperature of the heated surface never to exceed that required for nucleation. The liquid then passes through a let-down valve into a flash vessel at a sufficiently low pressure to flash off the required amount of vapour; this can be calculated from the fact that the enthalpy of the mixture after the valve is equal to that before the valve.

Some flash evaporation also occurs in a heated pipe as a result of the pressure drop. This may be important when there is boiling in upward flow in a vertical pipe. It is necessary to calculate the pressure drop due to friction plus that due to gravity, whence it is possible to determine the amount of vapour generated as a result of pressure drop (additional to that resulting from the input of heat).

CHAPTER 3

Types of vaporisers

In this chapter the many types of vaporiser used in the process industries are classified according to the form of the heating surface used. The large water-tube boilers used for nuclear and conventional power stations are outside the scope of this book. Sections 3.1 to 3.3 deal with equipment where there is forced convective boiling inside tubes, the later sections dealing with shellside vaporisation, falling film evaporators, the use of plate heat exchangers and direct contact heating. The suitability of each type is discussed in relation to the eight duties listed in Chapter 1. The problems that may arise with each type are described, and some account is given of how they might be dealt with.

The appropriate section of Chapter 5 should be consulted when selecting a type for a specific duty.

3.1 Upward flow inside vertical tubes (or ducts)

A wide range of equipment employs upward flow of the fluid to be vaporised inside vertical tubes, covering all the duties listed in Chapter 1. A high velocity in the tubes is needed, to achieve a high heat transfer coefficient and, in many applications, to reduce the fouling or scaling of the heated surface. Usually only a small fraction (5 to 20%) of the liquid is vaporised, the remainder being recirculated to the inlet of the tubes, with the object of enhancing heat transfer, as explained in section 2.7. With this arrangement, circulation may be natural, assisted or forced. Alternatively, the once-through system is sometimes used in vaporisers or evaporators; with very clean fluids complete vaporisation can be achieved, but with less clean fluids the outlet quality must be limited to 90–95% (quality is the ratio of the mass flow of vapour to the total mass flow rate).

3.1 Variation with depth of the boiling temperature of water at several pressures –
from equation [9.50]

Pressure (bar)	0.1	1	10	100
(Pa)	10^4	10^5	10^6	10^7
Increase in boiling temperature per unit increase in depth of liquid (°C/m)	18.9	2.6	0.38	0.05

Since the vapour pressure of a liquid increases with temperature, the
boiling temperature of a single-component liquid (or the bubble point of a
mixture) is greater at the bottom than at the top of a vertical tube. This
effect of hydrostatic pressure may significantly reduce the mean temperature
difference along a long vertical tube, the effect being particularly important
when operating under vacuum with a small temperature difference. To
illustrate this point, Table 3.1 gives the variation of boiling temperature
with depth for pure water at several pressures.

In a system using recirculation, the temperature of the liquid at the inlet
to the tubes is the same as the temperature at the surface; thus when the
liquid enters the tubes it is below its boiling temperature by an amount
equal to the depth of the bottom of the tubes below the surface multiplied
by the increase in boiling temperature per unit increase in depth. Thus
boiling does not occur in the lower part of the tubes, the temperature rising
due to sensible heat transfer until it reaches the boiling temperature
corresponding to the local pressure; thereafter the temperature follows the
boiling temperature (as further explained in section 10.2).

3.1.1 Water-tube boiler

Water-tube steam generators are used as fired boilers and as waste heat
boilers. Each consists of a water drum at a low level and a steam drum at a
high level. The drums are connected by an array of tubes that are vertical,
or inclined. Preheated water is fed into the water drum and rises through
heated boiler tubes of 30 to 60 mm internal diameter into the steam drum,
as shown in the figures that appear later. The steam generated is removed
from the top of the steam drum. Water that has not vaporised is returned
to the water drum either through external downcomers connected to the
ends of the drum or through tubes situated in a less intensely hot zone. An
alternative arrangement used in many modern boilers has a manifold in
place of the water drum, the feed water being supplied to the steam drum.

The steam and water are normally introduced into the steam drum
through several internal cyclone separators of the type described in section

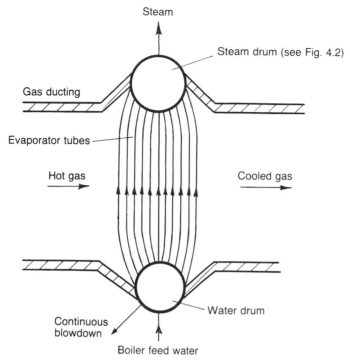

3.1   Vertical water-tube waste heat boiler (with external downcomers)

4.1.2 arranged in parallel along the length of the drum. Continuous removal of water from circulation prevents the build-up of too high a concentration of dissolved solids in the circulating water; this is known as 'blow-down'. The rate of feeding water into the boiler is controlled so that the steam drum is about half filled with water. Typical arrangements are shown in Figs 3.1, 3.2 and 3.3, and are discussed below.

The arrangements shown in Figs 3.1 to 3.3 are attractive because they work satisfactorily on natural circulation, thereby avoiding the expense of a high-pressure recirculating pump, which must be duplicated to cover the danger of a possible failure of a pump. It is also possible to operate at very high heat fluxes in the hottest tubes, as shown in section 8.2. They can be designed to cope with high pressure, being normally used for pressures in the range from 20 to 150 bar (2 to 15 MPa). It is possible to allow for the thermal expansion of the tubes without undue difficulty. Water-tube steam generators are used in the process industries both as fired boilers and as waste heat boilers. They may additionally contain superheater tubes, which superheat the steam taken from the steam drum, and economiser tubes, which preheat the water fed to the water drum. These boilers are designed and fabricated by specialist manufacturers.

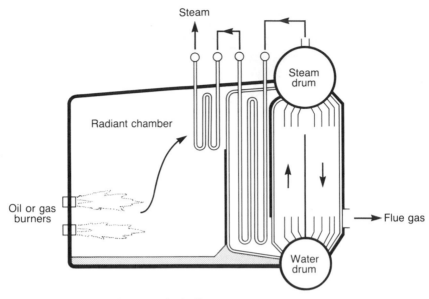

3.2 Oil- or gas-fired water-tube boiler

Fired water-tube boilers are heated by the combustion of solid, liquid, or gaseous fuel. Package water-tube boilers are now quite often used in the process industries for the supply of high-pressure steam, for driving steam turbines and for heating duties. Some of the steam may be taken from after the high or intermediate stages of the turbine for heating duties, the remainder passing to condensers. The waste heat boilers used in the process industry are designed according to the same principles, and are discussed first.

Figure 3.1 illustrates a water-tube waste heat boiler (without superheaters). This arrangement is compared with other types of waste heat boiler in section 5.2. If superheating is required, the steam taken from the top of the steam drum is usually.passed through a bank of vertical tubes placed in the gas duct on the upstream side of the evaporator. It is important to estimate the maximum temperature of the tubes in the first row of the superheater, as they may be overheated, because the gas is at its highest temperature here, there is enhanced convective heat transfer, and there is radiation from the gas and the wall of the duct to the leading edge of tubes in this row. Tubes may be connected in series so as to give co-current flow of steam and gas, thereby reducing the maximum temperature of the tubes in the superheater. The estimation of the temperature of tubes is dealt with in section 6.6. The water from the steam drum is returned to the water drum through external downcomers at the ends of the drums.

Figures 3.2 and 3.3 illustrate the essential features of typical fired water-tube boilers, but details vary according to the duty. Figure 3.2 shows a

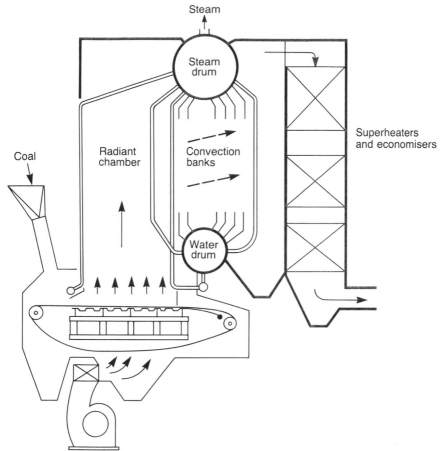

3.3 Stoker-fired water-tube boiler

package oil- or gas-fired boiler; Fig. 3.3 shows a chain-grate coal-fired boiler. The fuel is burned in the furnace, which acts as a radiant chamber. Its walls are lined with evaporator tubes, arranged close together so that the walls cannot be heated by direct radiation. The gas then passes through the convection section of the boiler, which contains more evaporator tubes and superheater tubes. The remarks in the previous paragraph about the danger of overheating superheater tubes apply also to fired boilers. Interested readers should consult the specialist manufacturers of boilers.

Information on the many practical problems in the design and operation of water-tube boilers is given by Goodall (1980), Part 5.

In addition to an array of cyclone separators introducing the steam and water into the steam drum, it may be necessary to install some form of mist separator opposite the vapour outlet pipe, e.g. as described in section 4.1.3. In small boilers without superheaters the free surface area in the drum may be sufficient to give an adequate separation of steam and water, in which case no separators are needed.

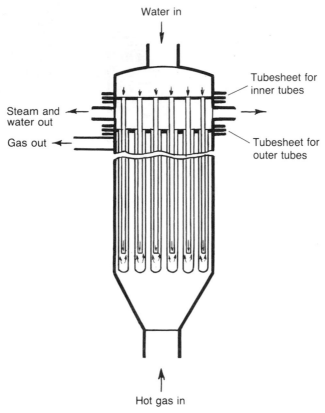

Water in

Tubesheet for
inner tubes

Steam and
water out

Gas out

Tubesheet for
outer tubes

Hot gas in

3.4 Vertical bayonet-tube waste heat boiler

Alternative types of water-tube waste heat boiler with vertical tubes are shown in Figs 3.4 and 3.5. In both cases the hot gas enters the bottom of a vertical shell and flows to the top in a single pass; there are two passes for the boiling water, which enters and leaves at the top. In the bayonet type, shown in Fig. 3.4, there are pairs of concentric tubes. A cap is attached to the lower end of each outer tube. The water flows downwards through the inner tubes to the bottom of the outer tubes; it boils as it flows up the annular space between the inner and the outer (heated) tube. There is no heat transfer in the smaller tubes, except that necessary to heat the water to its boiling temperature, which is slightly higher than that in the steam drum, due to the hydrostatic head. The upward-flow boiling in the annular space is free from the danger of static instability (see section 9.2.4). In the arrangement shown in Fig. 3.5, the bayonet tubes have been replaced by U-tubes, the arrangement being otherwise similar. A forced circulation is preferred with the U-tubes, as circulation would be slow in starting with natural circulation and there would be a risk that circulation might start in the wrong direction if the steam drum were filled to above the level of the

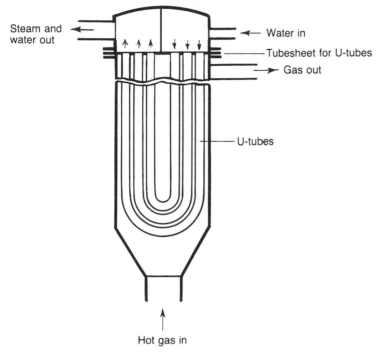

Steam and water out ←

← Water in

Tubesheet for U-tubes

→ Gas out

U-tubes

Hot gas in

3.5   Vertical U-tube waste heat boiler

riser pipe that circulates the steam and water to the cyclone separator (see section 4.1.2).

However, with the bayonet tubes, natural circulation may be used, because heat is applied only to the surface where flow is upwards; hence circulation begins as soon as heat is applied during start-up.

The advantages of the arrangements shown in Figs 3.4 and 3.5 are that there are no restraints on differential thermal expansion of the tubes and that the tubes are not attached to a tubesheet where the hot gas enters the shell. A high velocity through the tubes is essential to prevent the accumulation of solids at the lowest point in the water passage, i.e. where it passes from the inner to the outer tube of the bayonet arrangement, or at the bottom of U-tubes. There is a danger of failure due to dryout at and near to the bottom of the outer tubes with the bayonet arrangement. The former can be avoided by attaching protective covers of temperature resistant material to the bottom of each tube. It is difficult, but important, to locate the inner tube centrally inside the outer tube, especially at the bottom.

The most important factor controlling the heat transfer rate in a water-tube boiler is the heat transfer coefficient from the heating gas to the outer surface of the tubes; this is defined as the heat flux divided by the difference between the temperature of the gas and of the outer surface of the tubes.

The heat transfer due to forced convection can be estimated as described in section 6.2. The heat transfer due to radiation, which will contribute 5 to 15 % of the heat, can be estimated as described in section 6.5. Other points to be checked are: (1) that the critical heat flux is never approached (section 8.2 gives a method of calculating this and it is advisable to be always below about 70 % of the estimated critical heat flux); (2) that the water circulation rate is adequate. Section 9.6 gives a method of estimating the circulation rate from the heat load and the geometry of the boiler. The results of such calculations are expressed as the recirculation ratio, i.e. the water rate entering the tubes divided by the steam generation rate. It is not possible to estimate the minimum tolerable circulation ratio; this depends very much on the water treatment and the heat flux. It is usual to aim for a recirculation ratio of 10, although many boilers have run satisfactorily with a recirculation ratio of only 5. Due to maldistribution of water between the many tubes that provide parallel paths of flow, the tubes deprived of water operate at a recirculation ratio below the mean, and these will be the first to be damaged by any fall in the standard of water treatment. Usually a ratio of about 10 can be obtained without much increase in the capital expenditure over that required for a ratio of 5, and this reduces risk of damaging the boiler by maloperation.

3.1.2   Short-tube vertical evaporator and vapour generator

There are many industrial applications where a liquid boils in upward flow through vertical tubes at absolute operating pressures ranging from 0.1 to 10 bar. Heat is usually supplied by the condensation of steam on the outside of the tubes. Varieties considered in this and subsequent subsections are used as vapour generators, concentrating evaporators, crystallising evaporators and reboilers.

   When used as an evaporator, recirculation is usually through a central downcomer, as shown in Fig. 3.6, where an axial flow pump at the bottom of the downcomer provides assisted circulation. Often the pump is omitted, relying on natural circulation. Alternatively the whole of the shell might be filled with tubes, and downcomers provided around the outside of the shell for recirculation. The cross-sectional area for flow through the downcomer should be similar in magnitude to the total area through all of the tubes. With external downcomers, it is important to ensure that there is sufficient head of liquid above their inlet (see section 9.6 for information on flow in natural circulation). The vapour separates from the liquid in the disengagement space between the level of liquid above the top tubesheet and the top of the shell. The amount of liquid carried over by the vapour may be excessive, in which case an external separator must be installed.

   The tubes of short-tube vertical evaporators are about 2 m long and their internal diameter ranges from 25 to 60 mm. The larger sizes are used in

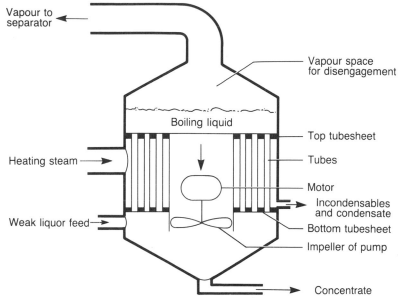

3.6 Short-tube vertical concentrating evaporator with assisted circulation

crystallising evaporators or if there is a significant tendency to fouling or scaling of the heated surface. Assisted circulation should be used when it is necessary to maintain a high velocity through the tubes, in order to achieve a high heat transfer coefficient, or to reduce the amount of scaling, or because the evaporator is required to grow large crystals.

When relying on natural circulation, an important consideration is that the circulation rate falls considerably when the heat transfer rate is reduced; therefore it is important to check that no troubles, such as scaling or fouling, will be introduced when operating at a reduced rate. It may be possible to improve the thermal performance by reducing the liquid level (as indicated by an external level gauge) from well above to well below the top tubeplate; this gives a greater enhancement of heat transfer coefficient due to boiling and increases the temperature difference (by reducing the saturation temperature) but the dangers of scaling are greatly increased.

With a crystallising evaporator, a device must be included for removing the crystals from the bottom of the evaporator, where they are allowed to settle and form a thick slurry. This slurry may be removed mechanically, e.g. by a slow-moving screw, or hydraulically, by allowing it to settle through a slowly upward flowing stream of liquor in a device known as an 'elutriator' and described in section 4.6. In the latter case, some of the liquor is recirculated up the elutriator to maintain the required low velocity there so as to ensure that only large crystals can fall through this stream and leave the evaporator.

The short-tube vertical arrangement may be used as a vapour generator, for example when a gas such as ammonia has been liquefied, to facilitate storage or transportation, and it is necessary to revaporise it. If the substance is noxious, a once-through system may be used, in order to minimise the amount of liquid contained in the shell. Under these circumstances, the vaporiser differs from that shown in Fig. 3.6 in that there is no internal downcomer, the whole of the shell being fitted with tubes. The feed rate must then be controlled so that evaporation is complete when the top of the tubes is reached. This leads to a reduction in the heat flux at the top of the tubes (see section 7.5). A spray separator may be needed (see section 4.1).

3.1.3   Basket-type evaporator

The basket-type evaporator, shown in Fig. 3.7, is a natural circulation short-tube evaporator in which the liquid recirculates through an annular downcomer formed by a gap between the heater and the shell of the evaporator. The cross-sectional area of the downcomer should exceed the

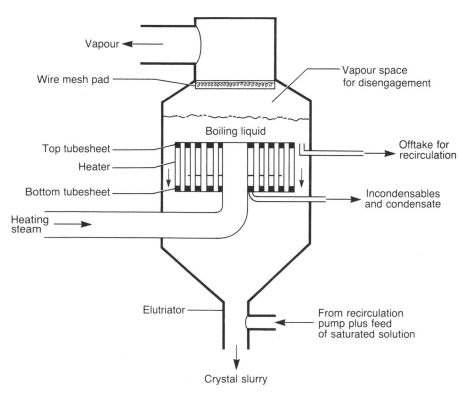

3.7   Basket-type crystallising evaporator with natural circulation

total cross-sectional area of all the tubes, to ensure an adequate circulation
ratio.

The uses of the basket-type evaporator are similar to those of the
conventional short-tube vertical evaporator, described in the previous
section. Its advantage is that the tubes may be removed for cleaning or
replacement; however, it must be supported from below or suspended from
the superstructure of the evaporator. Thus the basket-type is preferred to
the short-tube evaporator shown in Fig. 3.6 only when there is a possibility
of corrosion or heavy scaling of the tubes in the boiling zone or on top of
the upper tubeplate.

Basket-type evaporators may be used as crystallising evaporators if it is
required to produce crystals of a controlled size, e.g. ammonium sulphate
for agricultural uses. However, operation must always be at or near to the
full load, otherwise it produces crystals of reduced size. Removal of crystals
is as described in the previous section. Spray may be removed in an
external separator or by a pad of wire mesh supported on a metal grid as
shown in Fig. 3.7.

3.1.4 Long-tube vertical evaporator

Compared with the evaporators described in sections 3.1.2 and 3.1.3, the
long-tube vertical evaporator uses fewer tubes of somewhat smaller
diameter and of length 6 m or more. The downcomer is external, as shown
in Fig. 3.8. Forced circulation is usually preferred to natural to achieve a
high velocity in the tubes and hence obtain a high heat transfer coefficient.
The long-tube vertical evaporator is preferred when it is important to
achieve a high heat transfer coefficient and when the attendant
disadvantage of a higher boiling temperature, due to a greater hydrostatic
head, is not serious, and when the extra height required presents no
problems.

The long-tube vertical evaporator may be used as a concentrating or as a
crystallising evaporator. The concentrate or crystal slurry is removed from
a suitable point in the return line, as shown in Fig. 3.8. Alternatively the
heated liquid may pass from the evaporator first to a flash vessel and then
to a fluidised bed in a separate crystallising chamber, the vapour being
taken off the top of the flash vessel.

3.1.5 Climbing-film evaporator

By operating a long-tube vertical evaporator with a very low liquid level,
boiling begins soon after entry to the tubes and a large amount of vapour
is generated, producing a climbing film of liquid with a very high heat
transfer coefficient. This also eliminates the disadvantage of the hydrostatic

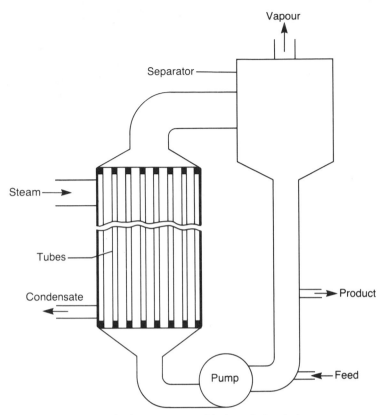

3.8 Long-tube vertical evaporator with forced circulation

head reducing the temperature difference. The climbing-film evaporator shown in Fig. 3.9 is a concentrating evaporator operating on the once-through system. The vapour and concentrated liquid pass immediately into a separator (usually a cyclone). Because of the high velocity, the time of contact with the heating surface is reduced. This arrangement is therefore ideal for concentrating a heat-sensitive liquid. The low temperature difference is an advantage for multiple-effect operation or vapour recompression, described in sections 5.4.5 and 5.4.6.

3.1.6 Vertical thermosyphon reboiler

The principle here is the same as in a long-tube vertical evaporator, except that the distillation column acts as the separator; a typical arrangement is shown in Fig. 3.10. The tubes are contained in a single-pass shell (a TEMA 'E' type shell); the heating medium, usually condensing steam, is on the shellside, entering at the top. The piping returning the two-phase mixture

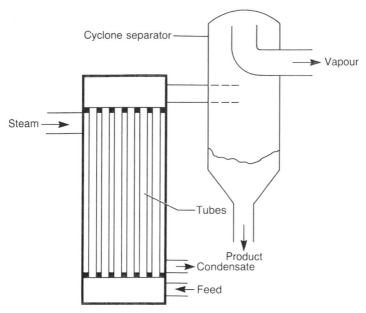

3.9   Climbing-film once-through concentrator

to the column should have a cross-sectional area at least as big as the total
area for flow up all the tubes, to avoid excessive pressure drop and to
ensure steady flow. Fluctuations in flow seriously upset the operation of
distillation columns. If the single-phase pressure drop in the inlet piping
exceeds the two-phase pressure drop in the outlet piping, flow will be
steady. It may be possible to achieve steady flow if the single-phase
pressure drop is only half the two-phase pressure drop, but accurate
prediction is difficult (see section 9.2.7). Hence it is good practice to provide
facilities for installing an orifice plate in the inlet piping to act as a
restriction to flow in the region of single-phase flow, if flow is found to be
unstable without it. The practice of providing a valve in the outlet piping is
often adopted but it is less satisfactory than the orifice restriction, because
its setting can be changed too easily. Any restriction to the circulation rate
may lead to a loss in temperature difference and increased fouling.

3.1.7   Spiral-plate vaporiser

All the types considered so far have been tubular heat exchangers in which
a fluid vaporises as it flows up vertical tubes. In this and the next section,
plate heat exchangers are considered in which a fluid vaporises as it flows
up ducts of constant cross-section (the use of various types of plate heat
exchanger to preheat a liquid is considered in section 3.6).
    The spiral-plate heat exchanger consists of two plain plates, normally

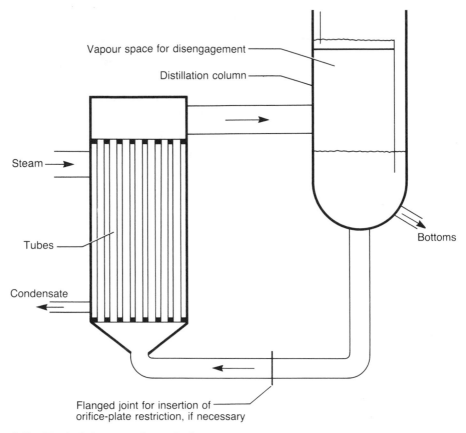

Vapour space for disengagement ———

Distillation column ———

Steam ——▶

Tubes ———

Condensate

Bottoms

Flanged joint for insertion of ———
orifice-plate restriction, if necessary

3.10   Vertical thermosyphon reboiler

made of stainless steel, wound round each other in a spiral, so as to form two separate flow passages. In single-phase heat exchangers the hot fluid enters at the centre of the unit and leaves at the periphery; the cold fluid flows in the countercurrent direction. In condensers the coolant flows from the outside to the centre through a spiral duct formed between adjacent plates sealed by welding at the top and bottom, as shown in the detail in Fig. 3.11. The vapour to be condensed enters the top of the unit and flows vertically downwards through the spiral gap between the spiral duct conveying the coolant. A similar arrangement is used for vaporisation, with the liquid to be vaporised entering at the bottom of the unit and flowing vertically upwards through the spiral gap between the spiral duct conveying the heating fluid, as shown in Fig. 3.11.

Further information on spiral-plate heat exchangers is given by Alfa-Laval (1969) and by Saunders (1987).

Spiral-plate heat exchangers operate very well for vaporisation, as shown in tests by Yilmaz et al. (1983). A spiral-plate vapour generator is shown in

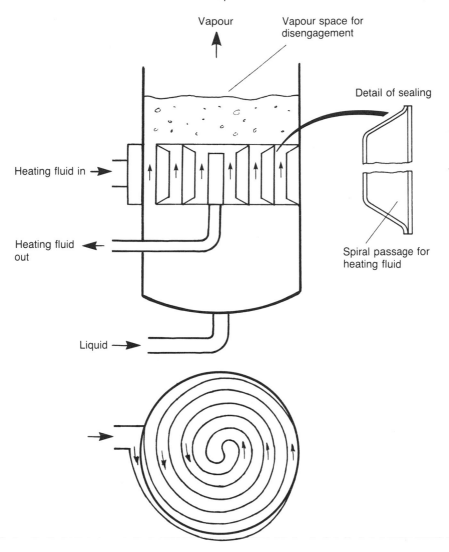

3.11 Spiral-plate heat exchanger with upward flow boiling

Fig. 3.11. It may also be used as an internal reboiler located in the base of a distillation column, or externally as a vertical thermosyphon reboiler.

The surface in contact with the vaporising liquid is easily accessible for cleaning. Mechanical cleaning of the spiral duct conveying the heating fluid is not possible, because it is sealed at the top and bottom by welded strips.

3.1.8 Plate-fin heat exchanger

The plate-fin heat exchanger is a form of compact heat exchanger built of alternate layers of flat plates and corrugated plates brazed together to form a block. The flat plates keep the fluids apart and form the primary heat transfer surface; they are 0.5 to 1 mm thick, except for the outer plates, which are considerably thicker to withstand the internal fluid pressure. The corrugated plates are 0.2 to 0.7 mm thick; together with their adjacent flat plates they provide channels for the flow of the fluids parallel to the direction of the corrugations, and they conduct heat to or from the flat plates, thus acting as fins (see section 6.4.4). The fluids are contained within each flow-channel by sealing bars located along the edges; there is a gap in the sealing bars where the fluid enters and where it leaves a channel. These channels are 3 to 12 mm deep. Further information on plate-fin heat exchangers is given by Saunders (1987).

Aluminium plate-fin heat exchangers are used extensively in industrial cryogenic plant as chillers. The vaporising refrigerant flows up the channels, arranged vertically, the gas to be cooled and condensed flowing downwards. Very high thermal effectiveness can be achieved with this countercurrent arrangement. With plain fins, the heat transfer coefficient and pressure gradient can be estimated as described in sections 6.2.8 and 9.1.1. However, complicated forms of plate-fins have been developed to give enhanced heat transfer (at the cost of greater pressure drop) by introducing serrated fins or fins in which there are regular changes in the direction of flow. Many of the available types are described by Kays and London (1984), who give the data necessary for the calculation of heat transfer and pressure drop in single-phase flow.

The plate-fin heat exchanger may also be used as an internal reboiler, arranged as shown in Fig. 3.12. The heating fluid, normally a condensing vapour, enters the top of the block and is conveyed across the top through distributor finning of the type shown in Fig. 3.12 (a), with a similar arrangement at the bottom for collecting and removing the heating fluid. The boiling fluid flows upwards through vertical channels, which are open at both ends, as shown in Fig. 3.12 (b). The liquid that has not vaporised falls into the liquid pool, the flow pattern being as shown by the arrows in Fig. 3.12 (a).

As the block is submerged in a pool of liquid in the base of the still, recirculation of liquid is through the space between the block and the still. These reboilers are used in oxygen plant and in the processing of hydrocarbon gases. However as they cannot be cleaned mechanically, they can be used only with clean fluids. They are useful in applications where there is only a small temperature difference between the streams, as is usually the case in cryogenic applications, but they are not suitable for high flow rates.

3.2   Flow inside horizontal tubes

Vaporisation inside horizontal tubes is less common than inside vertical

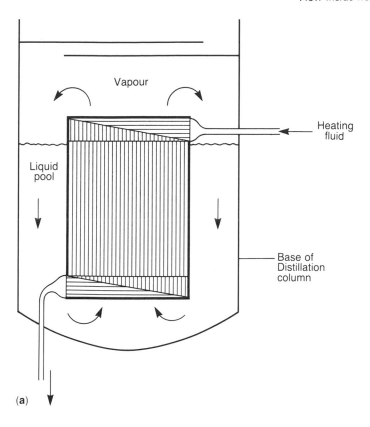

(a)

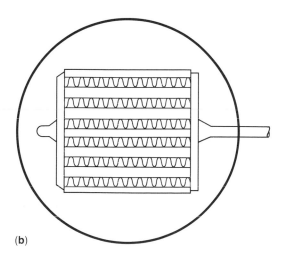

(b)

3.12 Plate-fin heat exchanger as an internal reboiler
    (a) Sectional elevation through a passage conveying the heating fluid
    (b) View looking down, showing the open ends of the passages conveying the
        boiling fluid

tubes with upward flow. Horizontal tubes are contained in a shell or duct, through which the heating fluid passes. Forced circulation of the boiling fluid is normally employed, because there is very little benefit from buoyancy forces, unlike upward flow in vertical tubes. Also a high velocity is needed to avoid dryout. On account of these difficulties, vertical tubes are preferred except where a horizontal layout is more convenient. Flow inside heated horizontal tubes may be used for preheating a liquid before it passes into a flash vessel.

When heating a fluid that tends to deposit dirt or form scale, if it flows through many tubes in parallel there is a danger of some tubes becoming choked. Thus if deposits begin to form preferentially in one tube, less fluid flows through that tube; consequently the rate of deposition in that tube increases, until it may become completely choked while other tubes are still quite clean. With vertical tubes there is some compensation, because a partially choked tube generates a higher quality, giving greater buoyancy, which tends to compensate for the reduced diameter. Orifice plates are sometimes installed as restrictions at the inlet to every tube to reduce the chance of scaling and to avoid unsteady flow, as described in section 9.2.7. This is particularly important in boilers, where the partial choking of a tube very easily leads to overheating of that tube.

When boiling occurs inside banks of tubes, it is advisable to check that there is no danger of static instability, in which some tubes take a much higher flow rate than others, as explained in section 9.2.4. This is liable to happen if the liquid at entry is below its boiling temperature.

3.2.1 Water-tube boiler

With horizontal tubes forced circulation is preferable, the water from the steam drum being recirculated through a pump to the inlet manifold that feeds the tubes. A spare pump is normally installed, on account of the serious damage to the tubes that results from a failure of the circulation.

Horizontal tubes are seldom used now for generating steam in fired boilers, the preferred arrangement being vertical or inclined tubes, as described in section 3.1.1. However, horizontal tubes are normally used in the 'economiser' where the heat in the gases leaving the furnace is used to preheat the boiler feed water. Economisers are usually designed to heat the water to its boiling temperature under normal operating conditions, but under abnormal conditions there may be some boiling inside the tubes near to the outlet. Similar devices for heat recovery may be installed inside the flue gas ducts of other types of furnace. When a certain amount of steam generation is intentional they are called 'steaming economisers'.

Horizontal water-tube boilers are used as waste heat boilers because they can be easily installed inside the duct conveying the hot process gases.

A possible arrangement is shown in Fig. 3.13 for a hot gas flowing

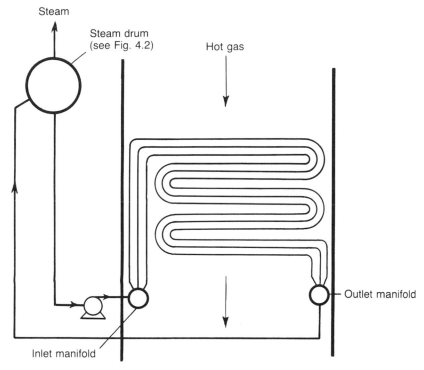

3.13 Section through one vertical layer of serpentine coils for waste heat boiler

downwards across an array of serpentine coils. Water is pumped into a horizontal manifold at the bottom of the duct and flows through many coils in parallel, installed in a vertical plane, to a similar outlet manifold. Each coil consists of several horizontal tubes in series connected by 180° bends.

For smaller applications, bundles of U-tubes are convenient because they may be installed immediately below a high-temperature reaction vessel, as shown in Fig. 3.14.

With the arrangement shown in Figs 3.13 and 3.14 there is a danger of overheating the top of the top row of tubes, where the heat flux is exceptionally high due to the enhancement at entry and radiation from the hot gases and the walls of the duct. It is also difficult to decide whether the general direction of flow of the boiling water should be upwards or downwards. The reason for the general direction being downwards in Fig. 3.13 is to provide water of low quality to the hottest tubes and thus reduce the risk of dryout. With such an arrangement, involving downflow boiling, it is important to check that there is no danger of static instability (see section 9.2.4). The upward direction was chosen for the waste heat boiler shown in Fig. 3.14 to avoid static instability and to reduce the pressure drop (much more steam is generated in the upper than in the lower half of

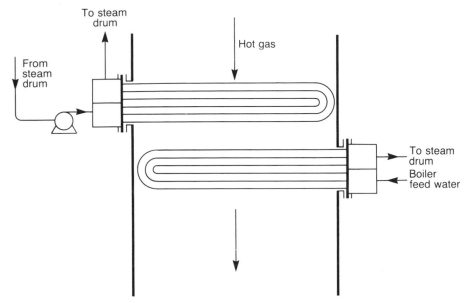

3.14 Section through centre of U-tube bundles for waste heat boiler and boiler feed water heater

a U-tube; thus the mean quality of the steam with respect to length of tube would have been greater with downward than with upward flow). With the arrangement in Fig. 3.14 there is a danger of dryout in the uppermost tubes, i.e. in the region of highest heat flux and highest quality.

Dryout can be avoided (see section 8.2.2) by using a very high velocity of the water and a high recirculation ratio, so that the quality at outlet is small (5–10%). Alternatively twisted tapes may be inserted in the uppermost tubes. With the arrangement shown in Fig. 3.14, twisted tapes would be inserted in the upper half of each U-tube in the upper bundle, and the direction of flow would then be downwards, in order to reduce the quality at the top, thus avoiding excessive pressure drop due to high quality two-phase flow through twisted tapes. Corrosion due to dryout is a serious problem, as discussed in section 12.2.2.

Finned tubes are sometimes used with the equipment described above, particularly towards the gas outlet, to compensate for the lower temperature difference and so maintain the heat flux. In such cases it is particularly important to check that the danger of dryout has been avoided.

3.2.2   Fired vapour generator

Gases, such as natural gas or ammonia, may be liquefied to facilitate storage and transport. When revaporising a liquefied gas for use in a process, a source of heat may be needed. Normally a hot process fluid or condensing steam is available as a source of heat. On sites where such a source can not be provided, e.g., at a storage depot, a fired vaporiser is needed. The vaporiser then consists of a small furnace with a circulating fan to circulate hot gases over the outside of the horizontal tubes in which the liquefied gas is being vaporised. A high gas recirculation ratio is needed to keep the temperature of the gas at inlet to a few hundred degrees and so avoid vapour blanketing of the inner surface, as described in section 2.5. The formation of a film of vapour inside the tubes would quickly lead to serious overheating of the tubes.

If the boiling temperature of the liquefied gas is substantially below 0 °C, there is a danger that the tubes will become so cold that ice forms on their outer surface, thus restricting the transfer of heat. The estimation of the temperature of the tube can be carried out as described in section 6.6.3.

Another use of the fired vaporiser is for the vaporisation of a high-temperature heat transfer fluid, such as the eutectic mixture of diphenyl ether and diphenyl (see footnote to Table 6.1). Such fluids are needed when it is required to heat a process fluid to a temperature higher than can be achieved with steam and it would not be safe to pass the process fluid through the furnace. Thus an intermediate heat transfer fluid must be employed, to be heated in a furnace and then transfer heat to the process fluid in a separate heat exchanger.

3.2.3 Evaporator

An arrangement of horizontal tubes such as that shown in Fig. 3.15 may be used as a concentrating evaporator. The heater is a standard shell-and-tube heat exchanger, the normal source of heat on the shellside being condensing steam. The floating-head bundle can be easily removed for cleaning. However, if it is unlikely that the inside of the tubes will need frequent cleaning and if they can be cleaned by washing *in situ*, then a U-tube bundle should be used, because this is appreciably cheaper than a floating-head bundle. Most of the heat supplied takes the form of sensible heat appearing as a rise in temperature of the circulating liquid; most of the vaporisation occurs in the flash vessel as a result of pressure drop.

Forced circulation is needed to give a high heat transfer coefficient and to avoid scaling or dryout. These problems must be investigated when considering the possibility of using a horizontal tubeside evaporator.

3.2.4 Chiller

A gas may be cooled by passing it across a bank of horizontal tubes through which a boiling refrigerant is being passed. If a liquid is condensing

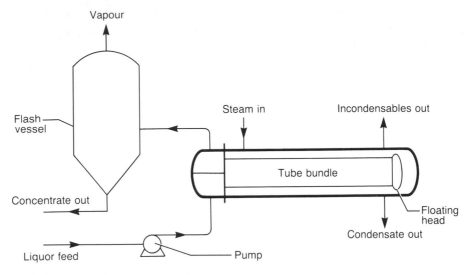

3.15 Horizontal in-tube evaporator

out of the gas, as when chilling atmospheric air, the gas should flow in a downward direction to aid the drainage of the condensate and the general direction of flow of the refrigerant between passes should be upwards. The saturation temperature of the refrigerant should be sufficiently high to avoid freezing of the condensate.

3.3 Vaporisation inside helical coils

Coiled-tube heat exchangers range in size from a single small coil inside a vessel to a multitubular unit with a surface area up to $20\,000\,\text{m}^2$. Large units are used in cryogenic applications for the liquefication of air and petroleum gases. They are used as steam generators in some nuclear power plants. Their chief advantage over shell-and-tube exchangers is that long tubes, with countercurrent flow, can be packed into a single shell, without any serious problems caused by differential thermal expansion, and without the need for large high-pressure tubesheets.

Details of multicoil units are given in section 3.4.5, which relates to their use as shellside vaporisers. Vaporisation inside helical coils is discussed below.

3.3.1 Coiled-tube boiler

Fired boilers are available with the water boiling in upward flow through helical coils. Heat is supplied by the combustion of fuel in a burner placed in the centre of the base of the unit. In a package boiler with coiled tubes it is common practice to evaporate to a quality of about 90 % in a once-through system, which may operate at high pressures, up to 80 bar (8 MPa), with tubes of 25 to 100 mm internal diameter. The steam passes through a separator and the water removed constitutes the continuous blowdown.

The chief advantages of coils over straight horizontal tubes for water-tube boilers are that higher single-phase heat transfer coefficients are achieved, facilitating once-through operation, and that a much higher heat flux may be used without risk of dryout, for a given pressure and mass flux. The price paid for these advantages is a higher pressure drop and hence a higher cost of running the pump needed for the forced circulation.

3.3.2   Vapour generator

Helical coils may be used to revaporise a liquefied gas or to obtain a pure vapour from an impure liquid, with upward flow of the liquid inside the tubes. This has the advantages described above for coiled-tube steam generators. The source of the heat may be waste heat in a process fluid, condensing steam or the combustion of a fuel.

3.4   Shellside vaporisation

This section deals with several different types of duties in which there is boiling in a shell, heat being supplied by a heating medium inside tubes. Circulation takes the form of: (1) forced external; (2) natural external; or (3) natural internal, i.e. inside the shell. Shellside evaporation from a falling film on the outside of helical coils is described in section 3.5.2.

The chief advantage of shellside over tubeside boiling is that there is no difficulty in obtaining a good distribution of the boiling fluid over the heated surface (maldistribution in tubeside boiling may be a serious problem, as explained in section 3.2). With liquids that are liable to boiling fouling, consideration must be given to the method to be used for cleaning the heated surface, when making a choice between tubeside and shellside boiling (see Ch. 5).

It is difficult to estimate the critical heat flux with boiling outside a bundle of horizontal tubes. It is known that this is considerably less than with boiling outside a single tube under otherwise identical conditions. Section 8.3.1 gives a method of estimating the critical heat flux for a single horizontal tube and gives a tentative suggestion for estimating the bundle correction factor, which is a function of the number and the pitch of the tubes.

When the temperature difference is only a few degrees, as at the outlet of a chiller, heat transfer is by convection, the contribution of nucleate boiling being negligible or non-existent. It is therefore important to ensure that there will be a significant velocity of flow in the shell where the temperature difference is small. This can be achieved either by using a high circulation rate or by arranging for the region of low temperature difference to be the region of high quality of the boiling fluid, so that the presence of the vapour increases the velocity of the liquid past the heated surface.

3.4.1 Fire-tube boiler

In a 'fire-tube' or 'shell' boiler, the boiling water is inside a shell, usually horizontal, and heat is supplied by hot gases passing through tubes. In a fired boiler the hot gases are produced by the combustion of a fuel in an adjacent combustion chamber; in a waste heat boiler the heat is taken from a hot process gas. A mixture of water and steam flows from the shell to a steam drum, where the steam is separated and exported, while the water, plus make-up water added to the drum, is circulated back to the shell; the steam drum is described in section 4.1.2. The various practical problems in the design and operation of shell boilers are discussed by Goodall (1980) (Part 4).

The shell boiler (like the vertical water-tube boiler) is attractive because it works satisfactorily on natural circulation, thereby avoiding the expense and complications of circulating pumps. It has the advantage over water-tube boilers that the dangers of overheating the tubes as a result of maldistribution of the water are considerably less. Shell boilers are less prone than water-tube boilers to failure due to debris in the water. It is important to ensure an adequate supply of water to the hot tubesheet.

Figure 3.16 shows a waste heat boiler (operating on natural circulation) for cooling a process gas entering at a temperature of 1000 °C or more. The hot tubesheet must be protected by an insulating sheet between it and the hot gas and by insulated ferrules inside the tubes. It is advisable to estimate the temperature of the tubes at inlet, to ensure that the weld is not overheated. With gases at lower temperatures, say 500 °C, it may not be necessary to use an insulating sheet or ferrules. A method of estimating the maximum temperature of the tubes is given in section 6.6.4.

Considerable stresses occur due to differential thermal expansion between the tubes and the shell. For the stress analysis calculations it is necessary to estimate the mean (with respect to length) temperature of the tubes, as described in section 6.6.1. The temperature of the inside of the hot tubesheet is very sensitive to the amount of dirt deposited on it by the boiling water; similarly the mean temperature of the tubes is sensitive to the amount of boiling fouling. Approximate information on fouling is given in section 6.4. However, inadequate water treatment will lead to more fouling

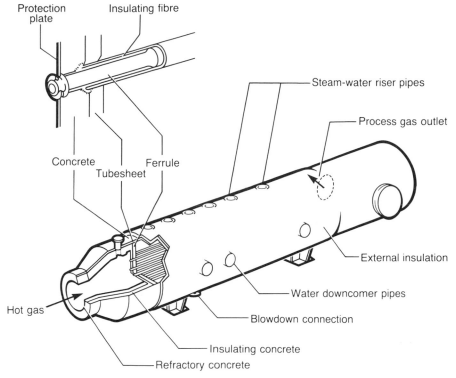

Protection plate

Insulating fibre

Steam-water riser pipes

Process gas outlet

Concrete

Ferrule

Tubesheet

External insulation

Hot gas

Water downcomer pipes

Blowdown connection

Insulating concrete

Refractory concrete

3.16 Natural circulation fire-tube waste heat boiler

than normal and a consequent increase in the temperature of the tubesheet and the tubes. Due to the need to insulate the channel at the hot end with an internal refractory lining, no tubes can be installed in the region of the tubesheet covered by this lining; consequently there is an unpierced annulus and the shell cannot be completely filled with tubes (see section 11.2.5A). From the point of view of the mechanical designer, this unpierced annulus has the advantage that it acts as a diaphragm and thus reduces the stresses due to differential thermal expansion.

It is important to return the circulating water to several points along the length of the shell; these return 'downcomers' should be arranged along the length of the shell at spacings such that equal amounts of steam are generated in each length of shell supplied. They should therefore be closer together at the hot end than at the cold end of the shell, to compensate for the greater vapour generation rate in the region of higher gas temperature. The 'risers' that collect the two-phase mixture must be at the top of the shell and spaced in a manner similar to the downcomers. Alternatively some designers incline the shell so that the hot end is slightly higher than the cold end, with a greater concentration of downcomers at the cold end, maintaining the greater concentration of risers at the hot end; however, this practice has not been much used in recent years.

Waste heat boilers of the type described above will operate satisfactorily in the generation of high-pressure steam, at *c*. 100 bar (10 MPa), provided that the various points have been taken into account at the design stage, and as long as there is no failure of water treatment (see Ch. 14).

Waste heat boilers for low-pressure steam (up to about 25 bar) may utilise the shell for separating the steam from the water, thus avoiding the considerable cost of a steam drum, as with the packaged boilers described in the next paragraph.

Small packaged fired shell boilers are used extensively in smaller factories in the chemical industry, generating steam in the region of 10 to 20 bar and at rates up to 8 kg/s. To avoid the need for a separate steam drum, there are no tubes in the top of the shell. A typical arrangement is shown in Fig. 3.17.

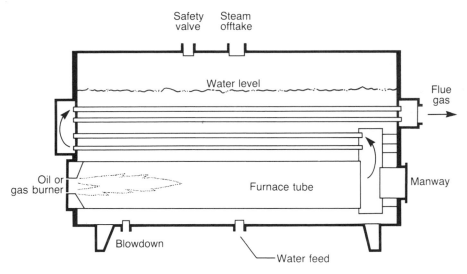

3.17 Fired shell boiler

The water level must be carefully controlled so that all of the tubes are always completely covered, but there must be sufficient space above the water level to permit the disengagement of the steam from the water without an undue amount of carry-over of liquid (see section 9.8). Section 8.3 should be consulted for estimating the critical heat flux. Further information is given by Goodall (1980), Part 4.

This type of boiler is often fired by a heavy fuel oil, which involves problems in the storage and preheating of the oil and in the design of the burners. These problems are dealt with by the manufacturers of the boilers and are outside the scope of this book.

3.4.2 Horizontal thermosyphon reboiler

Figure 3.18 shows a horizontal thermosyphon reboiler connected to a distillation column. The heating fluid (usually condensing steam) passes through the tubes, which may be contained in a TEMA X, G (as in the figure), H or possibly E shell, the different shells being illustrated by TEMA (1978). The natural circulation is provided by the difference in density between the liquid in the column and the two-phase mixture in the shell and outlet pipe of the reboiler. When designing the piping between the reboiler and the column, it is necessary to follow the procedure outlined in section 3.1.6 for vertical thermosyphon reboilers.

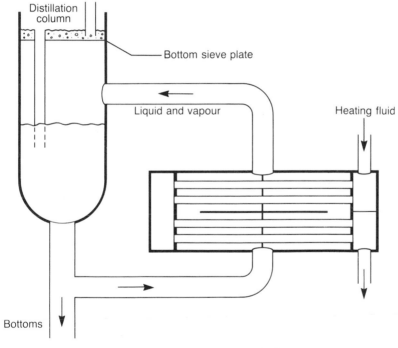

3.18 Horizontal thermosyphon reboiler with G shell

3.4.3 Kettle reboiler

In a kettle reboiler, illustrated in Fig. 3.19, the U-tubes containing the heating fluid are installed inside a shell whose diameter is usually one-and-a-half to two times the diameter of the tube bundle; this is known as a TEMA K shell. The liquid level is just above the top tubes. The level is usually controlled by a weir, as shown in Fig. 3.19; alternatively a level-control may be used. Recirculation occurs inside the shell. Thus the piping

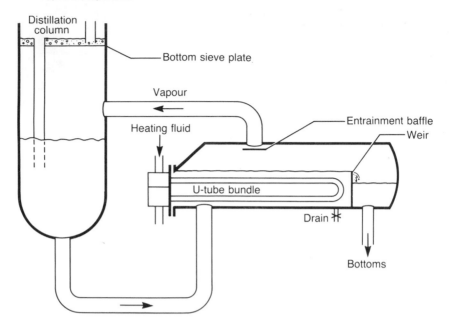

3.19 Kettle reboiler

between the reboiler and the column is smaller and there is no danger of unsteady flow, compared with the piping for vertical or horizontal thermosyphon reboilers described in sections 3.1.6 and 3.4.2. The bundle can easily be removed for cleaning.

It is difficult to calculate the internal circulation rate and the boiling heat transfer coefficient, which depends on both nucleate and convective boiling in most applications. This may be a problem with organic liquid, operating with a low temperature difference, when the shellside boiling coefficient may be the one that limits the heat transfer rate. If operating with a large temperature difference, it is advisable to estimate the critical heat flux – see section 8.3.1.

It may be necessary to check that there is not excessive carry-over of liquid droplets from the surface of the boiling liquid (see section 9.8). A small amount of carry-over does not often upset the performance of a distillation column, but it might cause a significant increase in the pressure drop in the vapour line from the reboiler to the column. An entrainment baffle, placed opposite the entrance to the vapour line, is a simple device for reducing the amount of liquid carried into the vapour line.

The kettle-type is susceptible to fouling. Non-volatile materials tend to collect in the shell, so the drain in the base of the shell must be opened periodically for purging.

3.4.4 Internal reboiler

This is the cheapest type of reboiler, consisting only of a bundle of horizontal tubes inserted into the reservoir of liquid at the bottom of a distillation column. Thus the expense of the shell and the interconnecting piping is eliminated, but in many applications it is physically impossible to accommodate the required amount of surface in the reservoir. The procedures discussed in section 3.4.3 for calculating the boiling coefficient and the critical heat flux in a kettle reboiler also apply to an internal reboiler. The bundle can easily be removed for cleaning, but fouling is likely to be less of a problem than with a kettle reboiler.

3.4.5 Boiling outside helical coils

Boiling inside helical coils has been discussed in section 3.3. This section is concerned with boiling in upward flow across the outside of helical coils. Downflow vaporisation outside helical coils has also been used, but as this is an example of evaporation from the surface of a falling film of liquid, it is dealt with in section 3.5.2.

Coiled-tube vaporisers are suitable for use with clean fluids when a high thermal effectiveness is required. They range in size from a single coil to the type used in the cryogenics industry and illustrated in Fig. 3.20, where the installation of a central core inside the coils concentrates the flow of the shellside fluid past the coiled tubes and gives strength to the exchanger. The heating fluid flows downwards through the tubes and condensation normally takes place. Alternate layers of coils are wound in opposite directions. They are separated from each other and the core by spacing strips. The coils in any layer are identical. The coils in different layers all have the same length of tube in order that they should all have approximately the same tubeside friction factor; consequently they all have the same height and the same helix angle, and there are more coils in the outer than in the inner layers. For example, in the case of the heat exchangers shown in Fig. 3.20, there are five layers of coils. The outermost layer contains seven coils, denoted by the letters A, B, C, D, E, F, G, the next layer contains six coils (H, I, J, K, L, M), the next five coils (N, P, Q, R, S), the next contains four coils (T, U, V, W) and the innermost layer contains three coils (X, Y, Z). Thus all coils have exactly the same length of tube, but those in the inner layers have a somewhat higher tubeside friction factor than those in the outer layers, due to their greater curvature. The geometrical calculations needed for the design of such a system of coils are described in section 6.2.7, which also describes the estimation of single-phase heat transfer coefficients. In certain cryogenic applications there might be several separate streams of fluids flowing down through the tubes, being cooled by a boiling refrigerant on the shellside (alternatively there

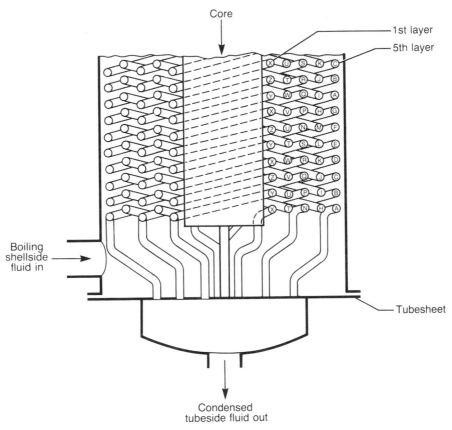

3.20 Coiled-tube heat exchanger. The figure shows five layers of coils, with three coils in the 1st (inner) layer, four in the 2nd, five in the 3rd, six in the 4th and seven in the 5th (outer) layer of coils.

might be several separate streams of cold fluid boiling in upward flow through the tubes and extracting heat from a fluid entering the top of the shell at atmospheric temperature).

Less complex arrangements of coiled tubes inside a vertical shell with a central core might be used for shellside boiling, using plain or finned tubes.

Helical coil heat exchangers are more expensive than shell-and-tube heat exchangers with the same number of tubes of the same length and diameter, because of the cost of coiling and because a larger shell is needed. Their advantages are compactness and the avoidance of serious problems due to differential thermal expansion.

3.4.6 Horizontal shellside evaporator

The TEMA type K shell, whose use as a reboiler has been described in section 3.4.3, may alternatively be used as a vapour generator or a concentrating evaporator. The normal arrangement is illustrated in Fig. 3.19. When it is used as a concentrating evaporator, the concentrate flows over the weir and out through the drain pipe used to remove bottoms in a reboiler application. This type of evaporator is not recommended when scaling is likely to occur, because the velocity of flow past the heated tubes is not sufficiently high to discourage the deposition of scale.

When it is used as a vapour generator, the liquid outlet may be used as a continuous purge, to limit the build-up of impurities in the boiling region. If the liquid that it to be vaporised is clean and free of dissolved solids, the weir and the liquid outlet may not be necessary. In the absence of a weir for continuous purge, the drain connection in the bottom of the shell may be used for occasional purging during operation.

The simplest form of evaporator consists of a horizontal bundle of U-tubes submerged in a large, partly filled, vessel, similar to the internal reboiler described in section 3.4.4.

3.4.7 Bayonet-tube vaporiser

The bayonet-tube heat exchanger has already been described for use as a waste heat boiler in section 3.1.1 and illustrated in Fig. 3.4. A similar arrangement of tubes may be used as a vaporiser, but with an inverted layout, i.e. with the tubes above the tubesheets. Heat is supplied by the condensation of steam, which is introduced at the bottom, rises up the inner tubes and condenses as it flows downwards through the annular space between the inner and outer tubes. The liquid is introduced into the bottom of the shell, where it vaporises, the vapour generated being taken off the top of the shell. The advantage of this arrangement is that the level of the liquid inside the shell is self-adjusting, providing a sufficiently large area for heat transfer so that the vapour generation rate equals the feed rate.

This arrangement has been used for the vaporisation of liquefied gases such as ammonia or carbon dioxide, where there is the extra advantage that if the shellside pressure is reduced, freezing of the condensate is avoided because it is self-draining and the inner tubes continue to be a heat source.

3.4.8 Chiller

A chiller for cooling a gas in flow across a bank of horizontal finned tubes was described in section 3.2.4. If the gas is at a high pressure, say more than 10 bar, it is better to put the gas through the tubes and to have the boiling refrigerant on the shellside. There may be one or two tubeside passes; U-tubes should be used for the latter if there is condensation, to avoid the maldistribution of the phases that would occur at the entry to the second pass. If some condensation has already occurred before entry to the chiller, it must be accepted that the liquid will flow preferentially in the tubes at the bottom of the chiller. An allowance for the loss in performance due to this may be made by assuming that the number of tubes cooling liquid will be such that the pressure drop of the liquid stream is the same as that of the gas stream. The shell may be TEMA type E, G, H or X.

It is normally thermodynamically desirable for a chiller to operate with a close approach, i.e., for the gas to be cooled to within a few degrees of the temperature of the boiling refrigerant. Thus at the gas outlet there is a very small temperature difference and it is likely that there will be no nucleate boiling or that the nucleate boiling coefficient will be low; this would lead to a serious reduction in heat flux, unless precautions were taken in design to ensure that the convective boiling coefficient there will be high. A high convective boiling coefficient at the gas outlet can be achieved by ensuring that at that end of the shell there will be either a high liquid velocity or a high quality.

3.5   Falling-film evaporator

This section is concerned with the description of equipment in which vaporisation occurs as a result of evaporation at the surface of a slightly superheated film of liquid flowing by gravity over a heated surface. The physical process has been described in section 2.8.

In a falling-film evaporator, a liquid to be concentrated flows down the inside of vertical tubes of 32–60 mm diameter. The heating fluid flows outside the tubes. This is described in more detail in section 3.5.1. Another type of equipment that comes under the same heading consists of helical coils, with their axes vertical; the heating fluid flows inside the coils and the liquid to be vaporised flows downwards over the outside of the tubes as a film, as discussed in section 3.5.2. Alternatively a falling film of evaporating liquid can be produced on the surface of vertical (heated) plates described in section 3.5.3.

3.5.1   Falling-film evaporation inside vertical tubes

The falling-film evaporator is used to concentrate a heat-sensitive liquid. The liquid to be concentrated is fed into a pool formed on top of the upper tubesheet of a vertical TEMA E-type shell; it flows by gravity as a thin film of liquid down the inner surface of the tubes. The heating medium flows across the outside of the tubes; it may be condensing steam, hot water, or a hot process fluid, steam being the most common heating medium. The

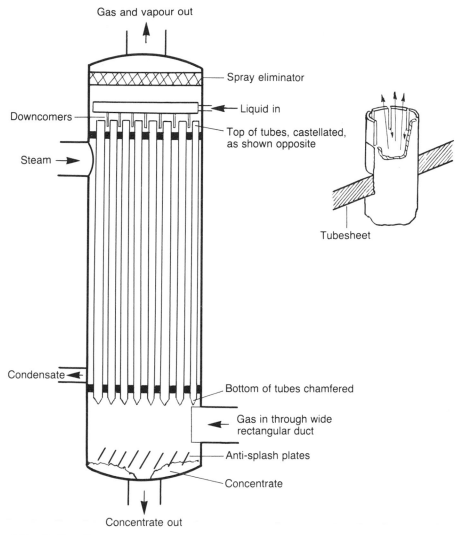

3.21 Falling-film evaporator with bottom entry of gas

concentrated liquid collects in the bottom of the shell, as shown in Fig.
3.21. The liquid flows quickly over the heated surface as a thin film, thus
reducing the danger of decomposition or scaling. It may be desirable in
extreme cases to evaporate at a temperature below the liquid's atmospheric
boiling temperature, but operation under vacuum may be excluded for fear
of the consequence of a leakage of air into the system. Under these
circumstances an inert gas is introduced to flow through the tubes. The gas
may be introduced into the upper channel, so that it and the vapour
generated flow co-currently with the liquid. Alternatively it may be
introduced into the lower channel so that gas and vapour flow

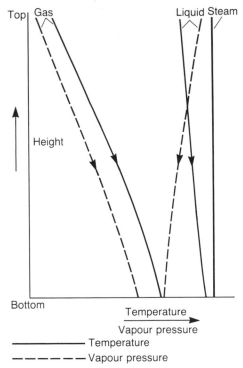

Top | Gas   Liquid Steam

Height

Bottom

Temperature

Vapour pressure

——————— Temperature

— — — — — Vapour pressure

3.22 Temperature and vapour-pressure profiles in a falling-film evaporator with gas entering at the top

countercurrent to the liquid. It is difficult to choose between these alternatives, their relative merits being discussed below.

If the gas is introduced into the top of the evaporator, the relationship between temperature and height is as shown in Fig. 3.22 by the full lines. The gas enters at a relatively low temperature, but it is heated by contact with the liquid. Similarly its vapour pressure, shown by the dotted line on the left, increases as it is mixed with vapour leaving the liquid. The temperature of the liquid increases towards the temperature of the heating steam as concentration proceeds. Consequently there is everywhere a good temperature difference for transferring heat from the surface of the liquid into the gas. However, the vapour pressure at the surface of the liquid, shown by the dotted line on the right, is likely to decrease slightly due to the greater concentration of solids (partly offset by the increase in liquid temperature). The consequent convergence of the dotted lines shows that there is little driving force for mass transfer at the bottom of the tubes and this may limit the amount of concentration achieved.

Alternatively if the gas is introduced into the bottom of the evaporator, the relationship between temperature and vapour pressure and height is as shown in Fig. 3.23. Here there is a large driving force for mass transfer at

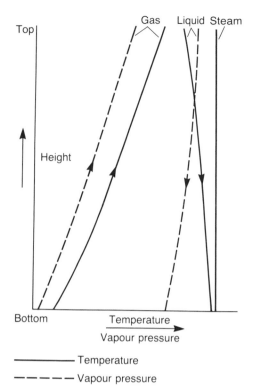

3.23 Temperature and vapour-pressure profiles in a falling-film evaporator with gas entering at the bottom

the bottom, where the concentrated liquid leaves, but the temperature difference at the top is reduced. With this arrangement it is also necessary to check that the flow of gas and vapour at the top of the tubes will not approach the limits to countercurrent flow (see section 9.10).

The moist gas that leaves the evaporator may be circulated through a cooler, to condense the vapour that it has picked up, and the returned to the evaporator.

It is important to establish an even falling film of liquid around the circumference at the top of every tube. This may be achieved by cutting six rectangular weirs at the top of each tube, as shown in the insert in Fig. 3.21. The purpose of the slots is to reduce the sensitivity of the distribution to slight errors in the levelling of the top of the tubes. Alternatively some form of insert may be fitted into the top of each tube to restrict flow and hence to force the level of liquid to rise, thus reducing the sensitivity to levelling. Great attention must be paid to the manufacture of the ends of the tubes. The shell must be set up so that the tops of the tubes are all at the same level. The tubes should be straight and vertical to within a fraction of their diameter.

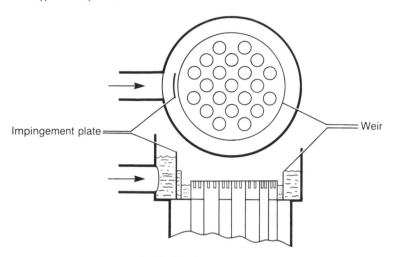

3.24 Simple distributor for falling-film evaporator

Careful attention must also be paid to the distribution of liquid over the top of the tubesheet, to minimise variations in the flow rate per tube. The ends of the tubes must protrude above the top of the tubesheet by an amount sufficient to avoid a significant drop in liquid level towards the centre of the tubesheet at the highest flow rate. Figure 3.24 shows a simple distributor with an impingement plate to destroy the momentum of the incoming liquid, followed by a weir around all the tubes, to give good distribution around the circumference of the tubesheet. The impingement plate and ducting must be carefully designed to avoid a build-up of liquid in the region diametrically opposite the liquid entry. In large evaporators, it is usually preferable to use a more complicated distribution system, such as that shown in Fig. 3.25, so that the liquid does not need to flow more than the short distances from the downcomers to the nearest tubes. The method of designing such a system is given in section 9.7.1.

It is necessary to operate a falling-film evaporator above a minimum wetting rate, otherwise the liquid stream will contract into rivulets, leaving dry patches between them. Waves on the surface of the film and turbulence in the film improve the transfer of heat and mass through the film to its surface. The hydrodynamics of a falling film are dealt with in section 9.11.

When introducing a gas into the bottom channel, it is important to introduce it through a wide duct, to ensure that it will be distributed evenly between the tubes. The flow rate of gas into the tubes should be well below the flooding rate (see section 9.10). The risk of flooding can be reduced by cutting a double chamfer at the bottom of each tube as shown in Fig. 3.21; the liquid then falls as drops from the two lowest points with minimum interference by the incoming gas. Inclined slats placed at the bottom of the channel avoid splash.

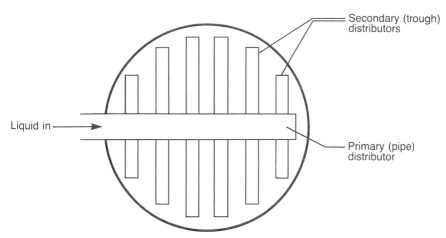

3.25 Trough distributor for falling-film evaporator. Downcomers take liquid from the primary to the secondary distributors from which it flows over weirs onto the tube sheet

3.5.2 Downward flow outside helical coils

The coiled-tube heat exchanger has been described in sections 3.3 and 3.4.5. Heat exchangers of this type have been used in cryogenic applications with the normal directions of flow reversed, i.e., with the condensing stream inside the tubes flowing upwards and the boiling stream outside the tubes flowing downwards. Downflow boiling of a multi-component fluid has been achieved by spraying the liquid on to the top of the coils, which are heated by a condensing fluid flowing upwards through the inside of the tubes. The shellside fluid flows by gravity down the outside of the tubes as a falling film of liquid, which evaporates. The vapour generated also flows downwards. This is therefore an example of a falling-film evaporator. With a multi-component fluid the design calculations are complicated by the fact that the rate of vaporisation is restricted by mass transfer resistance in both the liquid and the vapour phases.

3.5.3 Downward flow outside plates

Proprietary designs of plate heat exchanger are available in which pairs of plates are welded together and arranged vertically with the heating fluid passing through the space between the welded pairs. The liquid to be evaporated is sprayed on to the top of the plates and falls by gravity as a falling evaporating film. The vapour generated is taken off horizontally. The concentrated liquid is drained out of the bottom of the containing shell. The plates are normally made of stainless steel.

3.6   Plate heat exchangers for liquid heating

For a given rate of heat transfer, a gasketed plate heat exchanger costs less than a shell-and-tube exchanger. When using the arrangement of evaporator shown in Fig. 3.15, consideration should therefore be given to the possibility of replacing the horizontal shell by a gasketed plate heater. Various proprietary designs are available, as described, for example, by Saunders (1987). Plate exchangers may be used for vaporising or concentrating a clean liquid. With liquids that tend to foul or form scale on the heating surface, a plate exchanger should be used as a single-phase heater, boiling being suppressed by the insertion of an orifice-plate restriction in the outlet line from the heater just before the flash vessel; the size of the orifice must be calculated to ensure that the pressure inside the heater never falls below the saturation pressure (all vaporisation will then take place inside the flash vessel).

The use of plate heaters in the falling-film mode of evaporation has been described in section 3.5.3. Alternatively they may be used as climbing-film evaporators (see section 3.1.5).

3.7   Direct contact heating

For a given rate of heat transfer, direct contact heat exchangers have the lowest capital cost, due to the absence of heating surface, but they are limited to applications where one fluid is a gas and the other a liquid, where the two fluids are at the same pressure, and where there is no fear of contaminating one fluid with the other as a result of the inevitable mass transfer between fluids. Moreover it is difficult to establish a large interfacial area between the two phases, either as droplets or as bubbles. Direct contact condensers are used quite often in evaporative plant (they are described in section 4.2).

Submerged combustion is a form of direct contact heating which may be used for a concentrating evaporator or as a vapour generator to revaporise a liquefied gas, as described in sections 3.7.1 and 3.7.2.

Operating costs may be higher with direct contact than with indirect heating, because of the difficulty of heat recovery from the stack gases leaving the direct-contact vaporiser. Furthermore it is not possible to use multiple-effect evaporation. In plants where steam is available, it is likely to be more economical to use that steam as a source of heat, condensing it in an indirect heater, rather than to purchase fuel to burn in a direct-contact heater.

3.7.1 Submerged combustion concentrating evaporator

In this application, the burner for the combustion of gas or fuel oil employs downward flow and is placed in the centre of the vessel containing the liquid to be concentrated, as shown in Fig. 3.26. Ducting is provided so that the products of combustion flow through the liquid in the form of bubbles. The cooled combustion gases plus the vapour generated collect above the surface of the liquid and are taken away through a stack. Various devices are available for improving the contact between the gas and liquid and for increasing the time of contact. In some plants the burner itself is submerged, so that the liquid cools it and hence the necessity of using refractories or heat-resisting steels in the construction of the burner is avoided. More usually a conventional burner is located just above the level of liquid. It is essential to be able to use a clean fuel whose products of combustion will not contaminate the concentrated liquid. There may be problems due to corrosion of the stack.

Submerged combustion evaporators may be used for the concentration of acids, such as sulphuric acid (up to 80% by mass) or phosphoric acid (up to 75%). They are attractive for concentrating liquids that are likely to form scale on a heated surface. They may also be used for concentrating viscous liquids or aqueous slurries.

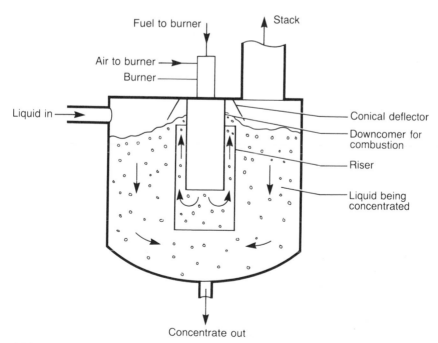

3.26 Submerged combustion concentrating evaporator

With oil fuels, a highly efficient burner must be used to avoid contaminating the concentrated liquid with unburnt fuel. A separator must be included to remove droplets from the stack gases; this may require elaborate and expensive equipment in plants for concentrating acids.

3.7.2 Submerged combustion vapour generator

The use of a fired in-tube vapour generator to revaporise liquefied gases is described in section 3.2.2. This is not suitable for revaporising a cryogen such as nitrogen or ethylene, due to the danger of ice forming on the outside of the tubes. An alternative device is shown in Fig. 3.27; here submerged combustion is used to heat water, which acts as a heat-transfer fluid and conveys heat indirectly to the cryogen, which flows through a helical coil submerged in the water bath. The water is maintained at 40–50 °C and a very high coefficient is achieved for the transfer of heat to the outside of the helical coil from the mixture of water and the gaseous products of combustion, thus reducing the danger of ice forming and leading to a compact design of coil. Water condensed from the products of combustion provides make-up. Materials must be chosen to withstand the water in the tank being saturated with carbon dioxide.

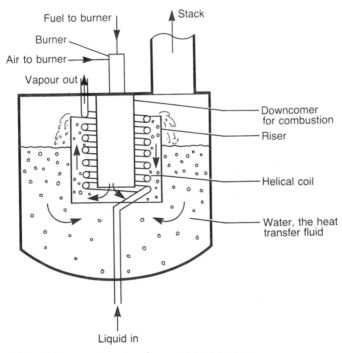

3.27 Submerged combustion vapour generator

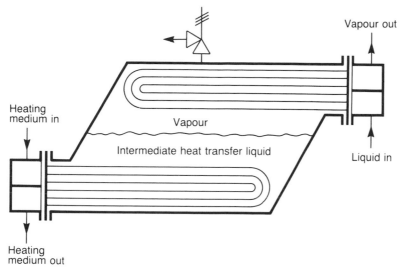

3.28 Double-bundle vaporiser

3.8 Double-bundle vaporiser

A device for conveying heat from a heating medium to a process fluid is the double-bundle vaporiser, described by Saunders (1987) in relation to its uses for single-phase duties. It may be used as a vaporiser, in the manner shown in Fig. 3.28.

The heating medium, usually condensing steam, passes through the lower bundle of U-tubes, which is submerged in the heat transfer liquid. This boils and then condenses on the upper bundle of U-tubes, through which the liquid flows and is vaporised.

This type of vaporiser is used to revaporise a cryogen that has liquefied if there is a danger that the heating medium would freeze if applied directly to the tubes containing the cryogen. If condensing steam is the heating medium, a heat transfer liquid must be chosen with an atmospheric boiling temperature intermediate between the temperature of the cryogen and that of the steam, and with a freezing temperature below the temperature of the cryogen. This arrangement responds rapidly to changes in demand.

CHAPTER 4

Ancillary equipment

There are several important items of ancillary equipment associated with vaporisers. They may be included in the purchase of the vaporiser or they may be obtained separately. They may be integral with, or separate from, the vaporiser. This chapter considers six separate items: (1) separators for removing drops of liquid from the vapour generated; (2) condensers; (3) equipment for producing vacuum; (4) desuperheaters; (5) devices for the introduction of the heating medium into a shell; (6) elutriators for removing large crystals from a crystallising evaporator.

4.1   Liquid/vapour separators

The vapour generated in a vaporiser contains some drops of entrained liquid. The type of separator required depends on the quantity of entrained liquid being removed from the vaporiser and the maximum tolerable amount of liquid in the exported vapour. The various types available are in three classes:

1. Impingement separator for removing large quantities of heavy carry-over (section 4.1.1).
2. Spray eliminators (sections 4.1.2 and 4.1.3).
3. Mist filters for small quantities of fine spray (section 4.1.4).

The first is necessary only when a gross amount of carry-over is likely and the third when an exceptionally pure vapour is required. The information in this section, and the design procedures in section 9.9, are taken from Stearman and Williamson (1972).

When vaporisation is the result of boiling at a submerged heated surface, the bubbles rise to the surface of the liquid, and as they break from the liquid, they carry drops of liquid with them. If the disengagement space is designed as described in section 9.8, most of the drops fall back into the liquid, but the amount entrained increases rapidly with increasing vapour rate. Most of the carry-over is spray, in the size range 100 to 500 μm; a small quantity is mist, in the size range 40 to 100 μm.

When a saturated vapour flows through a restriction, the fall in temperature resulting from the adiabatic expansion leads to the formation of fog. The droplets that constitute fog are mostly less than 40 μm in diameter; some of the larger droplets are removed by mist filters.

The spray eliminators described below all operate on the principle that when there is a change in the direction of mist flow, the vapour is easily deflected whilst the liquid drops tends to continue their motion in a straight line and impinge on a collecting surface. The velocity must be sufficiently high to produce this separation but not so high that vapour shear removes the liquid from the collecting surface. Design methods are given in section 9.9.

4.1.1   Impingement separator

The impingement separator removes heavy loading of coarse drops. It can take several forms, but consists essentially of a surface placed in the path of the flowing vapour and causing it to turn sharply through an angle of 90° to 180°. The simplest form is shown in Fig. 4.1.

With the arrangement shown in Fig. 4.1, the incoming vapour impinges

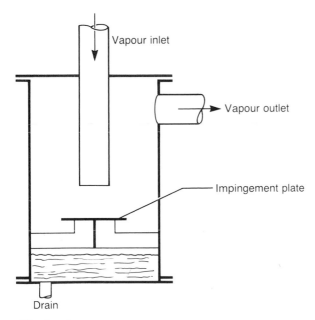

4.1   Impingement separator

on a flat plate. Alternatively a conical impingement plate may be used.
Such devices are used for separating vapour from a large quantity of liquid.

With shellside vaporisation, use is made of simple entrainment baffles,
placed opposite the outlet nozzles, as shown in Fig. 3.19; this prevents the
direct ejection of spray into the outlet pipes. There must be sufficient
clearance to avoid constriction of the flow; section 9.7.4 gives a method of
estimating the pressure drop in shellside nozzles.

4.1.2   Cyclone separator

If a fluid is introduced tangentially into a cylindrical vessel, all the fluid in
the vessel rotates about the centre-line of the vessel, forming a vortex. The
spinning speed must be high enough to throw the drops to the wall by
centrifugal action, but not so high that the liquid deposited on the wall is
re-entrained. The vapour must be removed from the centre in a manner
designed not to upset the flow pattern.

If steam generated in a water-tube boiler is to be superheated, it is
essential to provide efficient separation of liquid in the steam drum,
otherwise any liquid carried over will lead to scaling of the superheater
tubes, with consequent loss of superheat and the danger of overheating
these tubes. Therefore the steam drums shown in Figs 3.1, 3.2, 3.3 and 3.13
contain several cyclone separators, normally arranged in one horizontal
row along the length of the drum; occasionally two rows are used. Figure
4.2 shows a section through a drum containing one row of cyclones; these
cyclones are of a simple design. Consequently it may be desirable to install

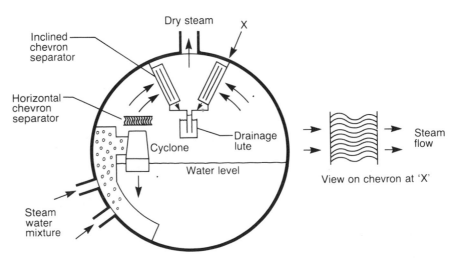

4.2   Section through steam drum showing cyclone separators and two stages of
chevron separation

an additional separator in the steam space for removal of spray from the steam before it leaves the drum. This may be a wire mesh separator, of the type described in the next section, placed immediately under the steam offtake pipe, or it may be in the form of two rows of chevrons running the length of the drum, as shown in Fig. 4.2. As the vapour flows along the sinusoidal path through the chevrons, drops are thrown by centrifugal action onto the surfaces, and flow by gravity off the surfaces into the drainage lute.

Cyclone separators are sometimes used with evaporation equipment, taking the form shown in Fig. 3.9. An alternative form of external cyclone is shown in Fig. 4.3. These external cyclones can be designed to give a higher efficiency than the simple internal cyclones shown in Fig. 4.2.

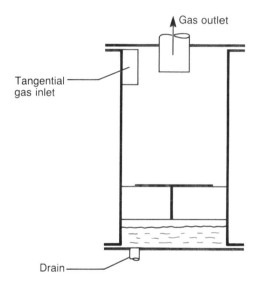

4.3   External cyclone separator

4.1.3   Knitted wire mesh separator

A pad of knitted wire of stainless steel (or other suitable corrosion-resistant material) is an efficient device for removing spray from a vapour. Such pads are normally 100 mm thick and they are typically packed to a voidage of 98 % with wire of about 0.33 mm diameter. Wire mesh separators may be installed internally, as shown in Fig. 3.7; alternatively they may be in a separate vessel, as shown in Fig. 4.4. If too much liquid is present in the vapour, a wire mesh separator becomes flooded and ceases to operate at all. Information on the maximum liquid loading is given in section 9.9. If there is a danger of flooding, the knitted wire mesh separator should be preceded by some simpler device to remove the bulk of the liquid, such as the crude

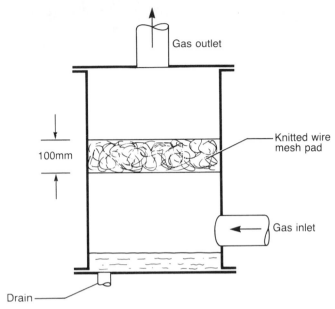

4.4 External knitted wire mesh separator

cyclone installed in a steam drum (described in section 4.1.2). In many applications, an impingement separator may be adequate for the first stage of separation. The impingement separator and wire mesh separator may be installed in one vessel, as shown in Fig. 4.5.

Care must be taken to install a pad of the correct cross-sectional area, because the performance of a wire mesh separator is sensitive to the momentum flux of the approaching vapour. The efficiency of separation increases with increasing momentum flux, up to the optimum momentum flux, after which liquid deposited on the wire is sheared off again and the removal efficiency falls considerably. Information on the optimum momentum flux is given in section 9.9.

4.1.4 Superfine glasswool filter

When it is necessary to remove fog or the last traces of spray of a noxious liquid, the devices described above must be followed by a mist filter, which is usually in the form of a superfine glasswool filter. The fibres have a diameter of only about two microns. Figure 4.6 shows a typical arrangement. Information on the design of these filters is given in section 9.9.4.

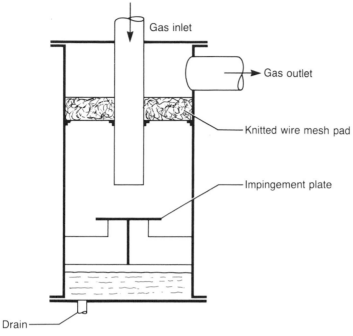

4.5   Combined impingement and wire mesh separator

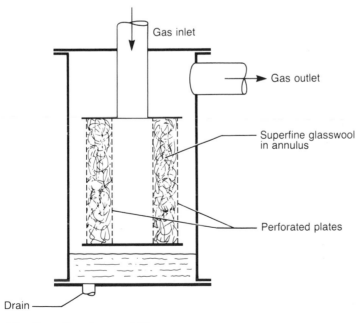

4.6   Superfine glasswool filter

4.1.5 Pressure drop in liquid/vapour separators

The pressure drop in the equipment described in the above section may restrict the performance of the vaporiser. In particular, pressure drop must be allowed for in the design of multiple-effect evaporators (described in section 5.4.5). Here efficient separation is important to avoid scaling of the heating surface of the next effect, but the pressure drop across the separator may significantly reduce the performance of the next effect.

The pressure drop across a liquid/vapour separator increases significantly with increasing liquid loading. It cannot be estimated accurately, but some information is contained in section 9.9.

4.2 Condensers

The vapour from an evaporator (or from the last effect of multiple-effect evaporators) usually passes to a condenser. Condensers are also part of electricity generating, distillation and refrigeration plants. The design of condensers is outside the scope of this book; a comprehensive treatment is being prepared, as part of this series of books. Further information is given by A. C. Mueller in Schlünder (1983). The choice of type of condenser for steam turbines is discussed by Goodall (1980) in his section 11.5. This section deals with the condensers that are specific to evaporators.

The cheapest form of condenser is that involving direct contact between the vapour and the coolant. These may be used only if:

1. The coolant is water.
2. The vapour and the coolant are at the same pressure.
3. There is no objection to mixing the condensate and the coolant.

Spray condensers are often used with evaporators operating under vacuum. They are described by the Heat Exchange Institute (1970). The vapour enters a spray chamber containing several sprays on a ring main. Various arrangements are available, one of the simplest being shown in Fig. 4.7. The sprays must be designed and located so that they produce a well-distributed shower of water (with as little as possible running down the wall of the chamber). As an alternative to spray chambers, direct contact condensation can be achieved in a vessel containing an arrangement of trays designed to produce sheets of falling water, the vapour flowing upwards. The various types of direct contact condenser are described by the Heat Exchange Institute (1970) and by How (1956). When operating under vacuum, condensers may be used with a barometric leg, as shown in Fig. 4.8, to eliminate the need for a vacuum pump to remove the condensate. This requires that the condenser should be placed in a building several floors above the ground. The cooling water and condensed steam are

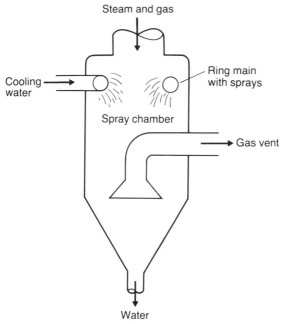

Steam and gas

Cooling water

Ring main with sprays

Spray chamber

Gas vent

Water

4.7 Spray condenser

removed by a downward inclined tail pipe into a hot well on the ground. The vertical drop in the tail pipe must be at least 10 m, so as to be able to maintain a deep vacuum in the body of the spray condenser (10 m of water pulls a vacuum of almost 1 bar). A further advantage of the direct contact vacuum condenser over a surface condenser is that a single inlet pipe may be used, whereas a surface condenser may require a large, expensive inlet manifold.

If surface condensers must be used, it is usual to put the vapour on the shellside. It is economic to use high vapour velocities at inlet, on account of the very low vapour density. It is important to check that the vapour velocity is not so high that the pressure drop would lead to a significant loss in temperature difference. It is also important to check that the high inlet velocity will not induce tube vibrations (see section 11.3). Many cases have been reported of damage to the tubes at the inlet to a condenser due to vibrations.

Adequate facilities must be provided for draining the condensate out of a shellside condenser. The drain connection must be located on the bottom of the shell and must be large enough to avoid flooding of the bottom tubes. The design of drains is dealt with in section 9.12. With direct-contact condensers, bigger drains are needed to remove the large volume of cooling water along with the condensate.

All condensers must be provided with good vents, first for purging during start-up and then for removing incondensable gases, which may have been

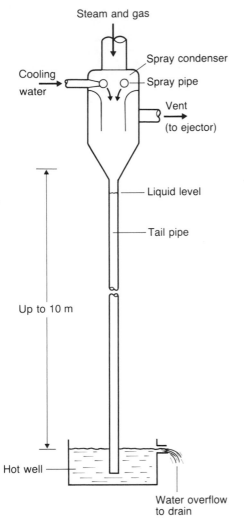

Steam and gas

Spray condenser

Cooling
water

Spray pipe

Vent
(to ejector)

Liquid level

Tail pipe

Up to 10 m

Hot well

Water overflow
to drain

4.8   Barometric leg. The level of water in the tail pipe above the level in the hot
well balances the vacuum in the spray condenser

introduced with the vapour or in solution in the liquid feed. In practice the
amount of gas to be vented may be considerably greater than had been
anticipated during design. It is difficult to estimate the amount of air that
will leak into equipment operating under vacuum. Gases removed from a
condenser are always saturated with vapour. Thus a compromise is needed
between operating with a high vent rate, with consequent high losses of
vapour, and designing for a low vent rate by installing a larger condenser
that will cope with the more arduous thermal duty.

4.3 Vacuum-producing equipment

With a vaporiser operating under vacuum, the vent from the condenser must be connected to vacuum-producing equipment. This usually consists of an ejector of the shape shown in Fig. 4.9. Here the vent enters from the left and is entrained by a jet of high-pressure driving steam entering through a standard steam nozzle. There is a substantial rise in pressure in the convergent mixing length and a further rise in pressure in the divergent diffuser. If a deep vacuum is required, two or three stages of ejection are needed. In such cases, the vent stream must pass through an interstage tubular condenser, so that the driving steam introduced in each ejector is condensed before the vent goes to the next ejector, to reduce the volume to be handled by that ejector.

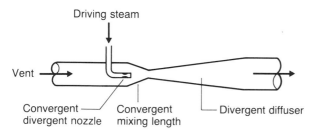

4.9   Steam-driven vacuum ejector

The ejection system must be designed for the maximum amount of gas that might leak into the system, plus the associated vapour. A 'hogging' ejector may also be installed. This has a higher output but lower efficiency than the normal ejector, the two being operated in parallel during start-up to reduce the time required to generate the required vacuum. Once a sufficiently high vacuum has been reached the inefficient hogging ejector is isolated.

For detailed information on the design and operation of ejectors, the reader should consult the manufacturers of this equipment. Valuable information on the generation of vacuum is given by Goodall (1980), section 13.2.

4.4   Desuperheaters

The steam that is used for heating a vaporiser is sometimes available at a pressure considerably greater than that required. The steam must pass through a let-down valve designed to control the pressure at the required lower value. During the process of reducing its pressure, the steam becomes superheated. The controversial question then arises of whether to install a desuperheater. The heat transfer rate depends on the saturation temperature, so is independent of the amount of superheat. The saturation temperature is chosen to give as high a temperature difference as possible without running into danger of excessive fouling, dryout or the decomposition of a heat-sensitive liquid. A decision to install a desuperheater may be taken for one or more of the following three reasons.

1. To reduce the cost of the steam main and vaporiser (a greater temperature requires a greater thickness of the wall).
2. To avoid trouble due to excessive thermal expansion, either of the steam main or of the shell of a tubeside vaporiser with fixed tubesheets.
3. To avoid overheating of the tubes of a tubeside vaporiser in the region of the steam inlet, due to the high local velocity of the steam.

In connection with the second item, the temperature of the wall of an efficiently insulated steam main carrying superheated steam can be calculated from the estimated heat transfer coefficients of the flowing steam and the insulation. However, it is difficult to estimate the mean temperature of a shell because there is no simple method of estimating the fall in steam temperature along the length of the shell; consequently a shell receiving superheated steam should be fitted either with expansion bellows, as shown in Fig. 2 of TEMA (1978), or with a vapour belt in a stepped shell, which is axially extensible, as explained in the next section.

A desuperheater normally consists of a spray nozzle installed in a steam main, as described in section 17.8 of Goodall (1980). It is essential to produce a fine spray that will evaporate in a few metres length of pipe; large drops take many seconds to evaporate and can form a pool running along the bottom of the main, despite the superheat of the steam above the water. Water in a superheated steam main may lead to erosion and thermal fatigue. As such desuperheaters are not expensive compared with the total cost of a vaporisation plant, they are often installed when there is some danger of any of the above troubles.

Some vaporisers normally heated by saturated steam are occasionally heated by superheated steam let down from a higher pressure. If there would be serious consequences of a failure of a spray valve on changing to superheated steam, it may be necessary to use a shell-and-tube desuperheater of the kettle type (see section 3.4.3). The steam to be used for heating passes at all times through the tubes. The shell contains a reservoir of water which will normally be at the saturation temperature and will start to boil immediately the change to superheated steam is made, thus guaranteeing that the steam will never be greatly superheated. The steam outlet from the shell is connected to the outlet pipe from the tubeside.

4.5   Introducing steam into a shell

The tubeside vaporisers shown in Figs 3.6, 3.8, 3.9, 3.10, 3.15 and 3.21 are all heated by steam introduced through a branch attached to the shell. TEMA (1978) requires that an impingement plate, or other means to protect the tube bundle against impinging fluids, shall be provided for all nominally saturated vapours. This is assumed to apply to all uses of steam for heating, because superheated steam may contain droplets of water entrained from a pool on the bottom of the main. If a simple impingement plate is installed opposite the end of the steam inlet branch, the space between the plate and the end of the branch must be sufficient to avoid a high steam velocity that might result in an excessive pressure drop or vibration of the nearby tubes (see section 11.3). Consequently it may not be possible to fill the shell completely with tubes. It is possible to fill the shell completely by introducing the steam through a vapour belt, as shown in Figs 4.10 or 4.11. With a long shell, the extra cost of the vapour belt will be offset by the saving due to being able to get a given number of tubes in

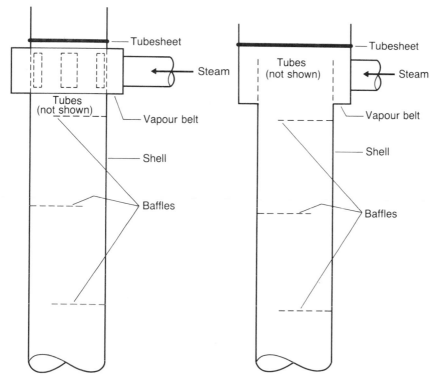

4.10   Vapour belt with stepped shell   **4.11**   Vapour belt with ported shell

a smaller shell. If the steam contains incondensable gas, the vapour belt has the advantage of giving a better distribution across the shell, thus avoiding the danger of high concentrations of gas collecting in stagnant regions. With the type of vapour belt shown in Fig. 4.10, the annular plate that joins the main part of the shell to the upper (larger) part acts as a diaphragm permitting extension of a shell with fixed tubesheets when the steam is introduced. This type has the disadvantage that it requires a large upper tubesheet. The type of vapour belt shown in Fig. 4.11 introduces the steam through ports cut in the shell, so does not require any increase in the diameter of the upper tubesheet, but the extensibility of the shell is lost, and it may be necessary to fit bellows in the shell.

4.6   Elutriator

When it is required to grow large crystals in an evaporator, an elutriator is attached to the base of the evaporator to allow crystals of an adequate size to leave the evaporator and to return smaller crystals so that they may grow further. It consists (see Fig. 3.7) of a vertical pipe through which liquor is pumped upwards at a controlled velocity such that only crystals of the required size can fall through the stream and pass into a collecting vessel below the elutriator.

CHAPTER 5

Selection of equipment

When selecting a vaporiser for use in a process plant on one of the duties listed in Chapter 1, a choice should be made between certain of the types of equipment described in Chapters 3 and 4. In the design of an evaporation plant, there is the additional problem of deciding whether to use multiple effects or vapour recompression; in the former case the optimum number of effects must also be decided upon. Some guidance is given in this chapter on selecting the best equipment for vaporisation.

5.1 Fired boiler

Steam is required in the process industries for one or several of the following uses:

1. To drive a turbine connected to an electric generator.
2. To drive a turbine connected to a compressor or pump.
3. As a process fluid.
4. For process heating.
5. For space heating.

In many plants steam is needed only for the last two of the above uses, its pressure being in the region of 10 bar (1 MPa).

In large plants steam may be needed for all of the above uses. The pressure of the steam generated for large process plants has, over the past 60 years, approached that used in coal-fired electricity-generating power stations, being over 120 bar in some modern plants.

When steam is required for power, it should be superheated, because the blades of steam turbines are liable to erosion by wet steam. The steam is usually wet when it reaches the last stages of the turbine, where the blades must therefore be made of corrosion-resistant material. The steam may be

exhausted direct to vacuum condensers, or it may be exhausted at some
intermediate pressure and used as a source of heat, condensing at pressures
between 1 and 10 bar.

A choice must be made between:

1. A water-tube boiler (see section 3.1.1 and Figs 3.2 and 3.3).
2. A coiled-tube boiler (see section 3.3.1).
3. A fire-tube boiler (see section 3.4.1 and Fig. 3.17).

The choice is determined by the pressure, the heat transfer rate required
and the relative costs.

For generating superheated steam for a small process plant, a package
fire-tube boiler may be less expensive than a package water-tube boiler at
pressures up to 20 bar and heat transfer rates up to 10 MW. Package
water-tube boilers are available for heat transfer rates up to 100 MW and
pressures up to 140 bar. For more arduous duties special designs of water-
tube boiler must be considered.

Package helical coiled-tube boilers may be available for some of the
range covered by package water-tube and fire-tube boilers. The attraction
of the coiled-tube boiler is that it is considerably cheaper than the other
types, but various mechanical troubles have been encountered. However,
with care, these can be overcome.

Many problems in design and operation are avoided in boilers burning
clean fuels but these are expensive, and much experience is available in
burning coal and heavy (residual) fuel oil. With coal and heavy fuel oil,
equipment must be provided for treating the fuel before it is taken to the
burners and for cleaning the effluent gas before it goes up the stack to
atmosphere. Furthermore, facilities must be provided for cleaning the heat
transfer surfaces. The design of such equipment is outside the scope of this
book. Another very important item of equipment in any boiler plant is the
water treatment plant, which is needed to prevent corrosion or fouling of
the heat transfer surfaces by the boiling water, and which is discussed in
Chapter 14.

5.2   Waste heat boiler

Many of the chemical reactions that take place in the process industries
are economical only when carried out at a high temperature; heat is
generated in some of these reactions (exothermic reactions). Thus waste
heat is available both in the flue gases from the preheating furnaces and in
the products leaving the reaction vessels. This heat may be recovered in
heat interchangers designed to preheat the incoming reactants or the air
supplied to the furnaces. Alternatively the heat may be used to generate
steam in a waste heat boiler. According to the temperature of the gas and the
requirements of the plant, the steam will be generated at a pressure from 10

to 150 bar. This reduces the amount of steam that must be generated in fired boilers. The operation of such plants becomes very complicated. Some generate more steam than they need, so become exporters of steam. Fired boilers are usually needed for the start-up of such plants, before their waste heat boilers are operational.

If several boilers (waste heat or fired) are operating at the same pressure, the steam from them may be taken to a common steam drum. If superheated steam is required, the steam from the drum may be heated by waste heat from the process gas or from flue gas or by direct firing.

A choice must be made between:

1. A vertical water-tube boiler in a horizontal duct (see section 3.1.1 and Fig. 3.1).
2. A bayonet-type boiler (see section 3.1.1 and Fig. 3.4).
3. A vertical U-tube boiler (see section 3.1.1 and Fig. 3.5).
4. A horizontal serpentine coil (see section 3.2.1 and Fig. 3.13).
5. A horizontal U-tube boiler in a vertical duct (see section 3.2.1 and Fig. 3.14).
6. A fire-tube boiler (see section 3.4.1 and Fig. 3.16).

The vertical water-tube boiler (1) is ideal for extracting heat from a flue gas, because it can easily be placed in the flue-gas duct, it operates satisfactorily on natural circulation and there is little danger of dryout, because the rate of heat transfer is limited by the low heat transfer coefficient of gas at atmospheric pressure and dryout heat fluxes are very high with vertical tubes. However, it is not suitable for a gas at high pressure, because of the difficulty of fitting it into a suitable duct. For a high pressure gas, the bayonet-type and the vertical U-tube boilers permit the use of a more convenient shape of container for vertical boiler tubes. However, there is a danger of dryout occurring at the lowest point of the water/steam path, namely the entry to the outer tubes of a bayonet-type (Fig. 3.4) or the 180° bends of the U-tube boiler (Fig. 3.5). The situation is made worse if debris collects at these low points during a shut-down or when operating the plant at a low output. It is possible to operate bayonet-type boilers on natural circulation, whereas forced circulation is recommended for U-tubes to ensure that circulation starts promptly on start-up.

With horizontal tubes, types (4) and (5), the main problem is to avoid dryout, as discussed in section 3.2.1; furthermore boilers of this type usually require forced circulation. However, a very neat arrangement can be used when the source of heat is an exothermic gas-phase reaction; the boiler tubes may be placed in a vertical duct located immediately underneath the reaction vessel, the gas flowing down through the catalyst and straight into the waste heat boiler.

The fire-tube boiler (6) is of a convenient shape for handling a gas at high pressure and it operates well on natural circulation. Compared with the types described above, it has the disadvantages that there is a tubesheet in the region of high temperature and there is little facility for differential expansion between tubes and shell. It must therefore be chosen only when there is no danger of appreciable shellside fouling or dryout. Detailed points about the design of this type are discussed in section 3.4.1.

Many operational troubles have been experienced with waste heat boilers in the process industries. In the past, some fire-tube waste heat boilers have been unreliable on account of failures of the tube-to-tubesheet joints at the hot end, due to overheating; this is now overcome by the protective treatment shown in Fig. 3.16, namely the refractory lining and the ferrules. Recently troubles have usually been traced to poor water treatment, or to dryout, or to a combination of the two; sometimes it was not clear which was to blame. Hinchley (1977) summarises the problems that have occurred in the waste heat boilers on steam reforming plants and the solutions adopted, stressing the serious financial consequences of a single failure. It is important to make sure during design that dryout cannot occur under any circumstances. During operation special care must be taken in water treatment and in avoiding the return of contaminated condensate to the boiler.

5.3   Reboiler

Distillation columns are used for the separation of chemical components. They need a source of heat at their base to vaporise and return to the column some of the liquid that collects in the bottom of the column; they need a condenser to reflux some of the vapour that reaches the top. This section deals with the choice of a suitable heat exchanger to be the reboiler at the base of the column. Further advice is given by J. W. Palen in section 3.6.1 of Schlünder (1983), leading to the following conclusions.

A choice must be made between:

1.   An internal reboiler (see section 3.4.4 and Fig. 5.1, or sections 3.1.7 or 3.1.8 for a plate exchanger).
2.   A vertical thermosyphon reboiler (see section 3.1.6 and Fig. 3.10, or section 3.1.7 for a spiral-plate type).
3.   A horizontal thermosyphon reboiler (see section 3.4.2 and Fig. 3.18).
4.   A kettle reboiler (see section 3.4.3 and Fig. 3.19).
5.   A forced-flow reboiler (see Fig. 5.2).

If the process fluid is very viscous, a forced-flow reboiler should be used. The heater should preferably take the form of a gasketed plate heat exchanger, as described in section 3.6. If, for some reason, a shell-and-tube

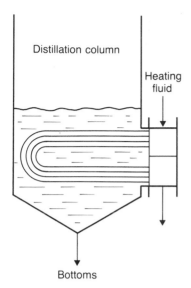

5.1 Internal reboiler

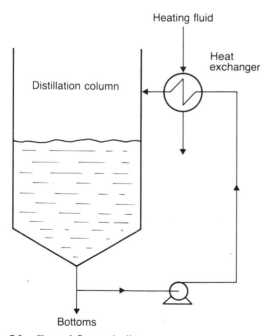

5.2 Forced-flow reboiler

exchanger is preferred, the viscous process fluid should be on the shellside (very low heat transfer coefficients are obtained with viscous flow inside long tubes). A restriction may be inserted in the return line to suppress boiling in the heater.

Similarly, if the process fluid is very dirty, the recommended type is a forced-flow reboiler, preferably in the form of a gasketed plate heat exchanger, as this can easily be dismantled for cleaning. If a shell-and-tube exchanger is used, it must be designed for easy cleaning.

For clean or not very dirty fluids of low to moderate viscosity, the choice is between the first four types listed above. The internal reboiler, sketched in Fig. 5.1, being the cheapest, should be considered first. If the space inside the reservoir of liquid is not adequate to accommodate the required amount of heating surface, the vertical thermosyphon is likely to be the best. As a result of the high velocity, fouling is reduced. Any dirt deposited inside the tubes is comparatively easy to remove. It is the best type for multi-component liquids with a wide range of boiling temperatures. Using a single-phase heating fluid, a pure countercurrent arrangement gives a favourable temperature profile. If headroom is limited, it may be difficult to design a vertical thermosyphon reboiler without exceeding the specified maximum length of tubes; under such circumstances, a horizontal might be preferable to a vertical thermosyphon. However, the horizontal thermosyphon reboiler is more prone to fouling or corrosion than the vertical, due to the danger of the formation of stagnant regions inside the shell.

The vertical thermosyphon reboiler is not suitable for operation under deep vacuum because there is a big variation of boiling temperature with depth, as explained in section 3.1. Similarly, with a small temperature difference at moderate pressure, the rise in temperature of the liquid before it starts to boil may produce a significant reduction in the mean temperature difference. In such circumstances a kettle reboiler may be preferable. At near-critical temperatures the difference between the densities of liquid and vapour is not sufficient to produce an adequate circulation in a thermosyphon, so a kettle reboiler should be used. Kettle reboilers are the most susceptible to fouling. Non-volatile materials tend to collect in the shell unless an adequate purge is maintained. There is a loss in effective temperature difference with wide-boiling-range liquids due to mixing in the shell.

5.4   Evaporator and vapour generator

Advice on the choice of an evaporator is given by Schlünder (1983), section 3.5.5, by Chilton and Perry (1973), pp. 11–27 to 11–31 and by Kern (1950), Chapters 14 and 15.

5.4.1   Vapour generator for revaporisation of a liquefied gas

For revaporising a gas that has been liquefied for storage or transportation, the choice is between:

1. A direct contact heating of a heat transfer fluid in a submerged combustion vapour generator (see section 3.7.2 and Fig. 3.27).
2. A fired tubular vapour generator with horizontal tubes (see section 3.2.2).
3. A once-through helical coil (see section 3.3.2 and Fig. 5.3).
4. A once-through short-tube vertical vaporiser (see section 3.1.2 and Fig. 5.3).
5. A kettle type (see section 3.4.6).
6. A bayonet-tube vaporiser (see section 3.4.7).
7. A double-bundle vaporiser (see section 3.8 and Fig. 3.28).

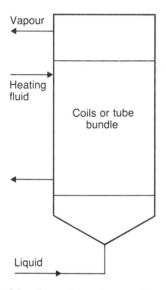

5.3 Once-through vaporiser

Type (1), (2) or (3) must be chosen when the only source of heat available is the combustion of a fuel. Type (3), (4), (5), (6) or (7) may be chosen when steam or a hot process stream is available for supplying heat.

The simplest device is a once-through arrangement type (3) or (4) in which the liquid flows inside heated tubes in the form of coils or a vertical bundle. The difficulties with this are that the level of liquid must be controlled so that the vapour is slightly superheated at outlet, where there is nevertheless still some mist present. The rate of heat transfer is considerably less in the mist flow region than in the region of wet-wall boiling. Also it is difficult to determine just where mist flow begins.

These troubles may be overcome by having shellside boiling, namely a kettle type or bayonet tubes, type (5) or (6). Furthermore the danger of unstable flow is eliminated. Kettle vapour generators are not popular for toxic or flammable liquids on account of the hazard of having a large volume of liquid in the shell. Bayonet-tube vaporisers have a smaller capacity of liquid, but they are more expensive due to the need for internal tubes that contribute nothing to the transfer of heat. The double-bundle vaporiser is expensive because two heat transfer surfaces are needed. Its advantages are as described in section 3.8.

5.4.2   Vapour generator for obtaining a pure vapour from an impure liquid

When a vaporiser is required to generate a pure vapour from a liquid that contains unwanted dissolved solids, as in a desalination plant, the choice is between:

1.   A short-tube vertical evaporator, with a purge at the bottom in place of the concentrate outlet (see section 3.1.2 and Fig. 3.6).
2.   A forced-flow flash vaporiser (see Fig. 5.4).

The short-tube vertical evaporator requires less heating surface area, and is therefore cheaper than the flash system, because the liquid is heated more with the latter and this reduces the temperature difference. However, the

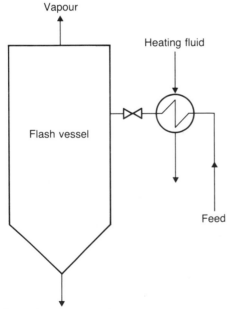

5.4   Forced-flow flash vaporiser or concentrator

forced-flow flash vaporiser, preferably with a gasketed plate heater, is used with dirty or viscous liquids; the remarks in section 5.3 on forced-flow reboilers also apply to forced-flow vaporisers.

5.4.3 Concentrating evaporator

When equipment is required to produce a concentrate from a dilute solution, a choice must be made between:

1. A submerged-combustion concentrating evaporator (see section 3.7.1 and Fig. 3.26).
2. A long-tube vertical evaporator, with natural circulation (see section 3.1.4 and Fig. 3.8 but without the pump).
3. A climbing-film evaporator (see section 3.1.5 and Fig. 3.9).
4. A falling-film evaporator (see section 3.5.1 and Fig. 3.21).
5. A long-tube vertical evaporator with forced circulation (see section 3.1.4 and Fig. 3.8).
6. A forced-flow flash evaporator (see sections 3.2.3 and 3.6 and Figs 3.15 and 5.4).

The first type is the cheapest, but is liable to the limitations due to direct contact heating, as described in section 3.7.1.

The long-tube vertical evaporator, with natural circulation, is the simplest and is the preferred type of tubular heat exchanger, except under the circumstances described below.

If the temperature rise in the non-boiling zone of a long-tube vertical natural circulation evaporator is appreciable compared with the overall temperature difference, the reduction in overall temperature difference may necessitate an uneconomical increase in heating surface; if so, it is preferable to use forced circulation or to use a climbing-film evaporator.

With foaming liquids, the climbing-film evaporator is best. With heat-sensitive liquids, either the climbing-film or the falling-film type should be used, so that the heat is supplied in a short time and the temperature of the heating surface does not rise much above the boiling temperature of the liquid. The temperature may be reduced by operating under vacuum or by introducing a gas into a falling-film evaporator, as described in section 3.5.1.

If a certain amount of fouling is expected, forced circulation should be used with a long-tube vertical evaporator, the forced circulation giving a high velocity and thus reducing the amount of fouling. With very dirty liquids, or with viscous liquids, a forced-flow flash evaporator should be used; the remarks in section 5.3 on forced-flow reboilers also apply to forced-flow concentrating evaporators.

5.4.4 Crystallising evaporator

The choice of the best type of evaporator for crystallisation depends very much on the nature of the crystals to be produced. The following types are available.

1. A short-tube vertical, with assisted circulation (see section 3.1.2 and Fig. 3.6 but with a device for removing crystal slurry in place of the concentrate outlet).
2. A short-tube vertical, with natural circulation (as above but without the axial-flow pump in the central downcomer).
3. A basket type (see section 3.1.3 and Fig. 3.7).
4. A long-tube vertical, with forced circulation (see Fig. 5.5).
5. A forced-flow flash evaporator (see Fig. 5.6).

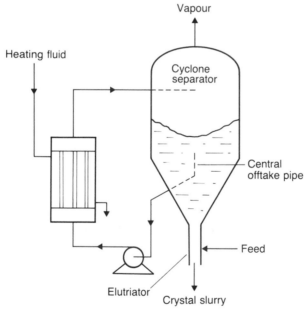

5.5 Long-tube vertical crystallising evaporator

In the first three types, small crystals are recirculated with the liquid passing up the tubes. The crystals grow as evaporation proceeds until they are large enough to fall to the bottom of the evaporator and not be recirculated. In the last type the crystals form in the flash vessel as the vapour is released. In long-tube vertical evaporators crystal growth occurs both in the tubes and in the separator. When there is crystallisation inside the tubes, large tubes are used, ranging in diameter from 40 to 80 mm.

The first type involves the pumping of liquid containing crystals; thus it

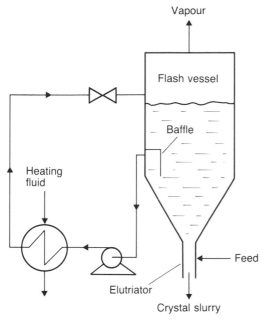

5.6   Flash crystallising evaporator

must not be used if the pump is liable to break the crystals or if the crystals are liable to abrade the pump. Under such circumstances it is better to use natural circulation, i.e. the second or third type. Alternatively, in the last two types, the pumps are handling a liquid with a low concentration of small crystals, so abrasion will be less likely.

In a crystallising evaporator, the size of the crystals is determined by the residence time; thus control is important and unstable flow must be avoided (see section 9.2). With natural circulation it is necessary for the velocity up the tubes to be sufficiently great, even when the output is reduced, for large crystals to be carried upwards and grow to the required size.

Short-tube and basket-type evaporators were frequently used in the past, but the long-tube vertical evaporator is the preferred type at present for many duties. However, the choice is very dependent on the method to be adopted for growing the crystals, and this is outside the scope of this book. Types (2) and (3) depend on natural circulation, so must be operated always at full throughput to obtain crystals of the required size.

The basket type is liable to unsteady flow, which also affects crystal size.

5.4.5   Multiple-effect evaporation

In multiple-effect evaporation, the steam that is generated in the evaporator that constitutes the first effect is used as the heating medium for the evaporator that is the second effect, which operates at a lower pressure; similarly the steam generated in the second effect heats the third effect, and so on. The steam generated in the last effect passes to the condenser. The consumption of steam is reduced by a factor approximately equal to the number of effects, but the surface area of each evaporator must be increased by a similar factor to compensate for the reduced temperature difference. Thus it is possible to determine an optimum number of effects by comparing the saving in fuel costs with the extra capital charges for operation in single-effect, double-effect, triple-effect, etc. Unfortunately the amount of steam generated in each effect is somewhat less than the amount generated in the previous effect, for two reasons: a small proportion of the heat load is devoted to preheating the feed to its boiling temperature; the specific latent heat of vaporisation increases with decreasing temperature. Consequently the amount of steam generated in an evaporator is less than the amount of condensing steam used for heating. Furthermore, in multiple-effect evaporation, it is necessary to provide extra surface area to compensate for the pressure drop of the vapour as it flows from one effect to the next. There is also a loss of temperature difference in concentrating and crystallising evaporators due to the elevation of the boiling temperature and the consequent generation of superheated vapour, which condenses at its saturation temperature in the next effect or in the condenser; this worsens the problem of designing a multiple-effect evaporator plant that will achieve a high heat transfer coefficient with a low temperature difference.

There are three principal systems for feeding the liquor to multiple-effect evaporators, known as 'forward feed', 'backward feed' and 'parallel feed'. The details of these systems and their relative merits are best understood by studying the example of triple-effect evaporation shown in the line diagrams and the temperature diagrams in Fig. 5.7.

For comparison the line diagram in Fig. 5.7 (a) shows three evaporators operating in single-effect; thus the feed enters the three shells through parallel paths and the steam enters the heating coils also through parallel paths. Line diagrams for each system of multiple-effect operation are shown in Figs 5.7 (b), (c) and (d). Underneath each line diagram is a temperature diagram in which the ordinate is temperature and the abscissa is distance along the heating coil. The effects are numbered 1 to 3 from left to right. The upper line in each diagram is the temperature of the steam that is being used for heating that effect and the lower line is the temperature of the liquid that is being evaporated. The saturation temperature of the steam supplied is denoted by T_s and of the feed by T_f. Temperature diagrams in Fig. 5.7 (a) are identical for each evaporator, with a large temperature difference, $T_s - T_f$, in each.

For each of the triple-effect systems, steam flows from left to right in the line diagrams (b), (c) and (d) of Fig. 5.7, the steam generated in the shell of the first effect passing to the coil of the second effect at a temperature and a

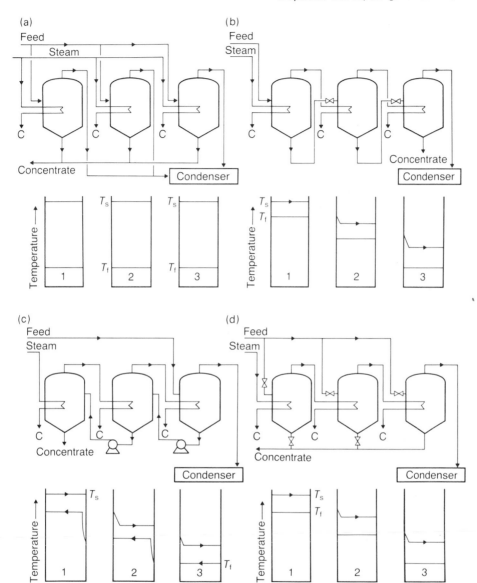

5.7 Examples of multiple-effect evaporation – four different ways of arranging
three evaporators. On each line diagram, C represents the outlet for
condensate and incondensables. The base of each temperature diagram is the
temperature in the condenser: the lower line is the temperature of the boiling
liquid, and the upper line is the temperature of the condensing steam
(a) three single-effect evaporators
(b) triple-effect, forward feed
(c) triple-effect, backward feed
(d) triple-effect, parallel feed

pressure equal to those of the liquid in the first effect. The amount of superheat depends on the concentration of dissolved solids. The same occurs from the second to the third effect. Thus the consumption of steam is only about a third of that in single-effect operation.

With forward feed the liquid flows in the same direction as the steam and passes through interstage let-down valves, and its temperature is reduced by flashing, thereby giving an adequate temperature difference in each effect. With this method a high feed temperature is needed.

With backward feed the flow of liquor is countercurrent to the steam, so interstage pumps are needed, as shown in Fig. 5.7 (c). Sensible heat transfer to the liquor is needed in the shells of the second and first effects, to bring it up to the higher boiling temperature in the new shell, where the pressure is higher. This arrangement is best for concentrating a feed that is cold or viscous, or if there is an appreciable elevation of boiling temperature with increasing concentration. However, it is not suitable if the maintenance of the interstage pumps would be difficult or if they might cause attrition of crystals in a crystallising evaporator.

In parallel feed, shown in Fig. 5.7 (d), fresh liquor is fed to each effect through parallel paths, but the pressure drop across the control valves is greater in the later effects because the operating pressure is less in these effects than in the first; there is a consequent reduction in temperature by flashing. Parallel feed is used when the concentration in each effect must be the same, or in a crystallising evaporator it may be used if several different feeds must be treated separately, but this requires separate offtakes for the different concentrates.

For further information see Chilton and Perry (1973), pp. 11–33.

5.4.6 Vapour recompression

As an alternative to multiple-effect operation, vapour recompression may be used to save fuel. The vapour generated in each evaporator is washed and recompressed and used as the main source of heating steam; a separate supply of steam is needed for start-up and to make up for the fact that less steam is evaporated than is condensed (as explained in the previous section). A simple arrangement is shown in Fig. 5.8. Alternatively steam injectors may be used to boost the vapour (in place of mechanical compression).

Vapour recompression is used much less frequently than multiple-effect operation, because mechanical compressors are expensive to buy, run and maintain and steam injectors are inefficient. It was developed for countries where hydroelectric power is cheap and plentiful. Also it may be used when it is necessary to operate over a restricted temperature range, thus limiting the number of effects that could be used. Alternatively if the steam supply is

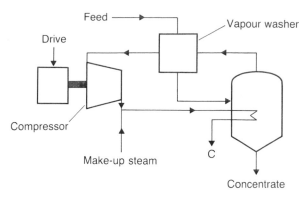

5.8   Evaporator using vapour recompression

at a pressure considerably higher than that required for heating, the steam
can be expanded in a turbine which is used to drive the compressor.

5.5   Chiller

For refrigeration duties where it is required to cool a fluid stream to
temperatures between about $-30\,°C$ and $0\,°C$, chillers are often horizontal
thermosyphons with the refrigerant boiling in the shell and the process fluid
flowing through the tubes in a single pass. If it is required to design a
chiller to cool the process fluid to within a few degrees of the boiling
temperature of a refrigerant, the benefits of nucleate boiling are likely to be
lost. It should therefore be arranged for the velocity of the refrigerant to be
high in this region, so that the convective component of the boiling
coefficient is high. This can be achieved by having co-current flow in a
TEMA E shell or by using a TEMA J shell with central inlet, so that the
process fluid leaves the tubes in a region where the quality of the refrigerant
is a maximum, i.e. where it leaves the shell.

    An alternative is to use a coiled-tube heat exchanger, as described in
section 3.4.5, and illustrated in Fig. 3.20. The boiling refrigerant flows up
through the shell. It may be advantageous for the process fluid also to flow
upwards, so that the closest approach is obtained at the top of the heat
exchanger where the turbulence on the shellside will be at a maximum and
compensate for the small temperature difference. This would not be
practicable in a gas cooler with a substantial amount of condensing
vapours, due to the tendency of the condensate to run backwards where the
shear of the gas is inadequate. In such cases it might be better to have the
gas and vapour flowing downwards through the tubes and for the
refrigerant to be introduced at the top as a falling film flowing down the

outside of the tubes (high coefficients are achieved with falling films with low temperature difference).

The complex heat exchangers used in the cryogenics industry for the liquefaction of air or natural gas are usually plate-fin heat exchangers. In many applications there are more than two streams; consequently the thermal calculations are very complex.

CHAPTER 6

Introduction to rating calculations

Chapters 6 to 10 describe the calculations in heat transfer and fluid flow that are required when designing a vaporiser, checking a design or analysing the results of tests on a vaporiser in operation. Basically the problem is to be able to predict the heat flux at any location in the heat exchanger. This is denoted by $\dot{q}$ and is defined as the ratio $\delta\dot{Q}/\delta A$, where $\delta\dot{Q}$ is the heat transfer rate (W) through a small element of the heated surface of area δA (m^2); thus $\dot{q}$ has the dimensions W/m^2.

The overall temperature difference required to produce a given heat flux is the difference between the local bulk temperature of the heating fluid and that of the vaporising fluid. It is equal to the sum of the temperature differences across some or all of five thermal resistances:

1. Heating fluid.
2. Dirt deposited by the heating fluid.
3. The wall of the exchanger, including any secondary surface (fins).
4. Dirt deposited by the vaporising fluid.
5. Vaporising fluid.

In a direct contact vaporiser, only the fluid resistances (1) and (5) are encountered. This book is mainly concerned with vaporisers in which the heating and the vaporising fluids are separated by a solid surface, usually in the form of metal tubes, alternatively in the form of plates; it is then necessary to be able to estimate the thermal resistance of the surface and any dirt deposited on it as a result of fouling. Account must also be taken of the thermal resistance of any permanent protective coating deposited on the surface during manufacture to prevent corrosion.

The estimation of the thermal resistance in non-vaporising situations is dealt with in this chapter and in vaporising situations in the next chapter. The various calculations that may be necessary in fluid flow are described in Chapter 9.

6.1 Heat transfer coefficients

When heat is transferred by conduction or forced convection without any change of phase, the required temperature difference is proportional to the heat flux. A useful concept in the design of heat exchangers is the heat transfer coefficient. The individual heat transfer coefficient, α, for any of the thermal resistances defined above, is given by:

$$\alpha = \dot{q}/\theta \qquad [6.1]$$

where $\dot{q}$ is the heat flux (W/m^2) and θ is the temperature difference (K) across an individual thermal resistance. Thus α has the dimensions W/m^2 K. The surface area (A) used in the estimation of $\dot{q}$ and α must always be defined.

The five individual heat transfer coefficients and their surface areas are defined by subscripts: v refers to the vaporising fluid, h to the heating fluid, d to the dirt and w to the wall. When plain tubes are used, the ratio of external to internal areas is equal to the ratio of the external to the internal diameters of the tubes. Surface area may be enhanced by the use of secondary surface, e.g. by fins wound round the outside of the tubes.

The local overall temperature difference, denoted by ΔT, is equal to $T_h - T_v$, where T_h is the local temperature of the heating fluid and T_v that of the fluid to be vaporised (K). Thus ΔT is the sum of the five values of θ for the five thermal resistances listed above. The overall heat transfer coefficient, U, is defined by

$$U = \frac{\dot{q}}{T_h - T_v} \qquad [6.2]$$

To determine the value of U from the values of α, the total temperature difference ($T_h - T_v$) determined from equation [6.2], can be equated to the sum of the five values of θ from equation [6.1]. The heat flux, $\dot{q}$, may not be the same for all the resistances, because the surface areas may not all be the same. Referring U to the surface area in contact with the vaporising fluid gives the following equation for calculating U:

$$\frac{1}{U} = \frac{1}{\alpha_v} + \frac{1}{\alpha_{dv}} + \frac{1}{\alpha_w}\frac{A_v}{A_w} + \left[\frac{1}{\alpha_{dh}} + \frac{1}{\alpha_h}\right]\frac{A_v}{A_h} \qquad [6.3]$$

where A_v is the total surface area in contact with the vaporising fluid (m^2), A_h is the total surface area in contact with the heating fluid, and A_w is the mean total surface area of the wall separating the two fluids. The coefficients for the two fluid streams, and their dirt, are referred to the surface area in contact with the fluid. Thus α_v and α_{dv} are calculated with reference to the surface area A_v; similarly α_h and α_{dh} are calculated with reference to A_h; α_w is referred to A_w.

Normally the assumption that the various values of α, defined in equation

[6.1], are *independent* of the temperature difference (θ) is valid, but *this is not true when there is nucleate boiling*. It follows from equation [6.1] that α_v is the slope of a graph of $\dot{q}$ plotted against θ_v. Thus Fig. 2.1 shows that α_v increases with increasing θ_v from the onset of nucleate boiling at the point B until the region C is approached, where film boiling begins and $\dot{q}$ reaches a maximum. Similarly in condensation, α_h is dependent on θ_h, but to a lesser extent. The values of α due to thermal conduction and forced convection are independent of θ. Hence in the design of a single-phase heat exchanger, the value of U calculated from equation [6.3] may be regarded as a constant, but the design of a two-phase heat exchanger may be complicated by the need to employ trial-and-error in the determination of local values of U.

The total surface area (A) of a vaporiser required to give a specified heat transfer rate of $\dot{Q}$ is obtained by integrating the reciprocal of the local heat flux, which is equal to $U(T_h - T_v)$; this must be expressed as a function of $\dot{Q}_x$, the heat transfer rate from the inlet to each of the various locations. Hence:

$$A = \int_0^{\dot{Q}} \frac{d\dot{Q}_x}{U(T_h - T_v)} \tag{6.4}$$

In single-phase heat exchangers, U may be assumed to be constant and T_h and T_v can be expressed as functions of $\dot{Q}_x$; hence analytical solutions to the integral are available, as given in section 10.1. When boiling is taking place it is necessary to define several locations in the boiling region by the quality and calculate the corresponding values of T_h, T_v, α_v and U. The integral in equation [6.4] must then be solved as described in section 10.3.

6.2   Heat transfer coefficient for a single-phase heating fluid

If the source of heat for a vaporiser is a hot liquid, or a hot gas that does not condense, the heat coefficient for the heating fluid (α_h in equation [6.3]) can be determined by the method described below for the geometry chosen; the arrangements dealt with are:

1. Flow inside plain straight tubes.
2. Flow inside coiled tubes.
3. Longitudinal flow outside tube bundles.
4. Longitudinal flow in annular spaces.
5. Flow across banks of straight tubes.
6. Flow across bundles of straight tubes in shell-and-tube exchangers.
7. Flow across banks of coiled tubes.
8. Flow between plates.

The method adopted in this section is to give simple approximate correlations for forced convection, neglecting any effects of free convection. The reader is given references to be consulted if a more accurate estimate of the heat transfer coefficient of the heating fluid is required. Sections 6.2.1 to 8 deal with the eight different geometries considered.

In the estimation of α_h, an important quantity is the *mass flux*, denoted by the symbol $\dot{m}$ and defined as the mass flow rate $(\dot{M})$ divided by the cross-sectional area for flow (S). Thus $\dot{m}$ has the dimensions kg/s m^2. Alternatively, the mass flux is the product of the fluid density (ρ) and the mean velocity of the fluid $(\bar{u})$.

If the source of heat is a hot gas containing vapour that will partially condense as it cools, the first step in estimating the heat transfer coefficient for the heating fluid is to determine the dry gas coefficient, assuming no condensation, using the appropriate method of calculation for the geometry chosen. Section 6.2.9 should then be consulted for the determination of the enhancement due to condensation.

Similarly, when estimating the heat transfer coefficient in two-phase convective boiling, it is first necessary to estimate the single-phase convective coefficient, as explained in section 7.2. The correlations given in this section may be used for these preliminaries to the calculation of two-phase convective boiling heat transfer coefficients.

When heat and mass are being exchanged between a flowing fluid and a stationary boundary, the rate of transfer increases as the shear stress between the fluid and the boundary is increased. The similarity between the rates of heat transfer and of mass transfer in forced convection was first discussed by Reynolds (1874). The so-called Reynolds analogy is discussed in many textbooks on heat transfer, e.g. McAdams (1954), pp. 208–10, and Coulson and Richardson (1977), section 10.4. It is shown in the latter that, for turbulent flow past a rough surface, the heat transfer coefficient is given by:

$$\frac{\alpha}{\dot{m}c_p} = \frac{\tau}{\dot{m}^2/\rho} \qquad [6.5]$$

where c_p is the specific heat capacity of the fluid (J/kg K), $\dot{m}$ the mass flux (kg/s m^2), α the heat transfer coefficient (W/m^2 K) and τ the shear stress at the surface (N/m^2). The estimation of the last is discussed in section 9.1.

Coulson and Richardson (1977), in their section 10.4, also summarise the efforts that have been made to improve on this simple analogy by allowing for the thermal resistance of the laminar sub-layer of the boundary layer and to predict the shear stress from measurements of the velocity profile. However, it has not been possible to produce a theoretical method of prediction. In practical applications, an empirical relationship is established, for the geometry under consideration, expressing Nu, the Nusselt number, as a function of Re, the Reynolds number, and Pr, the Prandtl number;

these dimensionless parameters are defined as:

$$Nu = \frac{\alpha d}{\lambda} \qquad\qquad [6.6]$$

$$Pr = \frac{\eta c_p}{\lambda} \qquad\qquad [6.7]$$

$$Re = \frac{\dot{m} d}{\eta} \qquad\qquad [6.8]$$

where d is the characteristic dimension of the arrangement (m), as defined in the appropriate sub-section, η is the dynamic viscosity of the fluid (N s/m^2, i.e. kg/s m), and λ is the thermal conductivity of the fluid (W/m K). The relationships given below are for surfaces of normal commercial roughness; increasing the roughness of the surface gives some increase in the Nusselt number, but this is usually neglected in design.

The author wishes to express his gratitude to Engineering Sciences Data Unit (ESDU) of 251–9 Regent Street, London, for permission to publish from their Data Items the following equations: [6.9] to [6.12], [6.19] and [6.20].

6.2.1 Convective heat transfer inside plain straight tubes

Based on an extensive literature survey, ESDU (1967) recommend the following correlation for the mean heat transfer coefficient in turbulent flow inside a straight tube of normal surface roughness, for use when $4000 < Re < 10^6$, $0.4 < Pr < 300$ and $L > 40f$, where L is the total length of the tube.

$$\ln Nu = -3.796 + 0.795 \ln Re + 0.495 \ln Pr - 0.0225 (\ln Pr)^2$$

or $\qquad\qquad\qquad\qquad\qquad\qquad\qquad\qquad\qquad\qquad\qquad$ [6.9]

$$Nu = 0.022\,46\, Re^{0.795}\, Pr^{(0.495 - 0.0225 \ln Pr)}$$

The characteristic dimension, d, for substitution in equation [6.6] and [6.8], is the internal diameter of the tube. Table A1 of ESDU (1967) gives the errors in equation [6.9] over the range of Re and Pr for which it was devised. Overall it was found to correlate with the experimental data with an overall r.m.s. error of 10.2%.

For ease of application to design calculations, equation [6.9] may be rearranged so that all the physical properties are combined in one term ϕ_{fci}, a physical properties parameter for turbulent flow inside tubes. This can be determined from the appropriate physical properties at the operating temperature. Hence, α_{fci}, the heat transfer coefficient for a gas or a nonmetallic liquid in turbulent flow through a tube is given by:

$$\alpha_{fci} = \phi_{fci} \frac{\dot{m}^{0.795}}{d^{0.205}} \qquad [6.10]$$

where $\phi_{fci} = 0.022\,46 \frac{\lambda}{\eta^{0.795}} \mathrm{Pr}^{(0.495 - 0.0225 \text{ in Pr})}$ \qquad [6.11]

A few typical values of ϕ_{fci} at various temperatures are given in Table 6.1 for several liquids and in Table 6.2 for some gases.

Methods of predicting the convective heat transfer coefficient in laminar flow (Re < 2000) and transitional flow (2000 < Re < 4000) are given by ESDU (1968a), which also deals with the minor effects of free convection.

At entry to a tube, the heat transfer coefficient exceeds the value calculated from equation [6.9] by a factor of two or more, but after about 10 diameters it falls to a constant value, i.e. the value given by the equation. Information on entry enhancement is given by ESDU [1968b].

Equation [6.9] is based on experimental data where there are no significant variations of physical properties, either radially or longitudinally. It follows from ESDU (1967) and ESDU (1968b) that when the heating

6.1 Internal forced convection properties parameter for some liquids – from equation [6.11]

Liquid	ϕ_{fci} at a temperature (°C) of						
	− 30	0	50	100	150	200	300
Ammonia	13.0	14.0	15.1	—	—	—	—
Diphenyl*	—	—	—	—	—	3.30	4.14
Methanol	—	—	4.81	6.15	—	—	—
Refrigerant R-12	2.07	2.14	2.14	—	—	—	—
Toluene	—	—	3.04	3.50	3.90	—	—
Water	—	—	10.2	13.3	15.5	17.0	—
Sea-water concentrated to 150 g/kg	—	4.8	8.1	10.7	12.5	—	—

* Eutectic mixture of 73.5% diphenyl ether and 26.5% diphenyl

6.2 Internal forced convection properties parameter for some gases – from equation [6.11]

Gas	ϕ_{fci} at a temperature (°C) of					
	0	200	400	600	800	1000
Nitrogen	2.89	3.22	3.49	3.71	3.87	3.98
Superheated steam	—	4.66	5.48	6.28	7.05	7.74
Saturated steam	—	6.12	—	—	—	—
Methane	5.18	7.52	9.42	10.98	12.20	13.08

fluid is a gas, the physical properties should be evaluated at the mean bulk temperature of the gas and the effect of the lower wall temperature may be neglected. If the hot gas contains polyatomic molecules, e.g. water or carbon dioxide, there may be an appreciable amount of radiation from the gas to the wall, especially at high pressure – see section 6.5.

When heating or cooling a viscous liquid, the heat transfer coefficient may differ appreciably from that calculated from equation [6.9], due to radial variations in viscosity. When heating a liquid, the viscosity at the wall is less than in the bulk, leading to an increase in heat transfer; conversely when cooling a liquid, the higher viscosity at the wall leads to a reduction in heat transfer. It is recommended by ESDU (1968b) that the values calculated from equation (6.9), using properties at the bulk temperature, should be corrected by multiplying the coefficient by a correction factor dependent on the ratio of the dynamic viscosity at the bulk temperature to that at the wall temperature, provided that $4000 < \mathrm{Re} < 4.6 \times 10^5$, $2 < \mathrm{Pr} < 157$ and $0.072 < \eta_b/\eta_w < 12.7$.

$$\left.\begin{array}{l} \text{When cooling a liquid, } \alpha = \alpha_b(\eta_b/\eta_w)^{0.30} \\[2mm] \text{When heating a liquid, } \alpha = \alpha_b(\eta_b/\eta_w)^{0.18} \end{array}\right\} \qquad [6.12]$$

where α is the corrected heat transfer coefficient, α_b is that calculated from the properties at the mean bulk temperature, η_b is the viscosity of the liquid at the mean bulk temperature and η_w is the viscosity of the liquid at the mean wall temperature, calculated as described in section 6.6.5.

It may happen that changes in physical properties along the length of the heat exchanger lead to a significant change in the heat transfer coefficient. This is likely with a viscous liquid, especially when there is a change from laminar to turbulent flow, or vice versa. It is then necessary to estimate α and $\dot{q}$ at several points along the length and to solve the integral in equation [6.4] by one of the methods given in section 10.3.

6.2.2 Convective heat transfer inside coiled tubes

The heat transfer coefficient in a coiled tube, such as those illustrated in Fig. 3.20, is greater than that in a straight tube under otherwise identical conditions. The subject of forced convective heat transfer inside a coiled tube is dealt with by ESDU (1978). The equation recommended by McAdams (1954) for turbulent flow is that of Jeschke (1925), derived from experiments with air. This gives:

$$\alpha_c = \alpha_{st}(1 + 3.5d/D) \qquad [6.13]$$

where α_c is the heat transfer coefficient for the coil and α_{st} is the coefficient for a straight pipe under otherwise identical conditions, d is the internal diameter of the tube and D the diameter of the helix.

This simple formula gives a safe method of estimating the heat transfer coefficient for liquids or gases being heated or cooled in coils with $D/d > 10$ and Reynolds number between 10^4 and 10^5.

6.2.3   Convective heat transfer in longitudinal flow outside tube bundles

It is sometimes advantageous to design a shell-and-tube heat exchanger so that the shellside fluid is in longitudinal flow, i.e. flowing parallel to the tubes, instead of the more usual cross-flow arrangement. Hence there are no baffles, but it is necessary to provide some support for the tubes. The lowest tubes in the bundle can be supported by a plate and the desired spacing between the tubes can be maintained either by using a system of rods or by having bands around the tubes at regular intervals. Very little has been published on the effects of the latter on performance, but it is believed that although the bands roughly double the pressure drop, they do not appreciably affect the heat transfer. Various patented systems of rod supports are available – see Saunders (1987).

It is expected that, in a long unbaffled shell-and-tube heat exchanger in which the tubes fill the shell, the parallel flow equation [6.9] may be used (with adequate accuracy) to calculate the external heat transfer coefficient. The cross-sectional area for flow (S, required for estimating $\dot{m}$, the mass flux) is given by:

$$S = (\pi/4)(D^2 - n_t d_o^2) \qquad [6.14]$$

where D is the internal diameter of the shell, d_o the external diameter of the tube and n_t the total number of tubes in the shell.

The characteristic dimension, d, required for estimating the Reynolds and Nusselt numbers, must be taken as the equivalent diameter, d_e, of the flow passage. The equivalent diameter is:

$$d_e = 4V/A \qquad [6.15]$$

where A is the surface area of the tubes (m^2) and V is the volume of the space inside the shell (m^3). In the case of an unbaffled exchanger:

$$d_e = \frac{D^2 - n_t d_o^2}{D + n_t d_o} \qquad [6.16]$$

When using longitudinal shellside flow, consideration should be given to the possibility of using tubes with longitudinal fins, i.e. several fins arranged radially round the circumference of each tube; a cross-section through a tube of this type is sketched in Fig. 6.1. This introduces two important improvements over plain tubes: (1) the external heat transfer surface area (A) is enhanced; (2) the fins considerably increase the strength of the tube and so reduce the number of intermediate supports needed.

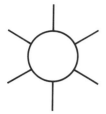

6.1 Cross-section of a tube with longitudinal fins

The parallel flow equation [6.9] may still be used to calculate the external heat transfer coefficient with longitudinal flow over longitudinal fins. The geometrical calculations are more complicated, but they follow the same lines as those described above for plain tubes. The cross-sectional area for flow is that given by equation [6.14] less the area blocked by the fins; the equivalent diameter must be calculated from equation [6.15]. It is also necessary to make an allowance for the thermal resistance of the fins, as described in section 6.4.4.

6.2.4 Convective heat transfer in annular spaces

The treatment in the previous section can be taken to the extreme of having only one tube per shell. This leads to the 'jacketed tube', or 'double pipe' heat exchanger, which consists of two concentric tubes; heat is transferred between a fluid flowing inside the inner tube and a fluid flowing in the annulus between the inner and outer tubes. Another example of the same geometry is the addition of a core rod in the centre of a tube to increase the velocity of the fluid flowing through the tube, and so to increase the heat transfer coefficient. In a jacketed tube, heat is transferred across the inner surface of the annulus; with a core rod, heat is transferred across the outer surface of the annulus. In either case, it is possible to use the parallel flow equation [6.9] to determine the heat transfer coefficient either to the inner or to the outer surface, as recommended by McAdams (1954), p. 243. Equations [6.14] and [6.16] simplify to:

$$S = (\pi/4)(D^2 - d_o^2) \tag{6.17}$$

$$d_e = D - d_o \tag{6.18}$$

where D is the inside diameter of the outer surface and d_o the outer diameter of the inner surface.

A more accurate treatment of heat transfer in annular spaces has recently been carried out by ESDU and made available in ESDU (1981b). This should be consulted if a more accurate estimate is required.

Longitudinal fins may be used on the outside of the inner tube of a double pipe (jacketed tube) heat exchanger. The remarks in the last two paragraphs of the previous section also apply to a single finned tube inside a larger tube.

6.2.5 Convective heat transfer across banks of straight tubes

Many studies of the data on convective heat transfer in flow across banks of straight parallel tubes have been published, the most recent being those of ESDU (1973c), which form the basis of the method recommended below, and which are adapted by J. Taborek in section 3.3.7 of Schlünder (1983). The ESDU method applies to flow at right-angles to banks containing at least 10 rows of tubes with at least 6 tubes per row, the length of the tubes exceeding five times their external diameter. The correlation for the heat transfer coefficient in flow across tube banks may be expressed in the form:

$$\mathrm{Nu} = 0.33 F_a F_p \, \mathrm{Re}^{0.60} \, \mathrm{Pr}^{0.34} \qquad [6.19]$$

The dimensionless parameters Nu, Re and Pr are as defined in equations [6.6], [6.7] and [6.8]. The characteristic dimension, d, in the Nusselt and Reynolds number is d_o, the external diameter of the tubes. The mass flux, $\dot{m}$, is the maximum, i.e. it is based on the minimum area for flow between the tubes. F_a is an arrangement factor; F_p allows for variations in fluid physical properties from the bulk to the wall. For heating or cooling liquids or gases, with any arrangement of tubes, F_p can be determined from the following equation, which relates it to the ratio of the Prandtl number at the mean bulk temperature to that at the wall temperature:

$$F_p = (\mathrm{Pr}_b / \mathrm{Pr}_w)^{0.26} \qquad [6.20]$$

where subscript b denotes the mean bulk temperature and w denotes the mean wall temperature, calculated as described in section 6.6.5.

The arrangement factor (F_a) depends on the Reynolds number, the tube pattern (see TEMA Figure R–2.4) and the ratio of the pitch of the tubes to their diameter (P/d_o). The last has only a minor effect, which is neglected in the approximate treatment given here. Values of F_a are tabulated below for various values of the Reynolds number and two arrangements: 'inline' denotes that the tubes are on a square pattern with one side of the square in line with the direction of flow; 'staggered' may denote a square pattern with the sides of the square at 45° to the direction of flow or an equilateral triangular pattern with a side of the triangle at 30° or 60° to the direction of flow.

Reynolds number (Re)	10	30	100	300	1000	10^4	10^5	10^6
Inline: $F_a =$	1.52	1.27	1.03	0.86	0.91	1.02	1.15	1.40
Staggered: $F_a =$	2.28	1.75	1.31	1.01	1.05	1.14	1.24	1.50

In the above treatment it is assumed that the tubes extend as far as the wall of the duct containing them, in order to prevent leakage between the outer tubes and the wall.

If the external heat transfer coefficient is low, consideration should be given to the possibility of using external transverse fins. Many proprietary designs of finned tubes are available; the manufacturers should be asked to provide data on the performance of their fins. It must be remembered that the addition of fins increases the temperature of the wall, so it is important to check that the maximum temperature of the wall, calculated as described in section 6.6, will not exceed the safe limit for the material of the tube.

6.2.6   Convective heat transfer on the shellside of a baffled shell-and-tube exchanger

Flow on the shellside of a baffled shell-and-tube heat exchanger is across the bundle of tubes, i.e. in a direction at right-angles to the tubes, except in the window zones, where flow is parallel to the tubes. The cross-flow equation [6.19] gives the heat transfer coefficient in the cross-flow zone, and the parallel flow equation [6.9] gives the heat transfer coefficient in the windows. Unfortunately it is difficult to estimate the amount of fluid flowing through these zones because there is a large amount of leakage between the baffles and the tubes and between the baffles and the shell; also fluid bypasses the bundle, flowing between the bundle and the shell and through any pass partition lanes between adjacent tubeside passes, if these are in the direction of the flow. It is a very difficult problem in fluid flow to estimate what fraction of the fluid entering the shell flows across the tubes and what fraction flows through the windows. It is beyond the scope of this book to give a reliable method of calculating these fractions by hand. The problem was first studied by Tinker (1951), but his method was not used until computers became available.

ESDU (1983b) presents an approximate method of estimating the flow fractions when flow is turbulent. The procedure can be carried out with a programmable calculator and may be applied to most industrial heat exchangers, except when the shellside fluid is a viscous liquid. The pressure loss can be calculated to within $\pm 30\%$ when flow is fully turbulent, which is a considerable advance on previous hand methods. In the transition region between laminar and turbulent flow, the method underpredicts pressure loss by up to 40%. (Section 11 of ESDU (1983b) contains an example on the use of this procedure.)

An empirical method of calculating heat transfer and pressure loss is available in Perry and Chilton (1973) on pp. 10–25 to 10–30, and a development of this method is given in Schlünder (1983) and in Chapter 12 of Saunders (1987). In this method, correction factors are given for application to the heat transfer coefficient and pressure loss calculated on the assumption that there is no leakage past the baffle and no bypassing around the bundle.

For an approximate estimation of shellside heat transfer coefficients, it may be assumed that only 50% of the fluid flowing along the shell is flowing across the tubes, the other 50% flowing through the leakage and bypass paths. This will normally give a slight under-estimate of the heat transfer coefficient in a well-designed shell. However, the flow fraction across the tubes may be even lower, if adequate care is not taken to minimise leakage and bypassing, or if the pitch of the baffles is less than a third of their diameter.

On account of the complexity of the topic, it is normal practice nowadays to use proprietary computer programs to estimate the various flow fractions and hence to estimate the shellside pressure drop and heat transfer coefficient. These programs are mentioned in section 10.7 of this book.

6.2.7 Convective heat transfer on the shellside of a coiled-tube exchanger

The multiple-coil heat exchanger has been described in section 3.4.5 and illustrated in Fig. 3.20. Geometrical calculations are necessary as a preliminary to the estimation of the mass flux, to permit the calculation of the Reynolds number and thence the external heat transfer coefficient. These geometrical calculations have been described by Smith (1964) and Abadzic (1974) and are summarised below. Figure 6.2 shows a general cross-section of part of an exchanger. To avoid maldistribution of the shellside fluid, adjacent layers are coiled in opposite directions.

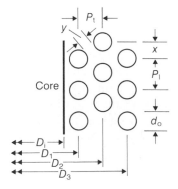

6.2 Part of coiled-tube exchanger: dimensions required for geometrical calculations. For the arrangement shown in Fig. 3.20, $D_i = 5P_t$, $D_1 = 6P_t$, $D_2 = 8P_t$, $D_3 = 10P_t$, $D_4 = 12P_t$, $D_5 = 14P_t$, $D_o = 15P_t$

The length of each coil (L) is

$$L = H \operatorname{cosec} \theta \qquad [6.21]$$

where H is the height of the coil and θ the helix angle. If n is the number of identical coils in a layer, the longitudinal pitch of the tubes (P_l) in a coil of diameter D is:

$$P_l = \frac{\pi D \tan \theta}{n} \qquad [6.22]$$

To avoid maldistribution of the tubeside fluid, the tubes in the different layers all have the same helix angle and the same length. Also the diameter of each layer must be in direct proportion to the number of tubes in that layer. If P_t is the transverse pitch and D_1 the diameter of the first layer, the diameter of the second layer is $D_1 + 2P_t$, of the third layer $D_1 + 4P_t$, etc. The annular space containing the coils should have an internal diameter, $D_i = D_1 - P_t$ and an outside diameter, $D_o = D_1 + (2m - 1)P_t$, where m is the number of layers, and D_i is the internal diameter of the space, i.e. the external diameter of the core.

To determine S, the cross-sectional area for flow on the shellside, it is first necessary to establish a relationship between y, the space between adjacent tubes, and x, the vertical displacement of the tubes of one layer relative to those in adjacent layers; these dimensions vary around the circumference, because adjacent layers are coiled in opposite directions. From Pythagoras's theorem:

$$(y + d_o)^2 = P_t^2 + x^2 \qquad [6.23]$$

where d_o is the external diameter of the tubes.

Thus y ranges from a minimum of $P_t - d_o$, when $x = 0$, to a maximum of $\sqrt{P_t^2 + (\tfrac{1}{2}P_l)^2} - d_o$, when $x = \tfrac{1}{2}P_l$. Integration with respect to x of the expression for y obtained from equation [6.23] gives the following equation for $\bar{y}$, the mean value of y:

$$\bar{y} = \tfrac{1}{2}\sqrt{P_t^2 + (\tfrac{1}{2}P_l)^2} + \frac{P_t^2}{P_l} \ln\left[\frac{\tfrac{1}{2}P_l + \sqrt{P_t^2 + (\tfrac{1}{2}P_l)^2}}{P_t}\right] - d_o \qquad [6.24]$$

The cross-sectional area for flow may then be estimated from:

$$S = \frac{\pi}{4}(D_o^2 - D_i^2)\frac{\bar{y}}{P_t} \qquad [6.25]$$

Abadzic (1974) gives equation [6.26]; this predicts the experimental data well, provided that P_t/d_o and P_l/d_o are in the range from 1.1 to 1.9 and P_t/P_l is in the range 0.9 to 1.1.

If $10^3 < \mathrm{Re} < 2 \times 10^4$, $\mathrm{Nu} = 0.332 \ \mathrm{Re}^{0.6} \ \mathrm{Pr}^{0.36}$

if $2 \times 10^4 < \mathrm{Re} < 2 \times 10^5$, $\mathrm{Nu} = 0.123 \ \mathrm{Re}^{0.7} \ \mathrm{Pr}^{0.36}$ $[6.26]$

if $2 \times 10^5 < \mathrm{Re} < 9 \times 10^5$, $\mathrm{Nu} = 0.036 \ \mathrm{Re}^{0.8} \ \mathrm{Pr}^{0.36}$

The dimensionless parameters Nu, Re and Pr are as defined in equations [6.6], [6.7] and [6.8]. The characteristic dimension, d, in the Nusselt and Reynolds number is d_o, the external diameter of the tubes.

6.2.8 Convective heat transfer in plate exchangers

The various uses of plate heat exchangers have been described in Chapter 3, section 3.1.7 dealing with the welded spiral-plate type, section 3.1.8 with the brazed corrugated plate-fin type, and section 3.6 with the gasketed plate type. The single-phase heat transfer coefficient for plain plates can be estimated from equation [6.9] for flow inside tubes if the internal diameter (d) is replaced by the equivalent diameter (d_e) defined by equation [6.15]. For parallel plates it follows that the equivalent diameter is twice the space between plates.

Usually the plates of gasketed plate exchangers are pressed into a complicated corrugated pattern to give a greater mechanical rigidity; this also enhances the heat transfer coefficient and considerably enhances the pressure drop. Similarly the corrugated plates of plate-fin exchangers are sometimes serrated and sometimes they take the fluid along a zigzag path; these arrangements produce similar enhancements to heat transfer and pressure drop, compared with plain fins. Most manufacturers of plate heat exchangers are prepared to give potential purchasers their data on heat transfer and pressure drop, for the type that they recommend for a particular duty. Alfa-Laval (1969) describe their range of plate exchangers and discuss in detail methods of calculating pressure drop and heat transfer. Graphical methods are given for the correct choice of heat exchanger design, together with alternatives.

Several manufacturers of plate heat exchangers have their own computer programs for estimating pressure drop and heat transfer in their own designs of exchanger, and these can also be used for design.

Various other forms of compact heat exchanger are available; heat transfer data for many have been published by Kays and London (1984).

6.2.9 Heating by a gas containing condensing vapour

The source of heat for a vaporiser may be a hot gas containing some vapour that will condense on the surface of the vaporiser as the gas is cooled. Even a small amount of condensation may provide the majority of the heat load. Under these circumstances the estimation of the local heat fluxes becomes very difficult. Methods of dealing with condensation have been published by Colburn and Hougen (1934) for a mixture of a vapour and an incondensable gas, and by Colburn and Drew (1937) for two condensing vapours. These involve trial-and-error calculations to determine the heat flux at several locations; the problem is further complicated when nucleate boiling contributes to the process of vaporisation. Furthermore, it is difficult to estimate the temperature and quality at locations after the inlet from the conditions at the previous location and the heat flux and mass flux to the wall in condensation. No attempt to write a computer program to deal with this situation has yet been successful.

An approximate method of dealing with multi-component condensation has been proposed by Silver (1947) and by Bell and Ghaly (1972). The method is applicable to any number of components of vapour and gas. Its basis is to assume that the incoming mixture is dry and saturated, and that the mass flux to the wall in condensation is always exactly that required to compensate for the heat flux, and so to maintain dry saturation. If the Reynolds analogy between heat and mass transfer holds, if the temperature difference is small, and if the concentration of vapour is small, then this assumption is valid. If the concentration of vapour is high, the mass flux to the wall will be greater than the method predicts, but under these circumstances the thermal resistance of the heating fluid is very low; consequently the rate of heat transfer is controlled by the other resistances, so the errors introduced by using the approximate method are small and on the side of underestimation. If the mass flux to the wall is less than that required to maintain dry saturation, the approximate method will overpredict the rate of heat transfer and fog will form in the vapour stream. The droplets of fog are very small and therefore difficult to remove. The serious consequences of fogging have been described by Steinmeyer (1972). Special demisters are required to remove the fog. Fogging can be reduced by reducing the temperature difference, but this will reduce the rate of heat transfer further.

In order to be able to use the approximate method of estimating the heat transfer coefficient in mixed condensation, it is necessary to have a cooling curve for the condensation of the mixture involved; this consists of a graph of temperature (T) plotted against the specific enthalpy (h) of the mixture, assumed to be dry and saturated at the temperature T. If α_{hg} is the heat transfer coefficient of the hot stream assuming that no condensation takes place, the true coefficient, α_h, is given by:

$$\alpha_h = \frac{\alpha_{hg}}{x_g c_{pg} \dfrac{dT}{dh}} \qquad\qquad [6.27]$$

where x_g is the quality, i.e. the ratio of the mass flow rate of gas and vapour to the total mass flow rate, c_{pg} is the specific heat capacity of the steam of gas and vapour, and dT/dh is the slope of the cooling curve, all measured at local conditions.

McNaught (1983) gives experimental validation of this method, and shows that there is little difference in the values of the overall heat transfer coefficients predicted by this method and by more accurate methods, for the binary and ternary mixtures used in the experiments.

6.3   Heat transfer coefficient for a condensing vapour

When the source of heat in a vaporiser is a condensing vapour or a hot gas containing a vapour that condenses, the thermal resistance of the heating fluid consists of two resistances in series:

1. Gas film resistance.
2. Condensate film resistance.

In the case of a pure vapour, e.g. when the heat source is condensing steam, it is not necessary to calculate the gas film resistance. It is known that the interface between the vapour and the film of condensate must equal the saturation temperature; therefore it is permitted to neglect the resistance of the vapour film when calculating the overall heat transfer coefficient (U), provided that the temperature of the hot stream is taken to be the saturation temperature. If the vapour is initially superheated, there will be some sensible heat transfer from the vapour to the interface, resulting in a reduction of the amount of superheat, but most of the heat transfer will be the result of mass transfer of vapour that condenses as soon as it reaches the interface; as the resistance to mass transfer is zero, the interface temperature cannot fall below the saturation temperature. Hence the heat flux to the vaporising fluid is equal to the reciprocal of the sum of all resistances from condensate to vaporising fluid multiplied by the difference between the temperature at the interface (the saturation temperature) and the temperature of the vaporising fluid. If there is a large amount of superheat, it may be advisable to determine whether the total heat flux could be provided by sensible heat transfer alone; if the wall temperature, calculated as described in section 6.6, assuming no condensation, is found to be above the saturation temperature, then no condensation can take place. It is important to appreciate that under these circumstances the heat flux is greater than that calculated on the assumption that there is condensation. Also the temperature of the vapour falls rapidly, so condensation begins after a short length (as soon as the temperature of the wall falls to saturation temperature).

  If an incondensable gas or more than one vapour is present, the interface temperature falls to below the saturation temperature, so it is necessary to estimate the gas film resistance. This has been discussed above in section 6.2.9.

  The remainder of this section is concerned with the estimation of α_c, the heat transfer coefficient of the condensate film. For further information on condensation, Chapter 2.6 of Schlünder (1983) should be consulted.

6.3.1   Theoretical heat transfer coefficient for laminar flow of condensate

Many textbooks on heat transfer give Nusselt's derivation of theoretical equations for the heat transfer coefficient in filmwise condensation; it is assumed that the flow of the condensate is laminar and due to gravity, the effects on the condensate of vapour shear being negligible. The local heat transfer coefficient is determined from the thermal conductivity of the condensate divided by the local thickness of the film, and the mean coefficient is obtained by integration of the local values, assuming a constant temperature difference along the length of the condenser. See Chapter 12 of Kern (1950) or Chapter 13 of McAdams (1954). Kern (1950) shows that the equations for condensation on a vertical surface and on the outside of a horizontal tube can be approximated to:

$$\frac{\bar{\alpha}_{cl}}{\lambda}\left[\frac{\eta^2}{\rho^2 g}\right]^{\frac{1}{3}} \qquad [6.28]$$

where g is the acceleration due to gravity (9.81 m/s^2), Re$_f$ is the film Reynolds number, defined below, $\bar{\alpha}_{cl}$ is the theoretical mean heat transfer coefficient of the condensate film (W/m^2 K), η is the dynamic viscosity of the condensate (Ns/m^2)), λ is the thermal conductivity of the condensate (W/m K) and ρ is the density of the condensate (kg/m^3). The Reynolds number of a falling film of liquid is:

$$\text{Re}_f = 4\Gamma/\eta \qquad [6.29]$$

where Γ is the mass flow rate of condensate per unit length (kg/s m).

To determine Γ it is first necessary to calculate $\dot{M}_c$, the total mass flow rate of condensate (kg/s), given $\dot{Q}$, the heat transfer rate for the heat exchanger (W) and Δh_v, the latent heat of vaporisation (J/kg), thus

$$\dot{M}_c = \dot{Q}/\Delta h_v \qquad [6.30]$$

With horizontal tubes and either shellside or tubeside condensation:

$$\Gamma = \dot{M}_c/Ln_t \qquad [6.31]$$

where L is the length of the tubes (m) and n_t is the total number of tubes.

With vertical tubes and tubeside condensation:

$$\Gamma = \frac{\dot{M}_c}{\pi d n_t} \qquad [6.32]$$

where d is the internal diameter of the tubes (m).

With vertical tubes and shellside condensation, it may be assumed that the baffles remove all the condensate formed in each inter-baffle space.

Hence:

$$\Gamma = \frac{\dot{M}_c}{\pi d_o n_t n_s} \qquad [6.33]$$

where d_o is the external diameter of the tubes and n_s the number of shellside passes.

By segregating the terms relating to the physical properties of the condensate and the acceleration due to gravity, equation [6.28] can be simplified to

$$\bar{\alpha}_{cl} = \phi_c \Gamma^{-\frac{1}{3}} \qquad\qquad [6.34]$$

where $\phi_c = 2.02\lambda(\rho^2/\eta)^{\frac{1}{3}}$ [6.35]

The values of the physical properties parameter ϕ_c for the condensate formed by the condensation of several vapours are plotted against temperature in Fig. 6.3. The properties of the liquid halogenated hydrocarbon refrigerants, designated by the letter 'R' and their specific

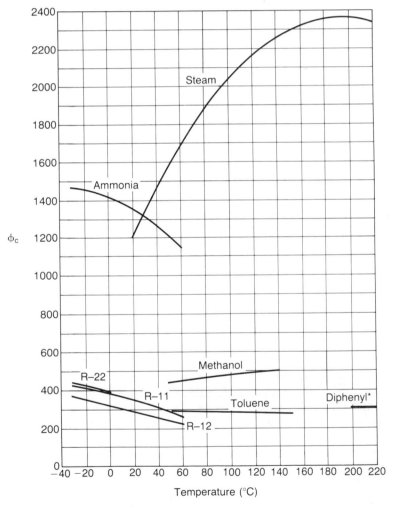

6.3 Physical properties parameter for condensing vapours (from equation [6.35])

number, were supplied by ICI, Mond Division, for whose help the author is grateful. The properties of the other liquids were obtained from ESDU (1980).

Research into condensation has shown that the heat transfer coefficient of a film of condensate does not always agree with the values calculated by laminar film theory, due to a variety of reasons. The following subsections deal with the usual arrangements; the reasons for deviation from theory are given and equations are recommended for design purposes.

6.3.2 Condensation inside horizontal tubes

The presence of condensate running along the bottom of a horizontal tube impedes the transfer of heat. Kern (1950), p. 269, recommends that to allow for this the condensate rate should be doubled; in other words, the heat transfer coefficient calculated from equations [6.31] and [6.34] should be multiplied by 0.79, i.e. $2^{-\frac{1}{3}}$. The correction would be nearer to unity if it were possible to install the tubes with a downward slope of a few degrees.

Near the entrance to a tube the thickness of the film of condensate may be appreciably reduced by vapour shear, giving an enhanced heat transfer coefficient. A simple method of estimating the local value of the coefficient in the region where vapour shear predominates is that of Boyko and Kruzhilin (1967). This gives:

$$\alpha_s = 0.021 \frac{\lambda_l}{d} \mathrm{Re}_{lo}^{0.8} \mathrm{Pr}_l^{0.43} \left[1 + x_g \left(\frac{\rho_l}{\rho_g} - 1 \right) \right]^{\frac{1}{2}} \qquad [6.36]$$

where d is the internal diameter of the tube (m), Pr_l is the Prandtl number of the condensate (from equation [6.7]), Re_{lo} is the Reynolds number for the total flow as liquid (see below), x_g is the local quality, i.e. the ratio of the mass flow rate of vapour to the total mass flow rate, λ_l is the thermal conductivity of the condensate (W/m K) and ρ_l is the density of the condensate and ρ_g that of the vapour (kg/m^3); Re_{lo} is obtained from equation [6.8] taking η_l as the viscosity of the condensate and $\dot{m}$ as the total mass flux, i.e. condensate plus vapour plus any incondensables.

The subject of condensation inside horizontal tubes is dealt with extensively in section 2.6.2D of Schlünder (1983). When the heating medium is condensing steam, the heat transfer coefficient of the condensate film is usually relatively high, and it is adequate to use the following approximate method, which is likely to err on the side of underpredicting the coefficient.

Calculate the mean coefficient in the gravity controlled region

$$\bar{\alpha}_{gr} = 0.79 \bar{\alpha}_{cl} \qquad [6.37]$$

where $\bar{\alpha}_{cl}$ is the theoretical coefficient, calculated as described in section 6.3.1.

Calculate the values of α_s at inlet and outlet, from equation [6.36], based on the values of x_g at inlet and outlet; denote these by $\alpha_{s\,in}$ and $\alpha_{s\,out}$ respectively.

Calculate the mean coefficient, $\bar{\alpha}_c$ as follows:

If $\bar{\alpha}_{gr} > \alpha_{s\,in}$, $\bar{\alpha}_c = \bar{\alpha}_{gr}$ [6.38]

If $\bar{\alpha}_{gr} < \alpha_{s\,out}$, $\bar{\alpha}_c = \frac{1}{2}(\alpha_{s\,in} + \alpha_{s\,out})$ [6.39]

If $\alpha_{s\,in} > \bar{\alpha}_{gr} > \alpha_{s\,out}$, calculate from equation [6.36] the value of x_g at which $\alpha_s = \bar{\alpha}_{gr}$; this can be regarded as the quality at which the change takes place from shear controlled to gravity controlled condensation. Divide the condenser into two separate condensers, the first being the shear controlled and the second the gravity controlled region. For the shear controlled region:

$$\bar{\alpha}_c = \frac{1}{2}(\alpha_{s\,in} + \bar{\alpha}_{gr})$$

For the gravity controlled region:

$$\bar{\alpha}_c = \bar{\alpha}_{gr} \qquad [6.40]$$

With condensing steam as the heating medium, it may be found that the first step gives a value of $\bar{\alpha}_{gr}$ very much greater than the other coefficients combined, in which case the enhancement due to vapour shear may be neglected; there is then no need to calculate α_s, the mean condensing coefficient being given by equation [6.40].

6.3.3 Condensation outside a bundle of horizontal tubes

Experiments have shown that equations [6.31] and [6.34] accurately predict the heat transfer coefficient for the film of condensate for a single tube, that is when $n_t = 1$. With a bundle of horizontal tubes, condensate formed on tubes in the upper layers drains onto tubes in lower layers and so increases the thickness of the film there and thus reduces the coefficient. The reduction in the mean coefficient has been found to be less severe than would be predicted if it were assumed that all condensate fell gently onto the tube immediately below. It is thought that this is due to splash and turbulence in the film induced by the liquid that has fallen onto it. No satisfactory method has yet been found for allowing for this inundation. Kern (1950) suggested an empirical treatment that gave good agreement between experimental results and predictions. Adopting this method gives the following equation for determining $\bar{\alpha}_c$, the mean heat transfer coefficient for the condensate film on a bundle of horizontal tubes, from $\bar{\alpha}_{cl}$, the theoretical value obtained from equations [6.28] and [6.31].

$$\bar{\alpha}_c = \bar{\alpha}_{cl}(n_v)^{-2/9} \qquad [6.41]$$

where n_v is the average number of tubes vertically above each other.

As the vapour flows past the tubes, the resulting shear stress reduces the thickness of the layer of condensate, and so improves the heat transfer. An allowance for this may be made by using the treatment given by ESDU (1984).

6.3.4 Condensation on vertical tubes

For a given heat load on a long tube, the Reynolds number is much greater if the tube is vertical than it would be if the tube were horizontal, as can be seen from equations [6.29] to [6.32]. Thus the theoretical heat transfer coefficient for the film of condensate (from equation [6.28]) is lower for a vertical than for a horizontal tube. However, as the Reynolds number is increased there is an enhancement above the theoretical value, first due to the formation of waves on the surface of the film and later due to the onset of turbulent flow of the condensate, as described by Butterworth in Section 2.6.2 of Schlünder (1983).

For laminar flow of condensate, Kutateladze (1963) has suggested that waves begin to affect the heat transfer coefficient in falling films when the film Reynolds number, from equation [6.29], reaches 30. At higher Reynolds numbers, up to about 1600, the local heat transfer coefficient is increased by a factor $0.687\,Re_f^{0.11}$. More recently Hirschburg and Florschuetz (1981) have analysed published data and developed a model to allow for the formation of waves.

If α_{cx} is the local value of the heat transfer coefficient at a point distant x (m) from inlet, and if $\bar{\alpha}_c$ is the mean coefficient from inlet ($x = 0$) to the location under consideration, it is recommended that these should be estimated from the following equations:

if $Re_f < 30$,

$$\alpha_{cx} = 0.75\bar{\alpha}_{cl}$$

and $\bar{\alpha}_c = \bar{\alpha}_{cl}$

[6.42]

if $30 < Re_f < 1600$

$$\alpha_{cx} = 0.504\,Re_f^{0.11}\,\bar{\alpha}_{cl}$$

and $\bar{\alpha}_c = \bar{\alpha}_{cl}\left[\dfrac{Re_f^{4/3}}{1.62\,Re_f^{1.22} - 7.8}\right]$

[6.43]

where $\bar{\alpha}_{cl}$ is the theoretical mean heat transfer coefficient from $x = 0$ to x, calculated from equation [6.28], and Re_f is the film Reynolds number at x, from equation [6.29]. The equations for $\bar{\alpha}_c$ have been derived by integrating the equations for α_{cx} for Reynolds numbers from 0 to Re_f, assuming a constant wall temperature. The experimental points used by Hirschburg and Florschuetz (1981) show good agreement with these equations over the range of Reynolds numbers, from 10 to 1000.

Whereas in laminar flow the local heat transfer coefficient decreases with increasing Reynolds number, a minimum is reached when Re_f is in the region of 1600 to 2000. Beyond these values the local coefficient increases steadily with increasing Reynolds number, due to the turbulent flow of the condensate. The following equation for α_{cx} in the turbulent region was obtained by Labuntsov (1957) from theoretical considerations confirmed by experiment. Assuming that transition from laminar to turbulent flow occurs at a Reynolds number of 1600, integration gave the following equation for $\bar{\alpha}_c$, the mean coefficient over the length of the tube, if $Re_f > 1600$.

$$\alpha_{cx} = 0.0153 \, Re_f^{0.58} \, Pr^{0.5} \, \bar{\alpha}_{cl}$$

$$\text{and} \quad \bar{\alpha}_c = \bar{\alpha}_{cl} \left[\frac{Re_f^{4/3}}{87 \, Pr^{-\frac{1}{2}} (Re^{\frac{3}{4}} - 253) + 13\,170} \right]$$

[6.44]

If condensing steam is the heating medium, the film of condensate will be in laminar flow, due to the high latent heat of vaporisation of water at the normally used pressures. However, with refrigerants condensing near to their critical pressure, the flow is likely to be turbulent and it will be necessary to use equation [6.44].

6.3.5 Enhancement of heat transfer in condensation outside horizontal tubes

Relatively high heat transfer coefficients are obtained with condensing steam, but with organic liquids the coefficients are many times lower, due to their lower values of latent heat of vaporisation and thermal conductivity; the former leads to a higher value of Γ and the latter to a lower value of ϕ_c in equation [6.34]. Methods of enhancing heat transfer in condensation are discussed in this section and in sections 6.3.6 and 6.3.7.

To enhance heat transfer in shellside condensation, tubes with low external transverse fins may be used if the shell is to be mounted horizontally. Beatty and Katz (1948) studied the available experimental data on single tubes and showed that the heat transfer coefficients averaged 5% less than those calculated by applying laminar flow theory for a vertical surface to the fins and for a horizontal tube to the primary surface between the fins. Rudy and Webb (1981) studied the retention of condensate at the bottom of the fins due to surface tension. They showed that the condensing heat transfer coefficient is less than predicted if there is a large amount of retention. The results were studied further by Owen et al. (1983), who showed that the fraction of the circumference of the fins flooded by condensate could be calculated from the dimensionless group ζ, defined by:

$$\zeta = \frac{\sigma}{\rho_l g d_f w}$$

[6.45]

where d_f is the external diameter of the fins (m), g is the gravitational acceleration (m/s²), w is the space between adjacent fins (m), ρ_l is the density of the condensate (kg/m³) and σ its surface tension (N/m). From theoretical considerations, confirmed by the experimental results, they showed that if $\zeta \geqslant 0.5$, the space between fins becomes completely filled with condensate. Otherwise, the fraction of the circumference flooded is:

$$x_c = \frac{\cos^{-1}(1 - 4\zeta)}{180} \qquad [6.46]$$

Owen *et al.* (1983) also showed that the method of prediction of Beatty and Katz (1948) gave good prediction of experimental results when $\zeta < 0.15$, which corresponds to $x_c < 37\%$. They also developed a more complex method of estimation, which gave better prediction when $\zeta > 0.2$ ($x_c > 44\%$). It is recommended that the method of Beatty and Katz, described below, be used if $\zeta < 0.2$, which will meet most industrial applications. Otherwise the reader should consult Owen *et al.* (1983) or one of the references mentioned in the following paragraphs.

Honda *et al.* (1983) describe experiments in which R-113 or methanol condensed at atmospheric pressure on a single horizontal tube. Tests on a plain tube were followed by tests on four different finned tubes. With either vapour condensing on the plain tube, the measured average heat transfer coefficient agreed well with that predicted from Nusselt's theoretical treatment – equation [6.28]. With either vapour condensing on a finned tube, the authors found that the amount of the circumference flooded agreed well with their theoretical equation – the same as equation [6.46]. With R-113 condensing on one of the finned tubes, 37% of the circumference was flooded, and the measured average heat transfer coefficients were 5 to 20% greater than those predicted by equation [6.47]; with methanol condensing on the same tube, 59% of the circumference was flooded, and the coefficients were only 43 to 67% of those predicted by equation [6.47], showing no enhancement compared with a plain tube. Honda *et al.* (1983) also showed that condensation on a horizontal finned tube is considerably augmented by attaching a porous drainage plate to the bottom of the tube, especially under conditions conducive to flooding.

Research is in progress at Queen Mary College, London, on condensation on the outside of a single horizontal tube. Several tubes have been tested; the space between fins ranged from 0.25 to 20 mm, the fins being 1.59 mm high, with a root diameter of 12.7 mm. The vapour, flowing downwards at low speed and near atmospheric pressure, was either steam or R-113. The experimentally determined values of the enhancement in heat transfer coefficient due to finning have been compared with predicted values from the various correlations. Yau *et al.* (1986) deal with the condensation of steam, including the effects of drainage strips; Masuda and Rose (1985) deal with the condensation of R-113. To apply the correlation of Beatty and Katz (1948), calculate, or take from the manufacturer's catalogue, the total external surface area per length, denoted by a_o (m²/m); this is the sum of a_r, the area of bare tube, and a_f, the area of the fins, per unit length of tube. Determine the condensate heat transfer coefficient for a single plain tube, $\bar{\alpha}_{cl}$, from equation [6.28]. In the absence of any information to the contrary, it will be assumed that equation [6.41], which allows for the effects of inundation with plain tubes, is also applicable to finned tubes. The mean condensate film coefficient is then given by:

$$\bar{\alpha}_c = 0.645 \left[\frac{a_r}{a_o} \frac{1}{d_o^{\frac{1}{4}}} + \frac{\eta_f a_f}{a_o} \frac{1.3}{H_f^{\frac{1}{4}}} \right]^{\frac{4}{3}} a_o^{\frac{1}{4}} \bar{\alpha}_{cl}(n_v)^{-2/9} \qquad [6.46]$$

where n_v is the average number of tubes vertically above each other, η_f is the fin efficiency (defined in section 6.4.4) and H_f is the mean height of each fin, given by:

$$H_f = \frac{\pi}{4 d_f} (d_f^2 - d_o^2) \qquad [6.48]$$

where d_f is the diameter of the fins and d_o of the outside of the tubes (m). In the derivation of equation [6.47], allowance has been made for the empirical fact that the coefficient is 5% less than the theoretical value. It gives the effective coefficient, allowing for the inefficiency of the fins, so $\bar{\alpha}_c$ is identical with α_h, the coefficient of the heating fluid, required for substitution in equation [6.3] to determine U, the overall heat transfer coefficient.

6.3.6 Longitudinal fins on vertical tubes

Thomas (1968) showed that longitudinal rectangular fins loosely clamped to vertical tubes increased the heat transfer coefficient of the film of condensate by a factor of up to nine. He also quoted results showing considerable enhancement in the condensate coefficient by the use of fluted surfaces, the vertical troughs in the surface helping the draining of the condensate and so reducing the mean thickness of the film.

6.3.7 Artificially roughened surfaces

Larkin (1981) studied the effects of cutting circumferential capillary grooves on the inside of the nearly horizontal tube used in an experimental two-phase closed thermosyphon operating as a heat pipe; he found no enhancement of the condensing coefficient when the fluid was water, but there was a five- to six-fold enhancement when the fluid was chlorodifluoromethane. From this it appears that although artificially roughening a surface may be an excellent way of enhancing an otherwise poor heat transfer coefficient in condensation, there is no method of predicting the amount of enhancement.

6.4 Thermal resistance of wall and dirt

This section is concerned with the estimation of the thermal resistance of the wall that is used to separate the heating fluid from the vaporising fluid. The use of extended surfaces is also considered. Methods are given for estimating the values of α_w, the heat transfer coefficient for substitution in equation [6.3], for the wall and any deposited dirt, so that the overall heat transfer coefficient (U) can be found. The heat transfer coefficient is the reciprocal of the thermal resistance. From the definition of thermal conductivity, the heat transfer coefficient for a plane wall is:

$$\alpha_w = \lambda_w / x_w \tag{6.49}$$

where x_w is the thickness of the wall (m), α_w is the heat transfer coefficient (W/m^2 K) and λ_w is the thermal conductivity of the material of the wall (W/m K).

6.4.1 Heat transfer coefficient for the wall

Equation [6.49] can be used to calculate the value of α_w, the heat transfer coefficient for the wall, from x_w, the thickness of the wall, and λ_w, its thermal conductivity. For substitution in equation [6.3] it is also necessary to know the area ratio (A_v/A_w). For plate heat exchangers this ratio is unity; for tubular heat exchangers the ratio equals the diameter ratio (d_v/d_w). With in-tube boiling, d_v is the internal diameter of the tubes; with shellside boiling d_v is the external diameter of the tubes. It is shown in textbooks on heat transfer, e.g. McAdams (1954), p. 13, and Coulson and Richardson (1977), p. 176, that for conduction through a cylindrical tube, the mean area is the logarithmic mean of the inside and outside areas. If d_h is the diameter of the surface in contact with the heating fluid

$$\frac{A_v}{A_w} = \frac{d_v}{d_w} = \frac{d_v}{d_v - d_h} \ln\left(\frac{d_v}{d_h}\right) \tag{6.50}$$

It is usually sufficiently accurate to use the arithmetic mean, rather than the logarithmic mean, in which case:

$$\frac{A_v}{A_w} = \frac{2d_v}{d_v + d_h} \tag{6.51}$$

If extended surfaces are used, it is still necessary to estimate the heat transfer coefficient of the primary surface in the way described above.

Allowance must then be made for the thermal resistance of the fins in the way described in section 6.4.4.

6.4.2   Fouling by the heating fluid

If the thickness of the deposit of dirt were known, the heat transfer coefficient could be estimated from equation [6.49]. Dirt coefficients are usually guessed from past experience; equation [6.49] is then used to calculate the thickness of the deposit, which is required for the calculation of pressure drop.

Very little information is available on fouling. The values published by TEMA (1978) are widely used, being the only published data. Otherwise the designer supplements this data with his own data obtained from past experience when heat exchangers have failed to meet their required performance and the deficiency has been attributed to excessive fouling. It is important to measure the performance of heat exchangers to check the assumptions made in design about fouling. If during overhaul it is found that a surface is badly fouled, a rough estimate should be made of the thickness of the deposit and the coefficient calculated from equation [6.49]; although very approximate, this may give a rough confirmation that the amount of fouling required to explain the poor performance is of similar magnitude to that observed.

The present state of knowledge on fouling in heat exchange equipment has been summarized in 12 papers edited by Chenoweth and Impagliazzo (1981) and presented at the 20th ASME/AIChE Heat Transfer Conference. Several papers relate to fouling by cooling-tower water and one to fouling by atmospheric air in aircoolers; none relate to fouling by the hot fluid.

Hawes and Garton (1967) have studied published work on heat transfer from hot gas suspensions to cold surfaces. They showed that precipitated particles adhere strongly to a cooler surface (but never to a hotter surface). Fouling increases with increasing temperature difference, with decreasing size of particle, and with decreasing gas velocity.

Condensing steam taken from a steam main gives very little fouling; a figure of $\alpha_{dh} = 10\,000\ W/m^2\ K$ may be used in design with safety. However, in multiple-effect evaporation, much lower values will be obtained in all but the first effect unless an efficient separator is installed after each effect. If boiler blowdown is the source of heat, TEMA (1978) suggest a figure of $\alpha_{dh} = 3000\ W/m^2\ K$. If a gas is the heating medium, the value of α_{dh} may range from 1000 to 4000, according to the cleanliness of the gas.

6.4.3   Boiling fouling

When the fluid to be vaporised is allowed to boil in the heater, the fouling resistance may be many times greater than it would have been if boiling had been suppressed (so that all vaporisation took place in a flash vessel); this is because normal fouling processes are reinforced and magnified by scaling from solids dissolved in the liquid. Very little information has been published on this topic.

Engelbrecht and Hunter (1974) report the results of a test on a climbing-film evaporator for the concentration of aluminium sulphate solutions, operating under vacuum. The tubes were 6.1 m long with internal and external diameters of 28.5 and 34.9 mm respectively. With the boiling fluid at an inlet pressure of 0.84 bar and with a true mean temperature difference of 75.5 K, the achieved overall heat transfer coefficient (referred to the inside surface area) was 791 W/m² K, giving a boiling fouling coefficient of 1360 W/m² K. In general it is not possible to say anything more precise than that in an evaporator the boiling fouling coefficient will range from 1000 W/m² K with heavy scaling to 10 000 W/m² K with light scaling. By contrast, when vaporising a fairly corrosive liquid, the amount of fouling may be negligible.

Oil used to lubricate the compressors of refrigeration plant frequently contaminates the refrigerant. In the chiller, contamination levels up to 10 % have been found. Inorganic refrigerants, e.g. ammonia, are generally immiscible with oil. They deposit a fairly stable film of oil on the heated surface. With shellside boiling of ammonia, values of the boiling fouling coefficients are expected to range from 5000 W/m² K, with fairly clean ammonia and a good purge, to 600 W/m² K with total evaporation of ammonia with traces of oil in the feed. Halogenated organic refrigerants are generally miscible with lubricating oil; thus the presence of oil introduces a mass transfer resistance. Studies by Stephan (1963) show a deterioration of heat transfer coefficient, compared with clean refrigerant, that is equivalent to a boiling fouling coefficient of 5000 W/m² K with an oil content of 6 %, and approximately 2000 W/m² K with an oil content of 9 %.

If a proprietary lining is used to reduce the amount of fouling, it must be remembered that this introduces an extra thermal resistance. The heat transfer coefficient for such a lining can be calculated from equation [6.49] if the thickness and thermal conductivity of the lining are known. For example, a lining 0.2 mm thick might introduce an extra coefficient of 5000 W/m² K.

6.4.4 Extended surfaces

If the heating fluid is a gas on the shellside of the vaporiser, or flowing in a duct across a bank of tubes, it may be economic to use tubes with external fins; these could be longitudinal or transverse to the tube. For a gas at roughly atmospheric pressure, high fins should be considered, if a high boiling coefficient is expected; for a high-pressure gas or for a liquid with only a moderate boiling coefficient, low transverse fins are preferable. Allowance must be made for the fact that the tip of the fins must be hotter than the base of the fins (in contact with the tube) in order that heat might flow along the fins from tip to base. If the fins are made of a material of low thermal conductivity, such as stainless steel, the temperature drop along each fin is large compared with the temperature difference between the heating fluid and the tubes, especially with high fins. By contrast, there is usually only a relatively small temperature drop along a high aluminium fin, the thermal conductivity of aluminium being nearly 13 times greater than that of stainless steel. To allow for the thermal resistance of fins, a term called 'fin efficiency' is introduced; this is defined as the ratio of the heat transferred along a real fin to the heat that would be transferred along an identical fin of infinite thermal conductivity. Thus the following expression must be used to calculate the value of A_h, the effective heating surface area, for substitution in equation [6.3].

$$A_h = A_r + \eta_f A_f \qquad [6.52]$$

where A_r is the primary surface area, i.e. the area of bare tube between fins (m²), A_f is the secondary surface area, i.e. the area of the fins (m²) and η_f is the fin efficiency. For a longitudinal fin of constant thickness, Coulson and Richardson (1977), p. 255, give the fin efficiency as:

$$\eta_f = \frac{\tanh mH_f}{mH_f} \qquad [6.53]$$

where H_f is the height of the fin (m) and m is given by:

$$m = \sqrt{\frac{2\alpha_{ch}}{\lambda_f x_f}} \qquad [6.54]$$

where x_f is the thickness of the fin (m), α_{ch} is the combined heat transfer coefficient of the heating fluid (α_h) and the dirt deposited by it (α_{dh}) (W/m² K) and λ_f is the thermal conductivity of the material of the fin (W/m K). The combined heat transfer coefficient is given by:

$$\frac{1}{\alpha_{ch}} = \frac{1}{\alpha_h} + \frac{1}{\alpha_{dh}} \qquad [6.55]$$

The above method of calculating the fin efficiency of longitudinal fins may be used for the approximate estimation of the fin efficiency of low transverse fins. Further information on fin efficiency is given by Kern (1950), Chapter 16, and by Kraus and Kern (1972).

It must be remembered that all theoretical work on fin efficiency is based on the assumption of a constant heat transfer coefficient between the fluid and the surface. However, the coefficient may be different at different locations, from the root to the tip of the fin and on the tube. This may not have a significant effect in boiling or in single-phase flow, especially with high theoretical values of the fin efficiency, but with condensation, the heat transfer coefficient is much higher at the tip of the fin than on the tube; in such circumstances the fin efficiency is less than that predicted.

It is important to check that the use of fins will not lead to overheating the tube – see section 6.6 for advice on how to calculate the temperature of a tube. Thus, when a hot flue gas is flowing across a bank of tubes in a duct, fins are often used on all but the first two rows of tubes where heat transfer is exceptionally high, due to high gas temperatures, enhanced heat transfer coefficients at entry, and radiation from the walls of the duct.

6.5   Radiant heat transfer

When a hot gas is the source of heat, radiation is significant compared with convection if the gas contains polyatomic molecules, such as steam, ammonia or carbon dioxide. This is usually neglected when estimating the performance of the vaporiser, but it may be important in the estimation of the critical heat flux or the maximum temperature of the tubes. Also, with a bank of tubes in a duct, radiation from the wall will significantly increase the heat flux at the leading edge of the tubes in the first row in the gas stream (and in the second row if the tubes are staggered). To allow for radiation, replace α_h, the gas convective heat transfer coefficient, by $\alpha_h + \alpha_R$, where α_R is the heat transfer coefficient due to radiation, estimated as described in this section. It may be necessary to allow for radiation when using equation [6.3] and the equations in section 6.6.

The coefficient of radiant heat transfer from a gas to a surface may be estimated from:

$$\alpha_R = \tfrac{1}{2}(1 + \varepsilon_s)\varepsilon_g\alpha_{bR} \qquad [6.56]$$

where α_{bR} is the black-body radiation coefficient (W/m^2), ε_s is the emissivity of the tube surface and ε_g the emissivity of the gas. The value of α_{bR} can be read from Fig. 6.4, where it is plotted against T_h, the temperature of the gas, for different values of T_s, the temperature on the surface of the wall; it is necessary to estimate T_s by trial and error. The curves in Fig. 6.4 have been calculated from the Stefan–Boltzmann law, which states that radiation is proportional to T^4, as explained in textbooks on heat transfer, e.g. Coulson and Richardson (1977), p. 208. α_{bR} has been calculated from the following equation, which may be used instead of the curves.

$$\alpha_{bR} = 5.67 \times 10^{-8}(T_h^3 + T_h^2 T_s + T_h T_s^2 + T_s^3) \qquad [6.57]$$

Some values of ε_s are given in Table 6.3 for commonly used tube materials.

The emissivity of a gas, ε_g, depends on its temperature, T_g, and on its partial pressure (expressed in bars) and on L, the effective path length (expressed in metres). The following approximate method of estimating L has been taken from McAdams (1954), pp. 88, 89.

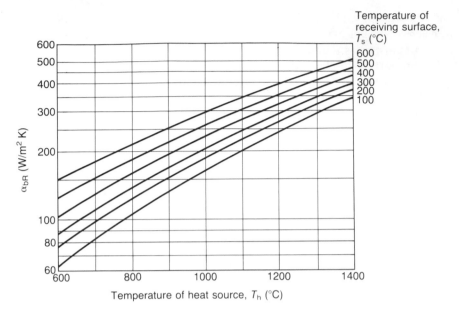

6.4 Coefficient of heat transfer for black-body radiation (α_{bR} from equation [6.57])

For the tubeside, $L = 0.9d_i$ [6.58]

For the shellside, $L = 3.6V/A_o$ [6.59]

where d_i is the internal diameter of the tube, V is the internal volume of the shell, excluding the volume of the tubes (m^3) and A_o is the external surface area of the tubes (m^2). The next step is to estimate the partial pressure of each radiating gas; this equals the total pressure multiplied by the molecular fraction of the gas.

In the case of radiation from a duct to the first row in a bank of tubes, ε_g may be taken as unity (see section 6.6.2). For the estimation of radiation from a gas inside a tube or in the midst of a tube bundle, the following procedure should be followed. If the gas consists only of monatomic and diatomic molecules, then $\varepsilon_g = 0$. Radiation from flue gases at nearly atmospheric pressure is dealt with thoroughly in the standard textbooks on heat transfer, e.g. Kern (1950), McAdams (1954). These may be used for dealing with radiation from a flue gas to a bundle of tubes. The only reference dealing with radiation from a gas at pressure is Wheatley (1972), who gives some experimental results for a flue gas at a pressure of 22 bar.

Figure 6.5 shows the product $\varepsilon_g . T_h$ (K) plotted against the product $(p_c + p_w)L$ (m bar), where p_c is the partial pressure of carbon dioxide and p_w is the partial pressure of water vapour (bar). The dotted line for a total pressure of 22 bar has been taken direct from Wheatley (1972). The full lines are for a total pressure of 1 bar; they have been calculated from the published data. This simplified method gives values for 1 bar that agree

6.3 Emissivity of surfaces

Surface	Emissivity (ε_s) at a temperature (°C) of			
	100	200	300	400
Aluminium				
Polished	0.09	—	—	—
Oxidised	0.11	0.12	0.13	0.14
Brass				
Polished	—	0.05	0.10	—
Oxidised	—	0.61	0.61	0.60
Graphite	0.44	0.47	0.49	0.51
Mild Steel				
Polished	0.12	0.17	—	—
Oxidised	0.80	0.80	0.80	0.80
Stainless Steel				
Oxidised	—	0.85	0.87	0.88
Titanium				
Polished	0.10	0.13	0.16	0.20
Oxidised	—	—	0.54	0.55

The values tabulated above have been taken from Hottel and Sarofim (1967) and Love (1968); these books should be consulted for further information

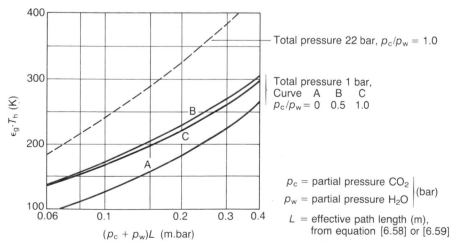

Total pressure 22 bar, $p_c/p_w = 1.0$

Total pressure 1 bar,

Curve	A	B	C
$p_c/p_w =$	0	0.5	1.0

p_c = partial pressure CO_2 ⎫
p_w = partial pressure H_2O ⎬ (bar)

L = effective path length (m), from equation [6.58] or [6.59]

6.5 Emissivity of gas containing carbon dioxide and water vapour

with the values calculated by the normal method to within 9% over the temperature range considered in the analysis (600 to 1100 °C). This graph may be used to estimate a rough value of ε_g by interpolation between curves. A comparison between the dotted curve and curve C shows that, with equal quantities of carbon dioxide and water vapour, increasing the total pressure from 1 to 22 bar increases the emissivity of the gas by 35 to 48%. A comparison between curves A, B and C shows the effect of changing the ratio of carbon dioxide to water vapour at atmospheric pressure.

6.6 Estimation of the temperature of tubes

In the normal course of rating calculations, the individual thermal resistances are estimated and added together to give the total resistance, which is equal to the reciprocal of U, the overall heat transfer coefficient in equation [6.3]; thus the interface temperatures between adjacent resistances are not given. For the purpose of mechanical design, it may be necessary to estimate certain temperatures of the tubes. The following problems are considered in this section.

1. When using a shell-and-tube heat exchanger with fixed tubesheets, it is necessary to know the difference between the mean temperature of the tubes and the mean temperature of the shell, because the tubesheets must be strong enough to cope with the differential thermal expansion between tubes and shell.

2. For the choice of a suitable material for the tubes, if operating in the region where the strength of the material falls substantially with increasing temperature, it is necessary to estimate the highest temperature that the tubes might reach at the hot end.

3. If the heating fluid is a gas containing a vapour, which, if it condensed on the surface, would produce a corrosive liquid, then it is necessary to estimate the lowest temperature that the tubes might reach at the cold end.

4. For the design of the tube-to-tubesheet joint, it may be necessary to determine the temperature distribution through the tubesheet.

In any of the calculations described in this section, it is essential to carry out the calculations for the worst possible operating conditions with respect to the problem under investigation. This may differ considerably from the flowsheet conditions used in thermal design; furthermore it may be difficult to estimate just what the worst conditions might be, as discussed in each of the following subsections.

6.6.1 Mean temperature difference between tubes and shell

This is the difference between the mean, with respect to both radius and length, of the temperature of the tubes and that of the shell. If the heat transfer coefficients do not change significantly along the length of the heat exchanger and if the boiling temperature is constant, the mean temperature difference is given by:

$$\bar{\theta}_{st} = U_o \, \Delta T_{lm}(1/\alpha_{fo} + 1/\alpha_{do} + 1/2\alpha_w) \qquad [6.60]$$

where $\bar{\theta}_{st}$ is the difference between the mean temperatures of the shell and tubes, ΔT_{lm} is the logarithmic mean of the temperature differences between fluids at inlet and outlet (see section 10.1), U_o is the overall heat transfer coefficient referred to the external surface area, α_{fo} and α_{do} are respectively the fluid and dirt coefficients for the fluid outside the tubes and α_w is the heat transfer coefficient for the wall (all temperatures being in K and coefficients in $W/m^2\,K$).

If the tubes and the shell are made of materials of significantly different coefficients of thermal expansion, e.g. stainless steel tubes in a mild steel shell, then it is necessary to calculate separate values of $\bar{\theta}_s$ and $\bar{\theta}_t$, defined respectively as the mean temperatures of the shell and the tubes above ambient temperature (see equations given by TEMA (1978) for equivalent differential expansion pressure with fixed tubesheets).

With boiling on the shellside:

$$\bar{\theta}_s = T_b - T_a \qquad \text{and} \qquad \bar{\theta}_t = \bar{\theta}_s + \bar{\theta}_{st} \qquad [6.61]$$

where T_b is the temperature of the boiling fluid and T_a is the ambient temperature (K). With boiling on the tubeside:

$$\bar{\theta}_s = T_b + \Delta T_{lm} - T_a \qquad \text{and} \qquad \bar{\theta}_t = \bar{\theta}_s - \bar{\theta}_{st} \qquad [6.62]$$

If the temperature of the boiling fluid changes significantly along the length of the heat exchanger, due to subcooling at inlet, superheating at outlet, changes in boiling temperature due to changes in composition or in hydrostatic pressure, the heat exchanger must be divided into lengths over which the changes are sufficiently small that an arithmetic mean may be taken of the values at the two ends of each length. If there are n such lengths, the overall mean values are given by:

$$\bar{\theta} = \left[\sum_{i=1}^{n} \theta_i L_i \right] \Big/ L \qquad [6.63]$$

where L_i is the length of the i th length and L is the total length (m).

If the heat source is condensing steam, the above equations may be simplified by putting $\Delta T_{lm} = T_s - T_b$, where T_s is the saturation temperature of the steam.

If boiling is taking place in vertical tubes, the effect of hydrostatic pressure is to increase the boiling temperature by an amount that is often significant in comparison with the temperature difference. If so the value of T_b to be substituted in the above equations must be the average with respect to length.

If the heating medium is superheated steam on the shellside and the shell is lagged, the temperature of the shell may be assumed to equal the temperature of the steam; the steam will travel some distance before losing its superheat. The temperature gradient may be estimated from:

$$-\frac{dT_h}{dz} = \frac{\alpha_f(T_h - T_s)}{\dot{M}c_p} \frac{A_o}{L} \qquad [6.64]$$

where T_h is the temperature of the superheated steam, T_s is the saturation temperature, A_o is the external surface area (m^2), c_p is the specific heat of superheated steam (J/kg K), L is the length of the tube (m), $\dot{M}$ is the mass flow rate of steam (kg/s), z is the distance along the tube (m) and α_f is the coefficient of heat transfer by convection from the steam to the tube, calculated as described in section 6.2.6, i.e. for single-phase flow (W/m^2 K). This requires step-wise integration. However, if it is assumed that over the important inlet region there is no significant change in the value of $\alpha_f/\dot{M}$, the mean shell temperature is given by:

$$\bar{\theta}_s = T_s - T_a + (T_{h\,in} - T_s)(1 - e^{-N})/N \qquad [6.65]$$

where $T_{h\,in}$ is the temperature of the superheated steam at inlet and N is given by:

$$N = \alpha_f A_o/\dot{M}c_p \qquad [6.66]$$

The values to be substituted in equation [6.66] should be the values at the inlet to the shell. The value of N is likely to rise along the length of the shell, but this means that the mean shell temperature will be somewhat less than that given by the above approximate method.

If the use of superheated steam leads to an unacceptably high shell temperature, it is necessary to put expansion bellows in the shell or use a vapour belt with a stepped shell, as shown in Fig. 4.10. Alternatively a desuperheater may be installed (see section 4.4).

When designing fixed tubesheets and using the methods described above to estimate the amount of differential thermal expansion between tubes and shell, it is difficult to decide what operating conditions will give the largest amount of differential expansion. The calculations should be carried out first at the flowsheet conditions, because the temperatures and coefficients for this will be given in the output of the computer program used in the thermal design. If it is found that differential expansion is likely to be a problem but it is desirable to persist with the intention to use fixed tubesheets without any expansion device, it may be necessary to estimate the amount of differential expansion under several different operating conditions before it is possible to determine the worst. As a first estimate, it should be assumed that the heat exchanger is operating with the maximum possible difference between inlet temperatures, with the inside of the tubes clean and the maximum amount of fouling on the shellside. Each attempt requires a rerun of the computer program to give the different temperatures and coefficients.

The above treatment is for steady-state heat transfer. If there is a sudden change in the temperature of either fluid, the temperature of the tubes will change to its new value in a few minutes, whereas the shell temperature will change more slowly. For approximate calculations of the stress induced by a sudden change in the inlet temperature of the tubeside fluid, because it might take several hours for the shell to reach its new temperature, it may be assumed that the tubes reach their new temperature before there is any significant change in the temperature of the shell.

6.6.2 Maximum tube temperature

The maximum tube wall temperature is given by:

$$T_{w\,max} = T_{h\,in} - (T_{h\,in} - T_v)U[(1/\hat{\alpha}_h + 1/\alpha_{dh})(A_v/A_h) + \tfrac{1}{2}\alpha_w(A_v/A_w)] \qquad [6.67]$$

where $T_{w\,max}$ is the maximum value of the temperature of the mid-plane of the wall, $T_{h\,in}$ is the inlet temperature of the heating fluid, T_v is the temperature of the fluid to be vaporised at the hot end, U is the overall heat transfer coefficient referred to the surface area in contact with the liquid to be vaporised, A is the surface area and α denotes the individual heat transfer coefficients, suffix v denoting the fluid to be vaporised, h the heating fluid, dh the dirt from the heating fluid and w the wall. This is derived from equation [6.3]. The superscript over $\hat{\alpha}_h$ denotes that this is the maximum value; it is given by

$$\hat{\alpha}_h = \alpha_h E + \alpha_R \qquad [6.68]$$

where α_h is the normal value of the convective heat transfer coefficient, E is the enhancement factor at entry, and α_R is the heat transfer coefficient due to radiation (if the heat source is a gas containing radiating molecules – otherwise $\alpha_R = 0$). The estimation of E is discussed below; the estimation of α_R has been dealt with in section 6.5.

With tubeside flow of the hot fluid, there are two factors producing a local enhancement in heat transfer coefficient at the entry to the tubes; firstly there is the extra turbulence due to the contraction in flow, which dies away during flow through the thickness of the tubesheet; secondly, as soon as the fluid reaches the region of heat transfer to the shellside fluid, there is an added enhancement due to the fact that the equilibrium temperature gradient has not yet been set up. ESDU (1968b) (in curve 2 of Fig. 4) shows that with a short calming section, corresponding to the thickness of the tubesheet, the local value of E is 1.9 after a length of $0.4d$, falling to 1.4 after a length of $2d$, where d is the internal diameter of the tube. Extrapolating the curve to the point of inlet suggests that $E = 2$ is a safe value to assume for the mean over a length of up to one internal diameter.

With shellside flow of the hot fluid the enhancement factor (E) can be determined from the work of Thomson *et al.* (1951) who carried out experiments on local heat transfer rates with a bank of staggered leading edge for the first six rows. For each tube, the coefficient was a maximum at the leading edge. Comparing each maximum with the mean coefficient for all the tubes at all angles gives $E = 1.35$ for the tubes in the first row and $E = 2.0$ for the tubes in the second row. The pitch of the tubes was twice their diameter; no information is available on more closely pitched tubes, as are normal, but it is expected that reducing the pitch would not greatly affect the enhancement at the first row but would increase it at the second row, due to the very high velocity there. As far as α_R, the radiation coefficient, is concerned, results by Ross (1967) show that at the normally used values of the pitch-to-diameter ratio, the ratio of the actual radiation to black-body radiation is 0.9 for the first row and 0.2 for the second row.

6.6.3 Minimum tube temperature

The minimum tube surface temperature in contact with the heating fluid is given by:

$$T_{w\,min} = T_{h\,out} - (T_{h\,out} - T_v)U(1/\alpha_h + 1/\alpha_{dh})(A_v/A_h) \qquad [6.69]$$

where $T_{w\,min}$ is the minimum value of the temperature of the surface of the wall in contact with the heating fluid, $T_{h\,out}$ is the outlet temperature of the heating fluid, the other symbols being as for equation [6.67]. This is derived from equation [6.3].

The calculation of the minimum surface temperature is important when the heating fluid is a gas containing a vapour which condenses to form a corrosive liquid and if the temperature of the fluid to be vaporised might be below the dew point. Before deciding to economise by using tubes that are not resistant to corrosion by the condensate, it is very important to ensure that all contingencies have been taken into account in the calculations showing that the temperature of the tubes could never fall below the dew point. Disastrous consequences have resulted from unexpectedly low boiling temperatures or heating gas outlet temperatures, or from unexpectedly high vapour contents or gas film coefficients.

In deriving equation [6.69] it is assumed that the heating fluid is perfectly distributed. Any maldistribution will increase the danger of condensation in regions where the velocity of the heating fluid is less than the mean. Thus with tubeside flow of the hot fluid, if a tube becomes partly choked the value of $T_{w\,min}$ for that tube will be less than average. With shellside flow of the hot fluid there is a grave danger of condensation in any stagnant zones.

6.6.4 Temperature distribution through a tubesheet

Figure 6.6 shows a section through a tubesheet and tube end; Figs 6.7 and 6.8 show an end view of part of a tubesheet with tubes on square and equilateral triangular pitch respectively. It is assumed that this is a problem in heat conduction in an axial direction only, i.e. that the temperature of the tubes at a distance z from the inlet is a function of z only, changes in temperature in a radial direction being negligible. Given T_s, the temperature of the shellside fluid, and $T_{h\,in}$, the temperature of the heating fluid entering the tubes, the objective is to estimate $T_{w\,max}$, the maximum tubesheet temperature (at entry to the tube).

Hot tubeside fluid at $T_{h\,in}$

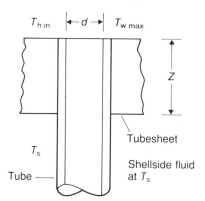

6.6 Section through a tubesheet

The first step is to calculate A_t, the tubesheet area per tube that is available for conducting heat from the hot fluid entering the tube through the tubesheet, to the vaporising fluid in the shell. If d is the internal diameter of the tubes and P_t is their pitch, from Fig. 6.7, for a square pattern:

$$A_t = P_t^2 - \frac{\pi}{4}d^2 \qquad\qquad [6.70]$$

from Fig. 6.8 for a triangular pattern

$$A_t = 0.866P_t^2 - \frac{\pi}{4}d^2 \qquad\qquad [6.71]$$

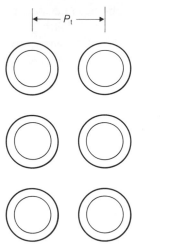

6.7   Tubes on a square pattern

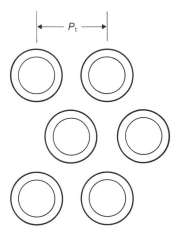

6.8   Tubes on a triangular pattern

It is also necessary to estimate the values of α_s, the heat transfer coefficient from the tubesheet to the boiling liquid (including any fouling) and of $\hat{\alpha}_h$, the heat transfer coefficient from the tubeside fluid to the tube in the tubesheet, including the entry enhancement, calculated as described in section 6.6.2, but excluding any possible fouling by the hot gas. Assuming that A_o is the effective area for α_s and that the heat transfer coefficient from the hot fluid to the outer face of the tubesheet is also equal to $\hat{\alpha}_h$, it can be shown that the temperature of the tube at inlet is:

$$T_{w\,max} = T_{h\,in} - \frac{(T_{h\,in} - T_s)L}{\sinh K + L\cosh K + \dfrac{\hat{\alpha}_h}{\alpha_s}L(\cosh K + L\sinh K)} \tag{6.72}$$

where $K = Z\sqrt{\dfrac{\pi d\hat{\alpha}_h}{A_t\lambda_w}}$ \qquad\qquad [6.73]

and $L = \sqrt{\dfrac{\pi d\lambda_w}{A_t\hat{\alpha}_h}}$ \qquad\qquad [6.74]

where Z is the thickness of the tubesheet (m) and λ_w its thermal conductivity (W/m K).

If it is found that $T_{w\,max}$ is too high, ferrules may be installed as a first step. If further protection is needed, the next step is to insulate the outside of the tubesheet, thus using the arrangement illustrated in Figs 3.16 and 6.9. The inclusion of ferrules necessitates the recalculation of α_h, the normal value of the convective heat transfer coefficient, which is increased owing to the greater velocity inside the ferrule (of diameter d_{fe}) compared with inside the tube (of diameter d). It must be remembered that the installation of

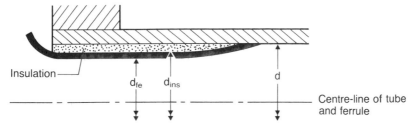

6.9 Details of a ferrule

ferrules also produces a considerable increase in the tubeside pressure drop. The new (lower) value of $\hat{\alpha}_h$ is given by equation [6.75], obtained from equations [6.3], [6.49], [6.50] and [6.68].

$$\frac{1}{\hat{\alpha}_h} = \frac{d}{2\lambda_{ins}} \ln\left(\frac{d}{d_{ins}}\right) + \frac{1}{\alpha_h E + \alpha_R} \frac{d}{d_{fe}}$$ [6.75]

where d_{fe} is the internal diameter of the ferrule, d_{ins} is the internal diameter of the thermal insulating material, and λ_{ins} is the thermal conductivity of the thermal insulating material in the ferrule. It is important to ensure that there is a seal between the ferrule and the tube; otherwise a leak of hot gas between the ferrule and the tube might result in an even higher temperature at the end of the tube. Assuming a perfect seal, and with no insulation of the outside of the tubesheet, the new (reduced) value of $T_{w\,max}$ can be determined by the procedure described at the beginning of this section, using the new value of $\hat{\alpha}_h$ in equations [6.72], [6.73] and [6.74], there being no change in the values of A_t, d or λ_w.

If it is decided to insulate the outside of the tubesheet, it may be assumed that the heat transferred from the hot fluid to the outside of the tubesheet is negligible, and equation [6.72] simplifies to:

$$T_{w\,max} = T_{h\,in} - \frac{T_{h\,in} - T_s}{\cosh K + \dfrac{\hat{\alpha}_h}{\alpha_s} L \sinh K}$$ [6.76]

The values of K and L are as calculated previously from the new, lower, value of $\hat{\alpha}_h$.

The equations given here tend to overestimate $T_{w\,max}$ because they neglect conduction along the tubes beyond the tubesheet, where the tubes act as fins. Also, with an uninsulated tubesheet $T_{w\,max}$ is likely to be overestimated, because the heat transfer coefficient from the hot fluid to the outer face of the tubesheet is less than $\hat{\alpha}_h$, except when there is intense radiation.

6.6.5 Mean wall temperature

Heat transfer coefficients are normally calculated from the physical properties evaluated at T_b, the mean bulk temperature of the fluid. However there may be a significant radial change of temperature between the bulk of the fluid and the wall. Some research workers have derived empirical factors to correct for the consequent radial changes in physical properties. Equations [6.12] and [6.20] give these factors for parallel flow and cross flow respectively. They require a knowledge of T_w, the mean with respect to length of the interface temperature between fluid and solid, usually known as 'the mean wall temperature'. For the hot fluid,

$$T_{wh} = T_{bh} - (T_{bh} - T_{bv})\frac{UA_v}{\alpha_h A_h} \qquad [6.77]$$

For the liquid to be vaporised

$$T_{wv} = T_{bv} + (T_{bh} - T_{bv})\frac{U}{\alpha_v} \qquad [6.78]$$

where A_h is the total surface area in contact with the hot fluid and A_v is the total area in contact with the fluid to be vaporised, U is the overall heat transfer coefficient, referred to A_v, and α_h and α_v are the coefficients for the hot and cold fluids respectively. The additional subscripts h and v relate to the temperatures of the hot fluid and the fluid to be vaporised respectively.

For a first approximation, the values of α_h or α_v must be evaluated from the physical properties at the mean bulk temperature, without any correction. It is then possible to use equations [6.12] or [6.20] to obtain a more accurate value of the coefficient. Iteration may be necessary.

CHAPTER 7

Calculation of heat transfer in vaporisation

This chapter is concerned with the estimation of the heat transfer coefficient from the heated surface to the liquid to be vaporised. Section 7.1 deals with single-phase heating of a liquid, i.e. where boiling has been suppressed by the pressure being greater than the saturation pressure and the temperature of the wall being below that required to initiate nucleate boiling (see Ch. 2 for an account of the physical processes involved in vaporisation). Sections 7.2 to 7.5 deal with two-phase boiling flow. Section 7.6 recommends a method of estimating the heat transfer coefficient in evaporation from the surface of a falling film of liquid. The correlations given in these sections apply only to single-component liquids, and it is known that significantly lower coefficients may be obtained with mixtures, as discussed in section 7.7. Proprietary devices for enhancing heat transfer in boiling are discussed in section 7.8.

7.1 Single-phase liquid convection

Single-phase heating of the liquid to be vaporised occurs when boiling is deliberately suppressed throughout. The object is to avoid scaling or fouling. This is achieved by the introduction of a restrictive device between the heater and the flash vessel where vapour and liquid are separated, as described in section 2.9. Single-phase heating also occurs at the base of heated vertical tubes due to the increase in the boiling temperature caused by the hydrostatic head (see section 3.1).

The methods described in section 6.2 for calculating the single-phase forced convection heat transfer coefficients of the heating fluid may also be applied to the heating of the fluid to be vaporised. Section 7.1.4 describes the estimation of single-phase heat transfer coefficients of a heated horizontal tube immersed in a liquid in an unstirred vessel, so that the transfer of heat is by free convection.

There are two limits to the single-phase heating of a liquid:

1. Two-phase convective boiling begins to occur in the bulk of the liquid, as described in section 7.1.1.
2. The wall becomes sufficiently superheated for the onset of nucleate boiling, the criteria for which are discussed in section 7.1.2.

7.1.1 The onset of two-phase convective boiling

As has already been mentioned in section 2.4, a liquid may be superheated above its saturation temperature if no nuclei are present to initiate boiling. If there has been subcooled nucleate boiling, the bubbles formed at the wall will grow into larger bubbles as soon as the bulk of the liquid reaches saturation temperature, thus preventing superheating of the liquid. Otherwise most industrial process liquids contain a sufficient quantity of nuclei in the form of solid particles and dissolved gases for the amount of superheat to be negligible. Bulk superheating of the liquid is dangerous, in that it will cause bumping and instability in natural convection vaporisers, as described in section 9.2.5. It can be avoided if the heat flux is sufficient to obtain nucleate boiling, as described in the next section. Except with especially clean liquids, it may be assumed that two-phase convective boiling starts as soon as the bulk of the liquid reaches the saturation temperature corresponding to the local pressure. Two-phase convective boiling is dealt with in section 7.2.

7.1.2 The onset of nucleate boiling

As the temperature of a heated wall is gradually increased above the saturation temperature of the liquid passing over it, nucleate boiling starts first at the larger nucleation sites and gradually extends to the smaller sites as the amount of superheat of the surface is increased. The situation is known as 'subcooled nucleate boiling' if the bulk of the liquid is still below saturation temperature; otherwise it is called 'saturated nucleate boiling' or 'two-phase convective boiling'. Accounts of how nucleate boiling begins have appeared in most of the textbooks on the subject, e.g. by J. G. Collier in section 2.7.3 of Schlünder (1983), in Section (8.3) of Bergles *et al.* (1981) and in Collier (1981) Chapter 5.

Butterworth and Shock (1982) recommend the correlation of Davis and Anderson (1966) for the determination of the temperature of the wall for the onset of nucleate boiling (T_{wONB}). This gives

$$T_{wONB} = T_s + \left[\frac{8\sigma \dot{q} T_s}{\lambda_l \, \Delta h_v \rho_g} \right]^{\frac{1}{2}} \qquad [7.1]$$

A similar correlation with the addition of an empirical correction factor, was put forward by Frost and Dzakowic (1967), who claimed to be able to cover a wider range of liquids; their correlation gives:

$$T_{wONB} = T_s + \left[\frac{8\sigma \dot{q} T_s}{\lambda_1 \Delta h_v \rho_g} \right]^{\frac{1}{2}} Pr_1 \qquad\qquad [7.2]$$

where T_{wONB} is the temperature of the wall at the onset of nucleate boiling, T_s is the saturation temperature of the liquid (K), $\dot{q}$ is the heat flux at the wall (W/m^2), Pr_1 is the Prandtl number of the liquid (as defined by equation 6.7), Δh_v is the latent heat of vaporisation of the liquid (J/kg), λ_1 is the thermal conductivity of the liquid (W/m K), ρ_g is the density of the vapour (kg/m^3) and σ is the surface tension of the liquid (N/m).

Equation [7.1] is likely to underpredict T_{wONB}; alternatives are suggested by Davis and Anderson (1966).

7.1.3 Example: onset of nucleate boiling

A domestic electric kettle with a power input of 2 kW has an element of 8 mm diameter, 0.7 m long. What is the temperature of the element when nucleate boiling begins? The properties of saturated water at atmospheric pressure are: $T = 373$ K, $\rho_g = 0.598$ kg/m^3, $\Delta h_v = 2257$ kJ/kg, $c_{pl} = 4222$ J/kg, $\eta_1 = 2.8 \times 10^{-4}$ Ns/m^2, $\lambda_1 = 0.6789$ W/m K, $\sigma = 0.0589$ N/m.

If the water is at 40 °C when nucleation begins, what is the heat transfer coefficient?

The heat flux is obtained by dividing the given heat input (2000 W) by the surface area of the heating element ($\pi d_o L$); thus

$$\dot{q} = \frac{2000}{\pi \times 0.008 \times 0.7} = 113\,680 \text{ W/m}^2$$

The Prandtl number of the liquid is obtained from equation [6.7]

$$Pr_1 = 2.8 \times 10^{-4} \times 4222/0.6789 = 1.741$$

From equation [7.1]

$$T_{wONB} = 373 + \left[\frac{8 \times 0.0589 \times 113\,680 \times 373}{0.6789 \times 2\,257\,000 \times 0.596} \right]^{\frac{1}{2}}$$

$$= 373 + 4.7 = 377.7 \text{ K } (104.7\,°C)$$

If equation [7.2] were used, ($T_{wONB} - T_s$) would be increased by a factor of Pr_1, compared with that given by equation [7.1], giving $T_{wONB} = 373 + 4.7 \times 1.741 = 381.2$ K (108.2 °C).

For water the first correlation is probably more accurate. It may therefore be concluded that nucleate boiling begins when the surface of the element reaches approximately 105 °C.

The temperature difference between the element and the water is

$\theta = 105 - 40 = 65\,°C$

hence the heat transfer coefficient is

$\alpha = 113\,680/65 = 1749\ \mathrm{W/m^2\,K}$

(This is somewhat greater than the value of $1560\ \mathrm{W/m^2\,K}$ that has been calculated by the method given in the next section.)

7.1.4   Free convection outside a horizontal tube

This section deals with the estimation of the heat transfer coefficient with free convection in single-phase flow, i.e. before the bulk liquid temperature has reached its boiling temperature and when the surface temperature is less than that for the onset of nucleate boiling.

Churchill and Chu (1975) showed that the heat transfer coefficient for free convection from a horizontal cylinder immersed in a fluid can be correlated by:

$$Nu^{\frac{1}{2}} = 0.60 + \left[\frac{Ra}{\{1 + (0.559/Pr)^{9/16}\}^{16/9}} \right]^{1/6} \times 0.387 \tag{7.3}$$

where Nu is the Nusselt number, defined by equation [6.6], d being the external diameter of the tube (m), for use in equation [6.6], Pr is the Prandtl number, defined by equation [6.7] and Ra is the Rayleigh number, defined by

$$Ra = \beta \frac{gd^3\rho^2\theta c_p}{\lambda\eta} \tag{7.4}$$

where c_p is the specific heat capacity of the fluid (J/kg K), g is the acceleration due to gravity ($9.81\ \mathrm{m/s^2}$), β is the coefficient of volumetric expansion of the fluid, i.e. the increase in volume per degree rise in temperature divided by the initial volume ($\mathrm{K^{-1}}$), θ is the temperature difference between the surface of the tube and the bulk of the fluid (K), λ is the thermal conductivity of the fluid (W/m K), η is the dynamic viscosity of the fluid ($\mathrm{N\,s/m^2}$) and ρ is the density of the fluid ($\mathrm{kg/m^3}$).

Equations [7.3] and [7.4] can be rearranged and combined with equation [6.6] to give α, the heat transfer coefficient ($\mathrm{W/m^2\,K}$).

$$\alpha = \lambda \left[\frac{0.60}{d^{\frac{1}{2}}} + \phi_f \theta^{1/6} \right]^2 \tag{7.5}$$

where

$$\phi_f = 0.387(\beta g\rho^2 c_p/\lambda\eta)^{1/6}[1 + (0.559/Pr)^{9/16}]^{-8/27} \tag{7.6}$$

Values of λ and ϕ_f are given in Tables 7.1 and 7.2 for water and ammonia

7.1   Properties of water for estimation of free convection

	Temperature (°C)			
	50	**100**	**150**	**200**
Thermal conductivity (W/m K)	0.642	0.679	0.683	0.663
Physical properties parameter ϕ_f (m$^{-\frac{1}{2}}$K$^{-\frac{1}{6}}$)	21.8	24.6	26.9	29.2

7.2   Properties of ammonia for estimation of free convection

	Temperature (°C)			
	− 30	**0**	**30**	**60**
Thermal conductivity (W/m K)	0.609	0.539	0.470	0.400
Physical properties parameter ϕ_f (m$^{-\frac{1}{2}}$K$^{-\frac{1}{6}}$)	26.7	29.1	31.7	34.8

at various temperatures. The physical properties were obtained from ESDU (1980).

7.1.5   Example: free convection heating

Water at 100 °C is being heated by the condensation of steam at 120 °C inside horizontal tubes submerged in the water. What are the convective coefficient and heat flux if the combined heat transfer coefficient for the condensing steam and the wall (including dirt) is 7200 W/m² K? The outside diameter of the tubes is 20 mm.

Let θ be the difference between the temperature of the outside of the tubes and the temperature of the water (100 °C). Then the heat flux from the steam to the outside of the tubes is given by:

$$\dot{q}_h = 7200\{120 - (100 + \theta)\} = 7200(20 - \theta) \qquad [7.7]$$

From Table 7.1 for water at 100 °C, $\lambda = 0.679$ and $\phi_f = 24.6$. Substituting in equation [7.5] gives:

$$\alpha = 0.679(0.60/0.020^{\frac{1}{2}} + 24.6 \times \theta^{1/6})^2$$

$$= 0.679(4.24 + 24.6\theta^{1/6})^2 \qquad [7.8]$$

The heat flux from the outside of the tubes to the water is given by

$$\dot{q}_c = \alpha\theta \qquad [7.9]$$

Under steady conditions of operation, the two heat fluxes must be equal. Thus a value of θ must be found by trial and error such that $\dot{q}_h = \dot{q}_c$.

Try $\theta = 17\,°C$. From equation [7.7]

$\dot{q}_h = 7200(20 - 17) = 21\,600\ \text{W/m}^2$

From equation [7.8]

$\alpha = 0.679(4.24 + 24.6 \times 17^{1/6})^2 = 1296\ \text{W/m}^2\,\text{K}$

From equation [7.9]

$\dot{q}_c = 1296 \times 17 = 22\,032\ \text{W/m}^2$

Repeating the calculations with $\theta = 16.95\,°C$ gives $\dot{q}_h = 21\,960\ \text{W/m}^2$, $\alpha = 1295$ and $\dot{q}_c = 21\,950\ \text{W/m}^2$. Hence the convective coefficient is $1295\ \text{W/m}^2\,\text{K}$ and the heat flux is $22.0\ \text{kW/m}^2$.

7.2   Two-phase convection, with wetted wall

The present state of knowledge on flow boiling has been summarised by Butterworth and Shock (1982). This section is concerned with the estimation of the local heat flux to a boiling liquid under the conditions listed below:

1.   The liquid to be vaporised is at its boiling temperature.
2.   The amount of liquid present is sufficient to wet the wall (see section 7.5).
3.   The wall temperature is below that for the onset of nucleate boiling.

The last condition is dealt with in section 7.1.2 and illustrated in Figs 7.1 and 7.2 below. These show the various temperatures plotted against z, the distance from the inlet of liquid to be vaporised. The boiling temperature (saturation temperature) is T_s; T_v is the temperature of the liquid to be vaporised, T_w of the wall and T_h of the heating fluid; T_{wONB} is the temperature of the wall for the onset of nucleate boiling, determined from equation [7.1]. The most important quantity in boiling is termed 'the quality', which is denoted by the symbol x_g and defined as the ratio of the mass flow rate of vapour at any location to the total mass flow rate.

At the entry to the preheating zone, the temperature of the liquid (T_v) rises rapidly, but the slope of the curve of T_v against z decreases as the heat flux $(\dot{q})$ decreases due to the reduction in the temperature difference $(T_h - T_v)$. Similarly there is a fall in T_{wONB} because, from equation [7.1], $(T_{wONB} - T_s)$ is proportional to $\dot{q}^{\frac{1}{2}}$. With a relatively small temperature difference between T_h and T_s, it is likely that the curve for T_{wONB} will never intersect the curve for T_w, as shown in Fig. 7.1, so nucleate boiling will never occur. When the heated liquid reaches its boiling temperature the liquid may become slightly superheated before bubbles begin to form. This is shown in Fig. 7.1 by the curve for T_v rising to above the line for T_s and

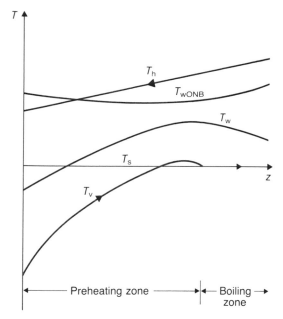

7.1 T/z diagram without nucleate boiling

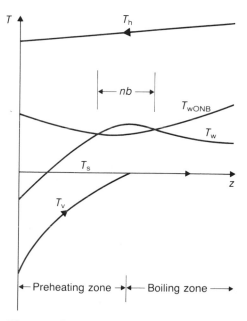

7.2 T/z diagram with a zone of nucleate boiling (nb)

then falling to T_s when bulk boiling is established. Industrial fluids normally contain sufficient impurities to avoid serious superheating, but with very clean liquids there is a danger of serious superheating, as discussed in section 7.1.1.

When T_h is very much greater than T_s (as in a fired or waste heat boiler) nucleate boiling will predominate. Figure 7.2 shows an intermediate condition, where there is nucleate boiling over a limited zone, marked 'nb'. Beyond this zone the convective coefficient has increased to such an extent that T_w has fallen significantly, and the resulting increase in $\dot{q}$ has caused T_{wONB} to increase and to exceed T_w, thus ending nucleation. Under the conditions shown in Fig. 7.2, the bubbles formed in the region of subcooled nucleate boiling provide a nucleus that grows when the bulk temperature reaches the saturation temperature, thus preventing gross superheating.

The steps in calculating the two-phase convective contribution to heat transfer in boiling are as follows:

1. Calculate the liquid convective heat transfer coefficient, α_{lo}, at zero quality, i.e. when the liquid has been heated to its saturation temperature but before any vapour has been established in the bulk of the fluid, using the appropriate correlation in section 6.2.

2. Calculate the enhancement factor (F_o) due to the presence of bubbles when the quality has reached x_g, applying the method in the appropriate section below to several values of x_g up to $x_{g\,out}$, the final (outlet) value.

3. Calculate the convective boiling heat transfer coefficient from

   $$\alpha_{cb} = F_o\alpha_{lo} \qquad\qquad [7.10]$$

4. Calculate U, the overall heat transfer coefficient, from equation [6.3], putting $\alpha_v = \alpha_{cb}$.

5. Calculate the heat flux from

   $$\dot{q} = U(T_h - T_s) \qquad\qquad [7.11]$$

6. If in doubt, check the validity of the assumption that there is no nucleate boiling by ensuring that $T_w - T_s$ ($= \dot{q}/\alpha_{cb}$) is less than $T_{wONB} - T_s$, determined from equation [7.1]. Otherwise use section 7.3 to determine the contribution to heat transfer of nucleation and section 7.4 to combine the two contributions.

7. If in doubt, consult Chapter 8 to check that the critical heat flux, $\dot{q}_{cr}$, exceeds $\dot{q}$ by a safe margin (say 50%).

7.2.1   Flow in tubes – the Chen method

Chen (1966) studies the flow of boiling saturated water and organic fluids in vertical channels with both forced convective boiling and nucleate boiling contributing to heat transfer. He assumed that the two could be combined by simple addition of the two contributions estimated separately. His procedure for predicting the convective component is given below: estimating the nucleate component is described in section 7.4.1. Chen based the latter on the analysis of Forster and Zuber (1955). He subtracted the nucleate component from experimentally determined values of the boiling heat transfer coefficient to determine the convective component. Most of the experiments were for flow up a tube, but results were also satisfactorily incorporated for the flow of water up an annulus and down a tube.

Chen used experimental data from different sources, obtained with water or one of several organic fluids, to determine experimental values of F, the convective boiling enhancement factor compared with the heat transfer coefficient for the liquid flowing alone (α_1); these were plotted against X, the parameter devised by Martinelli for correlating data on two-phase pressure drop – see Lockhart and Martinelli (1949). This parameter is defined as:

$$X = \sqrt{\frac{(dp/dz)_1}{(dp/dz)_g}} \qquad [7.12]$$

where dp/dz is the single-phase pressure gradient due to friction (see section 9.1). Subscript l denotes the value when the liquid is flowing alone and subscript g the value when the vapour is flowing alone. In two-phase boiling, the reciprocal of X is zero initially and this increases as the quality increases from its initial value of zero. In boiling flow either phase flowing alone is likely to be in turbulent flow, so X is estimated on this assumption and denoted by X_{tt}. Hence, assuming that the pressure drop in flow inside a tube is proportional to the velocity raised to the power of 1.8,

$$\frac{1}{X_{tt}} = \left(\frac{x_g}{1-x_g}\right)^{0.9}\left(\frac{\rho_1}{\rho_g}\right)^{0.5}\left(\frac{\eta_g}{\eta_1}\right)^{0.1} \qquad [7.13]$$

Table 7.3 gives some points from the curve recommended by Chen (1966) for determining the value of F from the value of $1/X_{tt}$ calculated from equation [7.13].

7.3 Chen's enhancement factor for flow in ducts

$1/X_{tt}$	0.2	0.5	1.0	2.0	5.0	10	30	100
F	1.2	1.8	2.7	4.2	8.0	13	29	70

The experimentally determined values of F lie within $\pm 30\%$ of the values recommended by Chen.

J. G. Collier in section 8.5.3 of Bergles *et al* (1981) and in section 2.7.3 of Schlünder (1983) gives the following equation to fit the curve proposed by Chen:

If $1/X_{tt} \leqslant 0.1$, $F = 1$;

if $1/X_{tt} > 0.1$,

$$F = 2.35(0.213 + 1/X_{tt})^{0.736} \qquad [7.14]$$

This equation gives excellent agreement with the values in Table 7.3.

The convective boiling heat transfer coefficient (α_{cb}) can then be found by multiplying the heat transfer coefficient for the liquid flowing alone (α_l) by the enhancement factor (F). As vaporisation takes place, the quantity of liquid flowing in the tube is reduced; consequently α_l is reduced. It is more convenient for calculations in boiling flow to relate α_{cb} to α_{lo}, the value of α_l when $x_g = 0$, because this is constant along the tube. Assuming that the heat transfer coefficient for the liquid flowing alone is proportional to the velocity raised to the power of 0.8, the value of F_o for substitution in equation [7.10] is related to F and x_g by:

$$F_o = F(1 - x_g)^{0.8} \qquad [7.15]$$

Chen gives no indication of the maximum value of $1/X_{tt}$ for which his method is valid; some of the points on his graph are up to a value of 100. However, Butterworth and Shock (1982) in their Fig. 6 show a steep fall in boiling heat transfer coefficient as $1/X_{tt}$ exceeds 70.

Above a critical value, the transition to mist flow takes place and the heated wall becomes dry, the heat transfer coefficient falling abruptly to the low value obtained in dry-wall convection. Section 8.2.6 suggests how it might be possible to estimate the quality at which the transition to mist flow takes place, in flow up a vertical tube. Section 7.5 deals with the prediction of the heat flux in dry-wall convection.

7.2.2 Flow in tubes – the Shah method

Shah (1976) considers the hundreds of correlations that have been proposed for estimating the heat transfer coefficients for boiling liquids and considers that most of them are not reliable beyond the range of data on which they were based, except for that of Chen for vertical flow. He developed a new correlation for saturated boiling at subcritical heat fluxes. His method, neglecting the benefits of nucleation, is described below, and the enhancement due to nucleation may be estimated as described in section 7.4.2. For horizontal tubes with low Froude numbers (defined below), the Shah method should be used in preference to the Chen method because it allows for the effects of stratification (see section 9.2.1). Otherwise it is difficult to say which method is better. Both have a similar approach and so would not be expected to give widely different answers. The safest approach is to use whichever gives the lower values for the enhancement factor F.

The first step in applying the Shah method to boiling in horizontal tubes is to determine Fr_l, the Froude number applied to all-liquid flow and defined by:

$$Fr_l = \frac{\dot{m}^2}{\rho_l^2 g d} \qquad [7.16]$$

If $Fr_l < 0.04$, gravity forces are significant and stratification occurs; then it is necessary to determine a correction factor from:

$$K_{Fr} = (25 \, Fr_l)^{-0.3} \qquad [7.17]$$

If $Fr_l > 0.04$ gravity forces are negligible in comparison with inertia forces, so $K_{Fr} = 1$ for horizontal or inclined pipes. For vertical pipes, $K_{Fr} = 1$ at all rates.

The next step is to determine the convection number (Co_h), modified to allow for stratification in horizontal pipes; this is defined as:

$$Co_h = K_{Fr} \left[\frac{1 - x_g}{x_g} \right]^{0.8} \left[\frac{\rho_g}{\rho_l} \right]^{0.5} \qquad [7.18]$$

The value of F for substitution in equation [7.15] is then equal to F_{cb} found from:

$$F_{cb} = 1.0 + 0.8 \exp(1 - \sqrt{Co_h}), \qquad \text{if } Co_h > 1.0 \qquad [7.19]$$

otherwise, $F_{cb} = 1.8 Co_h^{-0.8} \qquad [7.20]$

Shah is not able to give any precise advice on the estimation of the critical quality above which the amount of liquid present will not adequately wet the wall. However, the studies of Whiteway (1977) suggest that Shah's method may be safely used when $Co_h > 0.02$ (see section 8.2.6). The prediction of the heat flux in dry-wall convection is discussed in section 7.5.

7.2.3   Upward flow outside tube-bundles

With shellside vaporisation, flow is normally upwards with forced circulation and must be upwards with natural circulation. There is a high degree of turbulence and of mixing of vapour and liquid. The estimation of pressure gradient is discussed in sections 9.1.3 and 9.3.2.

J. W. Palen, in section 3.6.2 of Schlünder (1983) recommends the following equation for determining F, the enhancement factor

$$F = (\Delta p_{tp}/p_l)^{0.45} \qquad [7.21]$$

where Δp_{tp} is the two-phase pressure drop per row of tubes and Δp_l is the single-phase pressure drop when the liquid is flowing alone.

Alternatively, it may be adequate to take the following simplified approach. If it is assumed that the mixing of liquid and vapour is so efficient that a homogeneous mixture is produced, the density of the two-phase mixture (ρ_{tph}) is

$$\rho_{tph} = [x_g/\rho_g + (1 - x_g)/\rho_l]^{-1} \qquad\qquad [7.22]$$

In flow outside tube-bundles, the quality (x_g) is not likely to exceed 20%. As x_g increases from zero to its final value, the velocity of the mixture increases in proportion to the two-phase specific volume, which is equal to the reciprocal of ρ_{tph}. Assuming that the heat transfer coefficient depends on the properties of the liquid, i.e. assuming that the wall is completely wetted at all places by impingement, the heat transfer coefficient is given by the equation [7.23], assuming also that the heat transfer coefficient is proportional to the velocity raised to the power of 0.6, as in equation [6.19].

$$\alpha_{cb} = \alpha_{lo}[1 + (\rho_l/\rho_g - 1)x_g]^{0.6} \qquad\qquad [7.23]$$

where α_{lo} is the all-liquid convective heat transfer coefficient calculated from equation [6.19], using the physical properties of the liquid at the saturation temperature.

It is important to remember that there will always be a certain amount of slip, i.e. the vapour will move faster than the liquid; this results in the velocity and the pressure drop being less than the values calculated on the assumption of homogeneous flow, so the heat transfer coefficient will be less than that predicted by equation [7.23]. Slip is likely to be greater when the tubes are in line on a square pitch than with staggered arrangements, the vapour flowing fast up the channels between the tubes. On the other hand, heat transfer will be enhanced by the impingement of fast-moving drops on the tubes. Sections 7.4.3 and 7.4.6 explain how to combine this treatment of the convective component with the nucleate component of boiling. The procedure is evaluated by comparison with published experimental data.

With natural circulation it is difficult to estimate the flow rate through the bundle of a shellside vaporiser. Thus in the horizontal thermosyphon boiler shown in Fig. 3.18 it is necessary to estimate the external recirculation rate through the pipe back to the distillation column. In the kettle reboiler (Fig. 3.19) it is necessary to estimate the internal recirculation rate through the space between the bundle and the shell. The estimation of the flow rate in natural circulation is described in section 9.6.

7.2.4   Downward flow outside tube-bundles

Barbe et al. (1971) have shown in their experiments on the vaporisation of propane in downward flow outside a bundle of coiled tubes that the enhancement is given by:

$$F = \left[1 + \frac{1}{X_{tt}}\right]^{0.8} \qquad\qquad [7.24]$$

They also show how equation [7.24] can be deduced theoretically by assuming that the liquid and vapour flow as separate streams at different velocities such that the pressure drop of the liquid stream is equal to that of the vapour stream. The two-phase heat transfer coefficient is then given by

$$\alpha_{cb} = F\alpha_1 \qquad\qquad [7.25]$$

where α_1 is the heat transfer coefficient calculated from equation [6.26] for the liquid flowing alone.

7.2.5 Example: in-tube convective boiling

Ammonia boiling at $-20\,°C$ flows through horizontal tubes of internal diameter 13.6 mm at a rate of 580 kg/s m². Heat is supplied by a fluid at $-15.2\,°C$, the combined heat transfer coefficient for the heating fluid and wall (including dirt) being 8200 W/m² K. What is the heat flux when the quality is 0, 5%, 10%, 15% and 20% assuming that there is no nucleate boiling? Is this assumption justified? From ESDU (1980) the properties of ammonia at a saturation temperature of $-20\,°C$ are:

$p_s = 1.9010$ bar (190 100 Pa), $\qquad \rho_g = 1.6030 \qquad$ and $\qquad \rho_1 = 666$ kg/m³,

$\Delta h_v = 1329$ kJ/kg, $\qquad c_{pl} = 4493$ J/kg K, $\qquad \eta_g = 8.64 \qquad$ and

$\eta_1 = 236\,\mu$ Ns/m²

$\lambda_1 = 0.5857$ W/m K, $\qquad \sigma = 0.030\,73$ N/m.

From equation [6.7] the Prandtl number of the liquid is:

$$Pr_1 = 236 \times 10^{-6} \times 4493/0.5857 = 1.810$$

From equation [6.8] the Reynolds number of the liquid at zero quality is

$$Re_{lo} = 580 \times 0.0136/236 \times 10^{-6} = 33\,420$$

From equation [6.9] the Nusselt number at zero quality is given by

$$\ln Nu_{lo} = -3.796 + 0.795 \ln 33\,420 + 0.495 \ln 1.810 - 0.0225(\ln 1.810)^2$$
$$= 4.771$$

Hence $Nu_{lo} = 118.1$

From equation [6.6] the internal heat transfer coefficient for all liquid is given by:

$118.1 = \alpha_{lo} \times 0.0136/0.5857$, whence $\alpha_{lo} = 5085$ W/m² K.

To use the method of Shah, it is first necessary to determine whether stratification is significant. This depends on the Froude number, which is given by equation [7.16] as

$$Fr_1 = \frac{580^2}{666^2 \times 9.81 \times 0.0136} = 5.68$$

This is many times greater than the value of 0.040, below which stratification is significant. Hence either the Chen or the Shah method may be used. Both methods will be considered here.

In order to determine the internal heat transfer coefficient at various qualities by the Chen method, it is first necessary to calculate the Martinelli parameter, given by equation [7.13], as a function of the quality (x_g). Thus

$$\frac{1}{X_{tt}} = \left[\frac{x_g}{1-x_g}\right]^{0.9} \left[\frac{666}{1.6030}\right]^{0.5} \left[\frac{8.64}{236}\right]^{0.1}$$

$$= 14.64 \left[\frac{x_g}{1-x_g}\right]^{0.9} \tag{7.26}$$

Hence for each value of x_g it is possible to determine F from equation [7.14] and F_o from equation [7.15]. For example, when $x_g = 5\%$, equation [7.26] gives

$$\frac{1}{X_{tt}} = 14.64 \left[\frac{0.05}{1-0.05}\right]^{0.9} = 1.035$$

Equation [7.14] gives $F = 2.35(0.213 + 1.035)^{0.736} = 2.765$

Equation [7.15] gives $F_o = 2.765(1 - 0.05)^{0.8} = 2.654$

Equation [7.10] gives $\alpha_{cb} = 2.654 \times 5085 = 13\,500$

Equation [6.3] gives $U = (1/13\,500 + 1/8200)^{-1} = 5100\ \text{W/m}^2\ \text{K}$

Equation [6.2] gives $\dot{q} = 5100[(-15.2) - (-20.0)] = 24\,480\ \text{W/m}^2$

Carrying out similar calculations for all the qualities gives the results in Table 7.4.

In order to determine the internal heat transfer coefficient at various qualities by the Shah method, it is first necessary to calculate the parameter Co_h from equation [7.18]. On account of the high Froude number, $K_{Fr} = 1$. Thus when $x_g = 5\%$, equation [7.18] gives

$$Co_h = \left[\frac{1-0.05}{0.05}\right]^{0.8} \left[\frac{1.6030}{666}\right]^{0.5} = 0.517$$

Equation [7.20] gives $F_{cb} = 1.8 \times 0.517^{-0.8} = 3.050$.

This is somewhat more than the value of 2.765 obtained by the Chen method. Following the steps in the previous paragraph gives, for the Shah method, $F_o = 2.927$, $\alpha_{cb} = 14\,885$, $U = 5287$, $\dot{q} = 25\,380$. This is 4% more than the heat flux given by the Chen method. The results of the calculations for the Shah method are also given in Table 7.4. It is seen that at the maximum quality of 20%, the parameter Co_h has fallen to 0.1487, which is well above the minimum value of 0.020, recommended as the criterion for dry-wall conditions, so the assumption of wet-wall heat transfer is justified.

To determine whether there is nucleate boiling, it is necessary to follow

7.4 Summary of calculations for example in section 7.2.5

Symbol	[Equation]	$x_g =$					Units
		0	**0.05**	**0.10**	**0.15**	**0.20**	
Chen							
$1/X_{tt}$	[7.26]	0	1.035	2.026	3.074	4.205	—
F	[7.14]	1	2.765	4.254	5.641	7.014	—
α_{cb}	[7.10]	5085	13 500	19 880	25 190	29 830	W/m² K
U	[6.3]	3138	5100	5806	6168	6432	W/m² K
$\dot{q}$	[6.2]	15 070	24 480	27 870	29 690	30 870	W/m²
$T_{wONB} - T_s$	[7.27]	0.87	1.10	1.18	1.22	1.24	K
$T_w - T_s$	[7.28]	2.97	1.81	1.40	1.18	1.03	K
Shah							
Co_h	[7.18]	0	0.517	0.285	0.1965	0.1487	—
F	[7.20]	1	3.050	4.920	6.615	8.267	—
α_{cb}	[7.10]	5085	14 890	23 000	29 540	35 170	W/m² K
U	[6.3]	3138	5288	6045	6418	6650	W/m² K
$\dot{q}$	[6.2]	15 070	25 380	29 000	30 800	31 900	W/m²
$T_{wONB} - T_s$	[7.27]	0.87	1.12	1.20	1.24	1.26	K
$T_w - T_s$	[7.28]	2.97	1.70	1.26	1.04	0.91	K

the procedure in section 7.1.2. From equation [7.1]

$$T_{wONB} - T_s = \left[\frac{8 \times 0.030\,73\dot{q}(273 - 20)}{0.5857 \times 1\,329\,000 \times 1.6030} \right]^{\frac{1}{2}} = 0.007\,06\dot{q}^{\frac{1}{2}} \qquad [7.27]$$

Nucleate boiling will occur if this is less than the temperature difference across the boiling liquid, which is obtained from equation [6.1] as

$$T_w - T_s = \dot{q}/\alpha_{cb} \qquad [7.28]$$

The last two lines in Table 7.4 give these temperature differences. It will be seen that $T_{wONB} < T_w$ at qualities of 0, 5 and 10%. Hence the heat flux will be greater than estimated here at these qualities, as determined in the example in section 7.4.5.

7.3 Nucleate boiling

Many laboratory experiments have been carried out on nucleate pool boiling, using a small electrically heated element submerged in a saturated liquid, as has already been described in section 2.1. Conditions in the heater of an industrial vaporiser differ considerably from those in the laboratory. Apart from the much greater size, industrial equipment differs from laboratory equipment in the following important ways:

1.   Heat is supplied from a hot fluid at a given temperature (not electrically at a given heat transfer rate).

2.   Velocities past the heated surface are greater than those induced by natural circulation past the heated surface in experiments on pool boiling.

3.   There is likely to be a considerable amount of vapour with the liquid reaching the point in the heated surface under consideration.

In nucleate boiling, bubbles of vapour and associated liquid convey heat rapidly from the heated surface; consequently the boiling heat transfer coefficient increases with increasing heat flux. With convective boiling alone, the boiling heat transfer coefficient is independent of heat flux, and so it is possible to use the overall heat transfer coefficient in the normal manner, as illustrated in the example in section 7.2.5, but when nucleation also occurs, the overall heat transfer coefficient is a function of the heat flux. Hence the heat flux must be determined either by trial and error or by the graphical procedure which is illustrated in Fig. 7.3. Here the heat flux ($\dot{q}$) is plotted against $T_w - T_s$, the difference between the temperature of the heated wall at its interface with the boiling liquid and the saturation temperature of the bulk of the liquid.

The curve OB relates the heat flux absorbed by the boiling liquid to the wall temperature, showing how the heat flux rises from zero, when the temperature of the wall equals the saturation temperature, to a maximum, as in Fig. 2.1. In the region of nucleate boiling, from the onset of nucleate boiling at the point marked 'ONB' to the departure from nucleate boiling at the point marked 'DNB', the heat flux increases more rapidly than in proportion to the temperature difference.

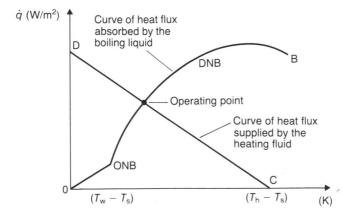

7.3   Estimation of heat flux in nucleate boiling (graphical method). Curves of heat flux against wall superheat

The line CD in Fig. 7.3 represents the source of heat and is determined as described below. If the wall in contact with the liquid were at T_h, the temperature of the source of heat, no heat would flow from the source, so the heat flux supplied would be zero, as shown by the point C. If the thermal resistance of the boiling liquid were zero, the interface temperature (T_w) would equal the saturation temperature (T_s). The temperature difference across the heat source would then be $(T_h - T_s)$, giving a heat flux supplied by the heating fluid of $U'(T_h - T_s)$, where U' is the approximate value of U obtained from equation [6.3] by putting $1/\alpha_v = 0$. This state is represented by the point D in Fig. 7.3. As U' may be assumed to be constant, CD must be a straight line with a negative slope equal to U'. The point at which the vaporiser will operate is that at which the heat flux supplied equals the heat flux absorbed by the boiling liquid, namely the point of intersection of the curve OB and the straight line CD. If U' is very high, the line CD is very steep. Under these circumstances, as T_h is increased, $\dot{q}$ increases to its maximum and then continues on the downward part of the curve OB. However at the inlet to a waste heat boiler operating with gas at a high temperature, the point C is well to the right of the diagram and the line CD is nearly horizontal, due to the high thermal resistance of the gas; these circumstances approximate to the state of constant heat flux obtained with electrical heating, and so there is a danger that an increase in gas temperature might lead to a sudden large increase in the operating temperature, the state of burnout described in sections 2.5 and 8.1. It is important to appreciate that the dramatic increase in temperature described as burnout cannot occur unless it is possible for the line CD to intersect the curve OB at three points as explained in section 2.1 and Fig. 2.1. This is likely with flow up a vertical tube, because then the downward slope at B is very steep.

The prediction of the curve OB in Fig. 7.3 between the onset of nucleate boiling and the departure from nucleate boiling is described below and in section 7.4. If there is a danger of operating to the right of the point marked DNB, Chapter 8 should be consulted.

7.3.1   Nucleate boiling of pure fluids inside ducts

This cannot be studied except in conjunction with studies of convective boiling. The methods devised by Chen and Shah should be used in conjunction with their correlations for convective boiling, described in sections 7.2.1 and 7.2.2. Their methods of determining the enhancements due to nucleation are described in sections 7.4.1 and 7.4.2. Mixture effects are discussed in section 7.7.

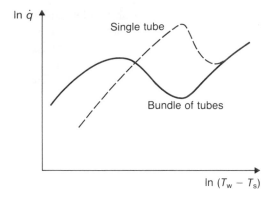

7.4   Heat flux curves for a bundle of tubes and a single tube (from Palen *et al.* (1972))

7.3.2   Nucleate boiling of pure fluids outside bundles of horizontal tubes

Measurements have been made of the mean overall heat transfer coefficient with boiling on the outside of bundles of horizontal tubes, by Palen and Taborek (1962), Palen and Small (1964), and Palen *et al.* (1972). Figure 7.4 shows a typical boiling curve for a tube bundle compared with that for a single tube under otherwise identical conditions. It will be seen that the bundle has an enhanced heat flux at low temperature differences, but the maximum heat flux is less with a bundle than with a single tube. It is thought that both of these changes are due to the presence of a large volume of vapour in the upper region of the bundle. A method is given below for estimating the nucleate boiling heat transfer coefficient for a single tube, and the enhancement for the bundle is discussed in section 7.4.3. The estimation of the critical heat flux of a bundle is discussed in section 8.3.

Heat transfer in pool boiling is very sensitive to the nucleation properties of the surface, which depend on its roughness, cleanliness and the characteristics of its wetting by the fluid that is being vaporised. The proposed correlation of Rohsenow (1952) allows for this by the inclusion of a coefficient which depends upon the nature of the combination of heating surface and fluid. Values of this constant are deduced from various experimental results for water or one of three organic liquids boiling on wires or plates of platinum, chromium or brass. It is not obvious how this method could be applied to industrial design, the values found for the constant ranging over a factor of five. Thus, for a given heat flux, the heat transfer coefficient could change by a factor of five according to the value selected for the constant. In view of the importance of being able to predict the order of magnitude of the heat transfer coefficient for nucleate boiling, with the wide range of materials used in industry, it is desirable to sacrifice accuracy and seek a simpler method.

Starczewski (1965) devised a graphical procedure to aid the design of flooded vaporisers, based on the generalised correlation of boiling heat transfer proposed by Mostinski (1963) for a single tube. This gives

$$\alpha_{nb1} = 0.106(p_{cr} \times 10^{-5})^{0.69}f(p_r)\dot{q}^{0.7} \qquad [7.29]$$

where α_{nb1} is the heat transfer coefficient for nucleate boiling outside a single horizontal tube (W/m^2 K), p_{cr} is the critical pressure of the liquid (Pa), $p_{cr} \times 10^{-5}$ being the critical pressure in bars, $\dot{q}$ is the heat flux (W/m^2) and $f(p_r)$ is a function, defined below, of the reduced pressure, which is defined by

$$p_r = p_s/p_{cr} \qquad [7.30]$$

where p_s is the saturation pressure, expressed in the same units as p_{cr}, the critical pressure. The required function of p_r is then found from

$$f(p_r) = 1.8p_r^{0.17} + 4p_r^{1.2} + 10p_r^{10} \qquad [7.31]$$

This method of estimating the nucleate boiling coefficient for a single tube is valid if $p_{cr} > 30$ bar and $0.0001 < p_r < 0.9$, with $\dot{q}$ up to 90% of the critical heat flux calculated as described in section 8.3.

J. W. Palen in section 3.6.2 of Schlünder et al. (1983) confirms that this method gives very reasonable results for a wide range of pure fluids, although the data near to the critical pressure are quite scattered. Thus the correlation may not be reliable as p_r approaches the limit of 0.9, but this is outside the normal industrial range.

The use of this method is illustrated in the example in section 7.3.3. Compared with other, more complex, methods of estimating heat transfer in nucleate boiling, the method proposed here is likely to predict lower rather than higher values of the coefficient. In many cases an approximate value is adequate, because the rate of heat transfer is controlled by the thermal resistance of the heating fluid, as in a gas-heated waste heat boiler. However, with boiling hydrocarbons and a low temperature difference, the boiling coefficient is likely to be the controlling thermal resistance.

A new approach to nucleate boiling is given by Cornwell et al. (1982) in their study of the effect of diameter on boiling on the outside of horizontal tubes. Their approach is more rational than that of Mostinski (1963) but it has not yet been developed to the extent that it may be offered as an alternative. It shows that the heat transfer coefficient increases with decreasing diameter below 25 mm, so that the coefficient is about doubled with a diameter of 8 mm.

Some suggestions for estimating the enhancement due to the presence of vapour towards the top of a bundle are suggested in section 7.4.3.

7.3.3   Example: shellside nucleate boiling

Gas at 920 °C enters the tubes of a waste heat boiler and is cooled by water boiling on the shellside at a pressure of 100 bar (10^7 Pa). The combined heat transfer coefficient for the gas, the dirt and the wall of the tubes is 450 W/m² K (including the effects of radiation and enhancement at entry). Calculate the heat flux. What is the external temperature of the tubes if all the dirt is on the inside? What is the external temperature of the tubes if the inlet temperature of the gas rises to 1100 °C and there is fouling on the outside of the tubes, giving a dirt coefficient of 2000 W/m² K? The critical pressure of water is 221 bar and its saturation temperature at 100 bar is 311 °C.

As a first approximation, neglect the temperature drop across the boiling film. From equation [6.2] the heat flux is:

$$\dot{q} = 450(920 - 311) = 274\,000 \text{ W/m}^2.$$

To estimate the nucleate boiling coefficient, first determine the reduced pressure from equation [7.30]. Thus:

$$p_r = 100/221 = 0.4525$$

This is well below the maximum of 0.9 recommended as the limit to the method. The function of reduced pressure is obtained from equation [7.31] as:

$$f(p_r) = 1.8 \times 0.4525^{0.17} + 4 \times 0.4525^{1.2} + 10 \times 0.4525^{10} = 3.121$$

From equation [7.29] the coefficient is:

$$\alpha_{nbl} = 0.106 \times 221^{0.69} \times 3.121 \times 274\,000^{0.7} = 87\,800 \text{ W/m}^2 \text{ K}.$$

Neglecting the contribution of convection within the shell, the overall heat transfer coefficient can be found from equation [6.3]:

$$U = (1/87\,800 + 1/450)^{-1} = 447.7 \text{ W/m}^2 \text{ K}$$

Thus a revised value of the heat flux is:

$$\dot{q} = 447.7(920 - 311) = 272\,600 \text{ W/m}^2$$

From equation [6.1] the temperature difference across the boiling film is:

$$\theta_b = 272\,600/87\,800 = 3.1 \text{ °C}$$

As the outside of the tubes is assumed to be clean, the external temperature is:

$$T_{to} = 311 + 3.1 = 314.1 \text{ °C}$$

If the gas temperature rises to 1100 °C and there is external fouling giving a dirt coefficient of 2000 W/m² K, the first approximation to the heat flux is modified to:

$$\dot{q} = (1/450 + 1/2000)^{-1}(1100 - 311) = 290\,000 \text{ W/m}^2$$

From equation [7.29] the new value for the coefficient is:

$$\alpha_{nb1} = 0.106 \times 221^{0.69} \times 3.121 \times 290\,000^{0.7} = 91\,400 \text{ W/m}^2 \text{ K}$$

The new value of the overall coefficient is:

$$U = (1/91\,400 + 1/450 + 1/2000)^{-1} = 365.9 \text{ W/m}^2 \text{ K}$$

Thus the revised heat flux is:

$$\dot{q} = 365.9(1100 - 311) = 289\,000 \text{ W/m}^2.$$

It is now possible to estimate the temperature difference across the boiling film and across the external fouling. Thus:

$$\theta_b = 289\,000/91\,400 = 3.2 \, ^\circ\text{C} \qquad \text{and}$$

$$\theta_{db} = 289\,000/2000 = 144.5 \, ^\circ\text{C}$$

Hence the external temperature of the tubes is:

$$T_{to} = 311 + 3.2 + 144.5 = 458.7 \, ^\circ\text{C}.$$

7.4 Combined convective and nucleate boiling

Having calculated α_{cb}, the convective boiling coefficient, using the appropriate method in section 7.2, and having established that nucleation may contribute significantly to heat transfer, it is necessary to estimate α_b, the combined boiling coefficient. Sections 7.4.1 and 7.4.2 deal with this for boiling inside tubes and section 7.4.3 for boiling outside bundles of tubes. Sections 7.4.5 and 7.4.6 are examples of such calculations for tubeside and shellside boiling respectively.

7.4.1 Upward flow inside ducts – the Chen method

The method of Chen (1966) has been recommended for dealing with boiling flow up vertical ducts, and possibly also for other orientations. One of the basic assumptions made in the derivation of this method was that the convective boiling coefficient (α_{cb}) could be added to the nucleate boiling coefficient (α_{nb}) to give the combined boiling coefficient (α_b). Thus α_{cb} should be calculated as described in section 7.2.1 and α_{nb} as described below. The combined coefficient is then given by:

$$\alpha_b = \alpha_{cb} + \alpha_{nb} \qquad\qquad [7.32]$$

Chen's equation for the nucleate component in $W/m^2\,K$ is:

$$\alpha_{nb} = 0.001\,22\,\frac{\lambda_l^{0.79}c_{pl}^{0.45}\rho_l^{0.49}S\theta_b^{0.24}\Delta p_v^{0.75}}{\sigma^{0.5}\eta_l^{0.29}(\Delta h_v\rho_g)^{0.24}} \qquad [7.33]$$

where c_{pl} is the specific heat capacity of the liquid (J/kg K), S is a suppression factor explained below, Δh_v is the latent heat of vaporisation (J/kg), Δp_v is the difference in vapour pressure (Pa) corresponding to θ_b, which is the wall superheat (K), η_l is the viscosity of the liquid (N s/m^2), λ_l is the thermal conductivity of the liquid (W/m K), ρ_l and ρ_g are the densities of liquid and vapour respectively (kg/m^3) and σ is the surface tension (N/m).

The empirical suppression factor (S) was introduced to allow for the fact that the shear stress at the wall due to flow renders the temperature difference less effective for nucleate boiling. Chen determined it as a function of the two-phase Reynolds number, which he defined as

$$\mathrm{Re}_{tp} = \frac{\dot{m}d}{\eta_l}(1-x_g)F^{1.25} \qquad [7.34]$$

where d is the internal diameter of the pipe (m), F is the Chen enhancement factor defined in section 7.2.1 for different values of x_g, which is the quality.

Table 7.5 gives some points from the curve recommended by Chen (1966) for determining the value of S from the value of Re_{tp} from equation [7.34]. This suppression factor is unity at very low flow rates and it falls asymptotically to zero as the flow rate is increased.

7.5 Chen's suppression factor for flow in ducts

$\mathrm{Re}_{tp} \times 10^{-3}$	15	30	50	100	200	400
S	0.83	0.69	0.55	0.36	0.21	0.11

J. G. Collier in section 8.5.3 of Bergles *et al.* (1981) and in section 2.7.3 of Schlünder (1983) describes the factor S as being the ratio of the mean superheat seen by the growing bubble to the wall superheat (θ_b). He gives the following equation to fit the curve proposed by Chen (1966):

$$S = (1 + 2.53 \times 10^{-6}\,\mathrm{Re}_{tp}^{1.17})^{-1} \qquad [7.35]$$

This equation predicts the values tabulated in Table 7.5 to ± 0.01.

Equation [7.33] is difficult to use because it is necessary to guess T_w, the wall temperature, determine the vapour pressure at T_w and T_s calculate Δp_v, the difference between these vapour pressures, and determine θ_b ($= T_w - T_s$), T_s being the saturation temperature. As the wall superheat in flow boiling inside ducts is normally only a few degrees, the procedure may be simplified by substituting for Δp_v from the Clapeyron equation, which is given in Chapter 9 as equation [9.48]. Replacing dp_v/dT_s by $\Delta p_v/\Delta T_s$ and neglecting $1/\rho_l$ in comparison with $1/\rho_g$ gives:

$$\Delta p_v = \Delta T_s \Delta h_v \rho_g / T_s \qquad [7.36]$$

Substituting this in equation [7.33] where the symbol θ_b is used instead of ΔT_s to denote wall superheat, and approximating the exponent of θ_b from 0.99 to unity, gives:

$$\alpha_{nb} = \phi_{nbi} S \theta_b \qquad [7.37]$$

where the parameter ϕ_{nbi} depends entirely on physical properties and is given by:

$$\phi_{nbi} = 0.001\,22 \frac{\lambda_l^{0.79} c_{pl}^{0.45} \rho_l^{0.49} (\Delta h_v \rho_g)^{0.51}}{\sigma^{0.5} \eta_l^{0.29} T_s^{0.75}} \qquad [7.38]$$

Substituting for α_{nb} from equation [7.37] in equation [7.32] and multiplying both sides of the equation by θ_b gives the heat flux $\dot{q}$ $(=\alpha_b \theta_b)$ as:

$$\dot{q} = \alpha_{cb} \theta_b + \phi_{nbi} S \theta_b^2 \qquad [7.39]$$

It is now possible to determine the heat flux $(\dot{q})$ at the several selected values of the quality (x_g) given the corresponding values of the temperature of the hot fluid (T_h) and of the heat transfer coefficient of the heat source (U_h), which is a combined coefficient for the heating fluid, the wall and the dirt, but excluding the boiling liquid. The temperature difference across the boiling liquid $(\theta_b = T_w - T_s)$ must be determined by trial and error so that the overall temperature difference required is equal to the available difference. Thus the value of $\dot{q}$ calculated from equation [7.39] must satisfy equation [7.40].

$$T_h - T_s = \dot{q}/U_h + \theta_b \qquad [7.40]$$

This procedure is illustrated in the example in section 7.4.5. It is quite simple when a computer or programmable calculator is available.

The confusing situation may arise in which neglecting nucleate boiling gives a value of T_w slightly greater than T_{wONB}, whereas following the procedure given in this section gives a value of T_w less than T_{wONB}. This means that nucleate boiling is incipient but not fully established. Under such circumstances the contribution of nucleate boiling is likely to be small and the heat flux should be taken as that calculated for convective boiling only.

Some typical values of ϕ_{nbi} are given in Table 7.6. The data for the calculations for the halogenated hydrocarbon refrigerants, designated by the letter 'R' and their specific number, were supplied by ICI, Mond Division, for whose help the author is grateful. The data for the other liquids were obtained from ESDU (1980).

7.6   Values of ϕ_{nbi} – from equation [7.38]

	Temperature (°C)						
	−30	**0**	**50**	**100**	**150**	**200**	**300**
Ammonia	1164	2146	4582	—	—	—	—
Diphenyl*	—	—	—	—	—	83.37	296.4
Methanol	—	—	269	756	—	—	—
Toluene	—	—	67.2	164.8	325	—	—
Water	—	—	240	772	1765	3317	9552
R-11	—	—	216	395	—	—	—
R-12	215	356	669	—	—	—	—
R-22	344	585	—	—	—	—	—

* Eutectic mixture of 73.5% diphenyl ether and 26.5% diphenyl

7.4.2   Flow inside horizontal ducts – the Shah method

The method of Shah (1976) has been recommended for dealing with boiling flow inside horizontal ducts. The steps for using the Shah method to estimate the local heat flux ($\dot{q}$) at specified values of the quality (x_g), taking account of the enhancement due to nucleation, are described below.

1.   Determine the convective boiling enhancement factor (F_{cb}) from equation [7.19] or [7.20] as described in section 7.2.2 and illustrated in the example in section 7.2.5.

2.   Determine the enhancement relative to conditions at zero quality (F_o) from equation [7.15] and substitute this in equation [7.10] to determine the convective boiling heat transfer coefficient (α_{cb}).

3.   Follow the steps in section 7.2 to determine $\dot{q}$ from equation [7.11]. This will be denoted by $\dot{q}'$ to indicate that it is approximate because it neglects the benefits due to nucleate boiling.

4.   To determine the enhancement due to nucleation, it is first necessary to determine Shah's boiling number Bo, which is defined by:

$$Bo = \frac{\dot{q}}{\dot{m}\,\Delta h_v} \qquad [7.41]$$

where $\dot{m}$ is the mass flux and Δh_v is the latent heat of vaporisation. If using the approximate heat flux $\dot{q}'$ gives a value of Bo less than 1.9×10^{-5}, there is no enhancement due to nucleation under any conditions, so the true heat flux is equal to the approximate value already calculated and no further calculations are needed. Otherwise proceed to the next step.

5.   Calculate F_{nb}, the value of F due to nucleation from

$$F_{nb} = 230 Bo^{\frac{1}{2}} \qquad [7.42]$$

6. Calculate the boiling coefficient at zero quality from

$$\alpha_{vo} = F_{nb}\alpha_{lo} \tag{7.43}$$

7. Determine F for each quality in the boiling zone as follows:

If $Co_h > 1.0$, put F equal to the value of F_{cb} given by equation [7.19] or equal to F_{nb} given by equation [7.42], whichever is greater. If F_{nb} is the greater, return to step (3) to obtain a more accurate value of $\dot{q}'$. If F_{cb} is the greater there is no enhancement due to nucleation and so $\dot{q} = \dot{q}'$. If $1.0 > Co_h > 0.02$, put F equal to the value of F_{cb} given by equation [7.20] or equal to F_{cnb} given by equation [7.44] whichever is greater.

$$F_{cnb} = F_{nb}(0.77 + 0.13F_{cb}) \tag{7.44}$$

If $F_{cnb} > F_{cb}$, return to step (3) to obtain a more accurate value of $\dot{q}'$. If F_{cb} is the greater, there is no enhancement due to boiling and $\dot{q} = \dot{q}'$.

   Note: Equation [7.44] has been developed empirically by reading the experimentally determined values of F_{cnb} from Fig. 2 of Shah (1976), dividing them by the appropriate value of F_{nb} calculated from equation [7.42] and plotting this ratio against F_{cb}, the value of F given by equation [7.20].

   If $Co_h < 0.02$, there is some doubt about whether there will be complete wetting of the wall, so there might be a fall-off in heat transfer coefficient due to dryout (see section 8.2.6). Under these circumstances, section 7.5 should be used for estimating the heat flux. This is likely to occur only in the high-quality region of a once-through vaporiser. Not being able to estimate accurately the quality at dryout does not seriously impair the accuracy of the estimation of the total area required for the exchanger, because the region of doubt occurs over quite a small proportion of the heat load.

7.4.3 Flow outside bundles of horizontal tubes

This section is concerned with combined convective and nucleate boiling outside horizontal tube bundles, with upward flow. The less usual arrangement of downward flow, described in section 7.2.4, is a case of evaporation from a falling film of liquid and nucleation is unlikely to occur.

   Having estimated α_{cb} (the convective component of boiling) from equation [7.23] and α_{nb} (the nucleate component) from equation [7.29], it is necessary to be able to combine the two. In a gas-heated steam generator, the nucleate component is very high and the convective component is negligible. In a chiller or vaporiser operating with a small temperature difference, the nucleate term may be non-existent or very small in comparison with the convective term.

   This section is concerned with intermediate circumstances where both components are of the same order of magnitude. A simple safe method, suggested by Taborek (1980), is to put α_b equal to α_{cb} or α_{nb}, whichever is the larger; this implies that the mechanism that gives the higher coefficient is the controlling mechanism and renders the other mechanism unimportant. Some allowance may be made for the benefits due to the smaller component by using the following method of interpolation due to Kutateladze (1961).

$$\alpha_b = \sqrt{\alpha_{cb}^2 + \alpha_{nb}^2} \qquad\qquad [7.45]$$

When using the equation, α_{nb} must be calculated at the prevailing heat flux, which must be determined by trial and error from the overall temperature difference.

Experiments to measure the contours of boiling heat transfer coefficients in an experimental kettle reboiler have been described by Brisbane *et al.* (1980), Leong and Cornwell (1979) and Cornwell *et al.* (1980). The experimental rig consists of a slice of a reboiler, 28 mm long containing 241 tubes of 19 mm diameter, with an inline arrangement on 25.4 mm pitch. The results are shown in Table 7.7, where they are compared with the values predicted by calculating the circulation rate by the method given in section 9.6 and then calculating the heat transfer coefficients from equations [7.23], [7.29] and [7.45]. The calculations for the heat flux of 20 kW/m² are shown in the examples in sections 9.6.5 (for flow rate) and 7.4.6 (for heat transfer). The physical properties of the refrigerant R-113 were supplied by ICI, Mond Division, for whose help the author is grateful.

It is seen from the top section of Table 7.7 that the measured heat transfer coefficients increase about fourfold from bottom to top of the bundle, as the amount of vapour present increases. At the entry to the bottom of the bundle, the boiling temperature is increased by 1.7 °C on account of the hydrostatic head. The length of the sub-cooled boiling zone was estimated as described in section 9.6.3 and found to range from 143 mm at the lowest heat flux to 25 mm at the highest. The predicted values of the combined convective and nucleate boiling coefficients have been calculated as described in section 7.4.4. These are given in the bottom section of Table 7.7. In the subcooled nucleate boiling zone at the bottom of the bundle, the calculated values agree quite well with the measured values of the heat transfer coefficient. In the saturated boiling zone, at the middle and top of the bundle, the measured heat transfer coefficients are considerably greater than the predicted values. The discrepancy would have been somewhat greater if equation [7.21] had been used instead of [7.23] to determine the convective component. Cornwell *et al.* (1980) in their Fig. 3 show a region of high voidage at the top and in the centre of the middle of the bundle, the extent of this region increasing with increasing heat flux. It is possible that here the heat transfer coefficient is enhanced by the impingement of fast moving drops of liquid on the tubes.

This analysis suggests that the method of calculation recommended in this section may be used to obtain a safe estimate of the rate of heat transfer in boiling outside a bundle of horizontal tubes. It can be shown that possible

7.7 Comparison between measured and predicted heat transfer coefficients in boiling of R-113 at atmospheric pressure in natural circulation outside a slice of a kettle reboiler

	Heat flux (kW/m^2)*		
	10	**20**	**50**
Measured heat transfer coefficient for the boiling liquid $(W/m^2\,K)$:			
at the top	4000–7800	4000–8500	6000–10 900
in the middle of the bundle	1400–3000	2000–4000	3500–6100
at the bottom	1000	1000–2000	3500–4000
Estimated mass flux between tubes $(kg/s\,m^2)$	504	446	271
Corresponding heat transfer coefficient for all liquid flow α_{lo} $(W/m^2\,K)$	886	804	596
Estimated quality, x_g:			
at the top	0.018 07	0.0409	0.1685
in the middle of the bundle	0.003 90	0.0150	0.0786
Predicted convective boiling coefficient, α_{cb}:			
at the top	2188	3073	5064
in the middle of the bundle	1230	1864	3266
Estimated nucleate boiling coefficient, α_{nbl}	800	1300	2456
Estimated combined boiling coefficient, α_b:			
at the top	2330	3340	5630
in the middle of the bundle	1470	2270	4090
at the bottom	1180	1530	2540

* Based on the heated length of 25.4 mm

inaccuracies in estimating the circulation rate have little effect on the estimated boiling coefficients, because they are controlled by nucleate boiling at the bottom of the bundle and by the high vapour velocity at the top.

7.4.4 Note on subcooled nucleate boiling

The methods of predicting boiling heat transfer coefficients given in this chapter are all based on experiments carried out with saturated boiling. With subcooled nucleate boiling the problem arises that as far as the liquid convective component is concerned the effective temperature difference is

$T_w - T_b$, whereas for the nucleate boiling component it is $T_w - T_s$, where T_w is the wall temperature, T_s is the saturation temperature and T_b is the (lower) bulk temperature. This presents a serious problem when dealing with boiling in a long vertical tube under vacuum. For the Chen method (recommended for upflow) Butterworth and Shock (1982) suggest that T_w must be determined by trial and error so that the following equations are both satisfied:

$$\dot{q} = U_h(T_h - T_w) \qquad\qquad [7.46]$$

$$\dot{q} = \alpha_{lo}(T_w - T_b) + \alpha_{nb}(T_w - T_s) \qquad\qquad [7.47]$$

where α_{lo} is determined for the liquid alone, as described in section 6.2.1, and α_{nb} is determined as described in section 7.4.1. Moles and Shaw (1972) found this method successful for water but to give low predictions of heat flux for other liquids; however, they compared data over only a narrow range of heat fluxes. Shah (1976) states that his method is not suitable for subcooled boiling, so it should not be used when there is appreciable subcooling. It is recommended mainly for flow in horizontal tubes where there is little or no hydrostatic head to suppress boiling, so no significant problem will arise in these circumstances.

The method of Mostinski (1963) and Starczewski (1965), which is recommended for the nucleate component in boiling outside tubes, is based on an exchange action between bubbles of vapour and liquid at saturation temperature. However, if it were applied to a slightly subcooled liquid, the interchange would be between bubbles and subcooled liquid, so it might be assumed in this method that the appropriate temperature difference is $T_w - T_b$. Hence the coefficients could be combined using equation [7.45] and multiplied by the difference between the wall and the bulk temperatures, to estimate the heat flux. Here it is especially important to ensure that $T_w > T_{wONB}$, determined as described in section 7.1.2; otherwise the heat transfer rate is no more than that due to natural convection (see section 7.1.4).

7.4.5 Example: tubeside combined convective and nucleate boiling

Recalculate the heat fluxes in the example in section 7.2.5 for convective boiling to allow for nucleation, where necessary.

In the results in Table 7.4 it is seen that T_{wONB} is less than T_w at qualities of 0, 5 and 10%, so the possibility should be investigated that some enhancement might occur at these qualities. Both the Chen and the Shah methods will be considered.

For the Chen method, the value of ϕ_{nbi} for ammonia at $-20\,°C$ can be found from Table 7.6 by interpolation to be 1490.

The Reynolds number at zero quality has already been calculated as

$Re_{lo} = 33\,420$. The Reynolds number at quality 5% can be found from equation [7.34], having already calculated that $F = 2.765$ when $x_g = 0.05$. Hence:

$$Re_{tp} = 33\,420(1 - 0.05)2.765^{1.25} = 113\,200$$

Equation [7.35] gives the suppression factor:

$$S = (1 + 2.53 \times 10^{-6} \times 113\,200^{1.17})^{-1} = 0.326.$$

The convective component (α_{cb}) has been estimated to be 13 500 W/m² K. Equation [7.39] gives:

$$\dot{q} = 13\,500\theta_b + 1490 \times 0.326\theta_b^2 \qquad [7.48]$$

The heat transfer coefficient of the heat source has been given as $U_h = 8200$ and the overall temperature difference as $T_h - T_s = 4.8$. Substituting in equation [7.40] gives:

$$4.8 = \dot{q}/8200 + \theta_b \qquad [7.49]$$

It was found by trying several values of θ_b that $\theta_b = 1.75$ satisfied both equations [7.48] and [7.49], giving $\dot{q} = 25\,110$ W/m². This shows an enhancement of $2\frac{1}{2}\%$ over the value of 24 480 given in Table 7.4 for convective boiling only.

Table 7.8 shows the results of this and similar calculations for qualities of 0 and 10%.

Turning to the Shah method, described in section 7.4.2, equation [7.41] gives the boiling number (Bo) in terms of the mass flux, given as

7.8 Summary of calculations for example in section 7.4.5

Symbol	Equation	x_g			Units
		0	**0.05**	**0.10**	
Chen					
F	[7.14]	1	2.765	4.254	—
α_{cb}	[7.10]	5085	13 500	19 880	—
Re_{tp}	[7.34]	33 420	113 200	183 760	—
S	[7.35]	0.334	0.326	0.215	—
θ_b	[7.48]	2.69	1.75	1.38	K
$\dot{q}$	[7.49]	17 280	25 110	28 040	W/m²
Shah					
Co_h	[7.18]	∞	0.517	0.285	—
$\dot{q}'$	[6.2]	15 070	23 380	29 000	W/m²
$Bo' \times 10^4$	[7.41]	0.1955	0.3293	0.3762	—
F_{nb}	[7.42]	1.017	1.319	1.411	—
F_{cb}	[7.19 or 20]	1.0	3.05	4.92	—
F		1.017	3.05	4.92	—
$\dot{q}$	[7.10, 6.3 and 6.2]	15 200	25 380	29 000	W/m²

580 kg/s m^2, and the latent heat of vaporisation, given as 1329 kJ/kg. Taking the approximate value at 5% quality of $\dot{q}' = 25\,380$ W/m^2 (from Table 7.4) gives Bo', the approximate value of Bo, as

$$Bo' = \frac{25\,380}{580 \times 1\,329\,000} = 0.329 \times 10^{-4}.$$

This is greater than the minimum value of 1.9×10^{-5}, so nucleation is possible. Equation [7.42] gives:

$$F_{nb} = 230 \times (0.329 \times 10^{-4})^{\frac{1}{2}} = 1.319$$

The results of this, and similar calculations for qualities of 0 and 10%, are given in Table 7.8.

The boiling coefficient at zero quality is given by equation [7.43]

$$\alpha_{vo} = 1.017 \times 5085 = 5171 \text{ W/m}^2 \text{ K}.$$

Hence $U = (1/5171 + 1/8200)^{-1} = 3171$ W/m^2 K and the heat flux is $\dot{q} = 3171 \times 4.8 = 15\,200$ W/m^2. This shows an increase of only 1% over the previous value of 15\,070 W/m^2.

The fourth and third from last lines of Table 7.8 show F_{nb} calculated in this section compared with F_{cb} calculated in Section 7.2.5 and given as F in Table 7.4.

It is seen that F_{cb} is very much greater than F_{nb} at qualities of 5 and 10% . At 5% equation [7.44] gives:

$$F_{cnb} = 1.319(0.77 + 0.13 \times 3.05) = 1.539,$$ which is less than F_{cb}. Thus there is no enhancement due to nucleation. The same applies to 10% quality.

The Shah method gives a lower heat flux when $x_g = 0$; otherwise it gives slightly higher results.

7.4.6 Example: shellside combined convective and nucleate boiling

Compare the measured with the predicted values of the boiling heat transfer coefficients for R-113 boiling on the outside of an experimental kettle reboiler operating at a constant heat flux of 20 kW/m^2, as described in section 7.4.3. The recirculation ratio has been calculated in the example in section 9.6.5 to be 24.4. The operating pressure is approximately 1.0 bar, at which the properties of R-113 are: relative molecular mass, $\tilde{M} = 187.39$, critical pressure, $p_c = 34.14$ bar, saturation temperature, $T_s = 47\,^\circ$C, $\rho_l = 1512$, $\rho_g = 7.38$ kg/m^3, $\lambda = 0.0699$ W/m K, $c_{pl} = 955$ J/kg K, $\Delta h_v = 145\,715$ J/kg, $1000\,\eta_l = 0.497$ g/s m, $1000\sigma = 14.9$ m N/m.

From the dimensions of the tubes given in section 7.4.3, the heat transfer surface area is:

$$A = 241 \times \pi \times 0.019\,05 \times 0.0254 = 0.3663 \text{ m}^2$$

The heat transfer rate is:

$\dot{Q} = 20 \times 10^3 \times 0.3663 = 7326 \, \text{W}$ and the vapour generation rate is:

$\dot{M}_g = 7326/145\,715 = 0.0503 \, \text{kg/s}$.

It is shown in the example in section 9.6.5 that the cross-sectional area for flow is 0.002 758. Hence the mass flux through the bundle is $\dot{m} = 0.0503 \times 24.4/0.002\,758 = 445 \, \text{kg/s} \, \text{m}^2$.

The Prandtl number of the liquid is given by equation [6.7]

$\text{Pr}_1 = 0.497 \times 10^{-3} \times 955/0.0699 = 6.79$.

For the onset of nucleate boiling, Davis and Anderson (1966) give equation [7.1] without the term Pr_1:

$$T_{\text{wONB}} - T_s = \left[\frac{8 \times 14.9 \times 10^{-3} \times 20 \times 10^3 (273.2 + 47)}{0.0699 \times 145\,715 \times 7.38} \right]^{\frac{1}{2}} = 3.19 \, ^\circ\text{C}$$

Frost and Dzakowic (1967) give equation [7.2], whence:

$T_{\text{wONB}} - T_s = 3.19 \times 6.79 = 21.66 \, ^\circ\text{C}$.

The heat flux is given as $20 \times 10^3 \, \text{W/m}^2$. Dividing this by the measured coefficients quoted in Table 7.7 shows that the measured temperature difference was 10 to 20 °C at the bottom, 5 to 10 °C in the middle and 2.4 to 5 °C at the top of the bundle. Thus according to the equation of Frost and Dzakowic there was no nucleate boiling anywhere, whereas according to the equation of Davis and Anderson there was nucleate boiling at the bottom and in the middle, but some suppression at the top of the bundle. Subsequent calculations support the latter conclusion.

The Reynolds number of the liquid entering the bottom of the bundle is obtained from equation [6.8] as:

$$\text{Re}_{\text{lo}} = \frac{445 \times 19.05 \times 10^{-3}}{0.497 \times 10^{-3}} = 17\,060$$

The Nusselt number (defined by equation [6.6]) can be calculated from equation [6.19] for the liquid in cross-flow. Neglecting the minor correction factors, F_a and F_p, in equation [6.19], this gives:

$\text{Nu} = 0.33 \times 17\,060^{0.60} \times 6.79^{0.34} = 219$

From equation [6.6]:

$\alpha_{\text{lo}} = 219 \times 0.0699/19.05 \times 10^{-3} = 804 \, \text{W/m}^2 \, \text{K}$

From equation [7.23] the convective boiling coefficient is given by:

$\alpha_{\text{cb}} = 804[1 + (1512/7.38 - 1)x_g]^{0.6} = 804(1 + 204x_g)^{0.6}$

It is now possible to estimate the convective coefficients at the bottom, the middle and the top from the values of the quality (x_g) at these locations.

Position in bundle	Bottom	Middle	Top
Quality, x_g	0	0.0150	0.0409
$1 + 204x_g$	1	4.06	9.344
$(1 + 204x_g)^{0.6}$	1	2.318	3.822
α_{cb}	804	1864	3073

These are only about half the measured values, suggesting that there was an appreciable amount of nucleate boiling. An approximate value of the nucleate boiling coefficient can be determined by the method given in section 7.3.2. This requires a knowledge of the reduced pressure, p_r, defined by equation [7.30]. At the operating pressure of 1.0 bar this is:

$$p_r = 1.0/34.14 = 0.029\,29$$

Equation [7.31]gives (neglecting the last term):

$$f(p_r) = 1.8 \times 0.02929^{0.17} + 4 \times 0.029\,29^{1.2} = 1.046$$

Equation [7.29] gives the nucleate boiling coefficient for a single tube:

$$\alpha_{nb1} = 0.106 \times 34.14^{0.69} \times 1.046 \times 20\,000^{0.7} = 1300 \; W/m^2 \; K.$$

Combining the convective and nucleate components at the bottom (from equation [7.45]):

$$\alpha_b = [804^2 + 1300^2]^{\frac{1}{2}} = 1529 \; W/m^2 \; K$$

If it is assumed that there is also nucleate boiling at the middle and top of the bundle, the coefficients there are increased to 2273 and 3337 W/m^2 K.

It appears from this example that nucleate boiling may occur sooner than is predicted by equation [7.2]. Equation [7.29] may underpredict the nucleate component in boiling, or the combined coefficient may be greater than that given by equation [7.45] (but not as great as that given by equation [7.32]). There may be enhancement due to impingement of drops against the heated tubes, as previously suggested. The recommended methods of calculation appear to give a safe estimate of shellside boiling coefficients.

It can be shown that possible inaccuracies in estimating the circulation rate have little effect on the estimated boiling coefficients, because these are controlled by nucleate boiling at the bottom and by the high vapour velocity at the top.

7.5 Dry-wall convection

The equations given in the previous sections of this chapter are for heat transfer from a heated wall to a vaporising fluid, the wall being wetted by liquid. If the heat flux is increased above a certain value, known as the 'critical heat flux', the wall becomes dry and the heat transfer coefficient falls very considerably. The estimation of the critical heat flux is dealt with in Chapter 8. In industrial heat exchangers, the critical heat flux is never intentionally exceeded, except in once-through vaporisers, where a dry wall becomes inevitable as the quality approaches unity. This section discusses the difficult problem of estimating the heat transfer coefficient when operating with a dry wall.

The mechanism whereby the supply of liquid to the wall fails depends on the flow region (see section 9.2). With subcooled nucleate boiling or with bubble-flow (at low quality), a film of vapour forms between the wall and the liquid, as has been described in section 2.5. Alternatively as a boiling fluid flows through a heated tube, a region of annular flow (at high quality) is usually reached before the end of the tube. If there is a further transition to mist flow there is a considerable danger of the wall becoming dry, as has been described in section 2.6.

Low-quality dryout is liable to occur in large high-pressure water-tube boilers. This is somewhat outside the scope of this book. The present state of knowledge on the subject has been described by J. G. Collier in Chapter 10 of Bergles *et al.* (1981), where there is a summary of the extensive research that has been done on post-dryout heat transfer at the operating conditions in a nuclear reactor. Correlations are available for determining the heat transfer coefficient to water boiling at high pressures.

No satisfactory data are available for the prediction of high-quality dry-wall heat transfer coefficients obtained with various fluids boiling at pressures somewhat above atmospheric, as is encountered in once-through vaporisers. The main mechanism of heat transfer is likely to be by convection from the heated wall to the superheated vapour. This can be estimated by the normal method of calculating single-phase heat transfer (see section 6.2.1), if the amount of superheat is assumed to be negligible. This is the method normally adopted for a rough estimate, but it is liable to overestimate the heat flux, because there is usually an appreciable amount of superheat of the vapour, the amount being difficult to estimate. However, there is also transfer of heat by radiation from the wall to the droplets and to the vapour, and also by the droplets which impinge on the wall and are heated by direct contact.

7.6   Evaporation from a falling film of liquid

The mechanism of heat transfer by conduction across a falling film of superheated liquid has already been described in section 2.8. At the interface between the liquid and the vapour, both are at the saturation temperature. The amount of superheat at the wall is usually not sufficient for the onset of nucleate boiling. With a pure liquid in laminar flow, heat is transferred by conduction only across the thin film of liquid. With mixtures of liquids and with aqueous solutions, there is also a resistance to the mass transfer of the more volatile component to the interface. This section describes how the heat transfer coefficient may be calculated for a falling film of pure liquid; mixture effects are discussed in the next section.

Chun and Seban (1971) and Chiesa *et al.* (1974) give experimental data on the heat transfer coefficient of water evaporating from a heated falling film. The Reynolds number ranged from 300 to 20 000 and only a small fraction of the water evaporated as it flowed down the heated tube. Hence the experimental heat transfer coefficient is a local value. The measured values of α_{cx}, the local heat transfer coefficient, were from 0 to 24% above the values predicted from equation [6.43] for the laminar region and from 7 to 50% below the values predicted from equation [6.44] for the turbulent region. It is therefore recommended that the equations given in section 6.3.4 for the local values of the heat transfer coefficient in filmwise condensation be used to estimate the local heat transfer coefficients in the partial evaporation of a falling film, provided that the Reynolds number does not fall below 100 (see section 9.11.1). The effects of vapour shear are considered in section 9.11.2. If the Reynolds number changes significantly, the mean coefficient may be determined as described in the next paragraph.

Determine Re_{in}, the value of Re_f at the top (inlet) of the tube and Re_{out}, the value at the bottom. Using the procedure in section 6.3.4, calculate $\bar{\alpha}_{c\,in}$ and $\bar{\alpha}_{c\,out}$, the value of the mean coefficient for complete vaporisation from Reynolds number Re_{in} and Re_{out} respectively. Calculate $\bar{\alpha}_c$, the true mean coefficient for the range of Reynolds numbers from Re_{in} to Re_{out} from:

$$\bar{\alpha}_c = \frac{Re_{in} - Re_{out}}{Re_{in}/\bar{\alpha}_{c\,in} - Re_{out}/\bar{\alpha}_{c\,out}} \qquad [7.50]$$

Whitt (1966) gives results on falling-film heating and falling-film vaporisation of water and single-component organic liquids. He showed that the heat transfer coefficients for evaporation are similar to those for condensation.

7.7 Mixture effects

The correlations for heat transfer coefficients in boiling given in this chapter apply only to pure single-component liquids. Many reboilers deal with mixtures of liquids (multicomponent liquids); evaporators are required to vaporise water from aqueous solutions of solids. This section is concerned with warnings of how mixture effects may significantly reduce the heat flux in boiling, either due to changes in physical properties or due to mass transfer resistance leading to a reduction in the temperature difference that is available, with nucleate boiling, convective boiling, or evaporation from a falling film. Further information is given by Collier (1981), Chapter 12, and by R. A. W. Shock in Chapter 3 of Hewitt *et al.* (1982).

7.7.1 Physical properties of mixtures

The boiling heat transfer coefficient of a mixture may be less than that of the pure components due to mixture effects leading to a reduction in the specific heat capacity or in the thermal conductivity or to an increase in the dynamic viscosity. When designing an evaporator, it is important to remember that the presence of dissolved solids in a saturated aqueous solution considerably increases the boiling temperature and reduces the heat transfer coefficient, compared with pure water at the same saturation pressure. For example, R. A. Smith, in section 3.5.7 of Schlünder (1983), states that typically the coefficient for a saturated aqueous solution of an inorganic salt will be 75–80% of that for water at the same temperature. However, Happel and Stephan (1974) have shown that in pool boiling of an azeotropic mixture of ethanol and benzene, the heat transfer coefficient is equal to that obtained by interpolation, on the molecular composition, between the values obtained for the two pure components.

7.7.2 Vapour–liquid equilibrium effects on pool boiling heat flux

With mixtures other than azeotropes, the composition of the vapour generated differs from that of the liquid in equilibrium with it. In nucleate boiling, the liquid around a growing bubble is deficient in the lighter component because there is more of that component in the vapour than in the liquid; this leads to an increase in the boiling temperature of the liquid at the interface and thence to a reduction in the temperature difference. The experiments mentioned in the previous section by Happel and Stephan (1974) showed that the nucleate boiling coefficient for ethanol–benzene mixtures ranged from 70 to 100% of the values obtained by linear interpolation on a molecular basis from the values for the two pure components; with benzene–toluene mixtures, the range was 80–100%.
. R. A. W. Shock in Chapter 3 of Hewitt *et al.* (1982) summarises the considerable amount of published research carried out in the last few decades on the pool boiling of binary mixtures. His Figs 4 to 7 show the variations with composition of the amount of wall superheat required for nucleate boiling, with mixtures of *n*-pentane and *n*-hexane, benzene and cyclohexane, ethanol and water, and ethanol and benzene. His Table 5 summarises measurements of the nucleate pool boiling heat transfer coefficients published by the authors of 31 papers, dealing with many binary mixtures and two ternary mixtures. The boiling surface, in most cases, was a smooth plate, a small tube or a fine wire. The pressures ranged

from 0.2 to 35 bar. In most cases, correlations were proposed. Shock, in section 6 of his Chapter 3, describes the work done on measuring the critical heat flux of binary mixtures, summarised at the end of section 8.1 of this book.

J. W. Palen, in section 3.6.2 of Schlünder (1983), recommends an approximate empirical correction factor to be applied to the nucleate boiling coefficient, in shellside boiling, to allow for the loss in heat flux with mixtures; he has been able to correlate the correction factor with the boiling range of the mixture, which is defined as the difference between the dew point and the bubble point. Thus the nucleate boiling coefficient for a mixture is given by:

$$\alpha_{nb} = \alpha'_{nb} \times F_c \qquad\qquad [7.51]$$

where α_{nb} is the true coefficient for the mixture, α'_{nb} is the mean value calculated by neglecting mixture effects and F_c is the correction factor calculated from the boiling range (BR) from

$$F_c = \exp(-0.027\text{BR}) \qquad \text{or} \qquad\qquad [7.52]$$

$$F_c = 0.1, \qquad\qquad \text{whichever is larger.}$$

7.7.3   Convective boiling

Information on the convective boiling of mixtures is very limited. It is always neglected in the design of evaporators for concentrating or crystallising aqueous solutions, and there is no record of this ever having led to appreciable underdesign. There is no distinction between pure components and mixtures in the values of the boiling heat transfer coefficients quoted by Goodall (1980) in his Table 12.2.

With convective boiling of mixtures there is also a resistance to mass transfer in the vapour phase (as in multicomponent condensation). Bennett (1976), working with a mixture of ethylene glycol and water, found that mixture effects were significant not only for the nucleate boiling coefficient but also for the convective coefficient, using the method of Chen described in section 7.2.1.

7.7.4   Falling-film evaporation

When concentrating an aqueous solution in a falling-film evaporator, there is a resistance to the diffusion of water to the interface to replace the water that has vaporised. As a result of this mass transfer resistance, the concentration of dissolved solids at the interface is greater than it would otherwise have been. This produces an increase in the temperature at the interface, leading to a loss of temperature difference and a consequent loss of heat flux. Experiments by Chun and Seban (1972) on evaporation from falling films of water and brine have shown that heat transfer coefficients with brine, initially at a concentration of 14%, were about 10% less than those with water at the same flow rate and with the same temperature difference between the wall and the liquid. If a gas is introduced, as described in section 3.5.1, there is also a resistance to mass transfer in the gas phase.

7.8   Enhancement of heat transfer in boiling

Relatively low heat transfer coefficients may be obtained, especially with organic liquids, if the velocity of flow is not sufficient to give a high convective boiling coefficient and the overall temperature difference is not sufficient to produce nucleate boiling. Under such circumstances, consideration should be given to the possible use of one of the three methods of enhancing boiling heat transfer, namely: (1) increasing the surface area by introducing some secondary surface, such as fins, (2) using surfaces artificially made porous so as to enhance the nucleate component, (3) using devices that will enhance the convective component. The first two may be used to enhance boiling on the outside of tubes; the third may be used to enhance tubeside boiling, as discussed below.

7.8.1   Enhanced boiling on the outside of horizontal tubes

Proprietary treatments are available for coating the outside of tubes by brazing metal powder to the surface, thus forming a porous structure with a porosity of 50% or more. Bergles and Chyu (1981) have tested a single copper tube thus treated. They mounted it horizontally, testing it with boiling water and Refrigerant-113; they also tested a plain tube of the same size under the same conditions. For a given heat flux in boiling, they found that the temperature difference with a treated surface was many times less than with a plain surface; however, the boiling heat flux curves had large-scale hysteresis, being small initially until the heat flux had been increased to more than $10 \, kW/m^2$ with water or $3 \, kW/m^2$ with R-113.

Yilmaz et al. (1981) reported tests on two types of enhanced tube surfaces with re-entrant nucleation sites and on a plain surface, both as single tubes and as bundles of U-tubes in a steam-heated internal reboiler. The performance of the plain bundle was better than that of single plain tubes. Treated tubes gave similar performances whether tested singly or in a bundle, the heat transfer coefficient being about double that with a plain bundle, when boiling p-xylene (1,4-dimethyl benzene).

Carnavos (1981) measured heat transfer rates with Refrigerant-11 boiling on the outside of single horizontal tubes, in order to compare a plain tube with tubes enhanced in several ways, both on the inside, where heat was supplied by flowing hot water, and on the outside. The tubes were 1.3 m long and their external diameters (including any augmentation) ranged from 18 to 20 mm. In his Fig. 7 he compared the boiling performance at 35 °C of various tubes. The heat flux ranged from 12 to 55 kW/m², the area being based on the external diameter, neglecting any augmentation of area in the case of finned tubes. With plain tubes, the mean temperature difference across the boiling film ranged from 12 to 17 °C. The measured values of the temperature difference were 9 to 12% greater than the values predicted by the Mostinski correlation given in equations [7.29] to [7.31]. The best treated tube, with an applied metallurgically bonded type of porous surface, showed a 12-fold reduction in the temperature difference across the boiling film, for a given heat flux, compared with the plain tube. Other treatments gave 9-fold and 4-fold reductions. Tubes with integral fins showed a 2- to 3-fold reduction in boiling temperature difference; no details are given of the fins, but this is likely to be equal to the enhancement in external surface area, so it may be concluded that the use of fins does not significantly improve the heat transfer coefficient referred to the true external surface area. Myers and Katz (1952) showed that boiling heat transfer coefficients on the outside of externally finned tubes were slightly lower than those obtained with a smooth surface. Bondurant and Westwater (1971) have studied boiling from high fins and give information on minimum pitch.

Stephan and Mitrovic (1981) measured heat transfer in natural convective boiling of refrigerants in a bundle of seven tubes with proprietary T-shaped fins. Heat transfer coefficients were measured for the three tubes one above each other in the centre of the bundle. All tubes gave coefficients many times greater than those obtained with a smooth tube by Povolockaja, whose results were 30 to 60% greater than the values predicted by the Mostinski correlation (equation [7.29] above). The increase with T-shaped fins is explained by the vapour flow and bubble formation inside the T-shaped channels. The liquid is prevented from flowing freely past the primary surface by the small size of the slit connecting the liquid inside a channel to the outside liquid; hence the small effect of the position of the tube in the bundle.

Czikk et al. (1981) present data to show the effect of mixture composition on the performance of porous metal surfaces in boiling. A method is given for estimating the extra thermal resistance due to mixture effects.

7.8.2 Enhanced boiling inside tubes

Czikk et al. (1981) describe experiments in which boiling oxygen was flowing up a vertical (electrically heated) tube of 19.7 mm bore and containing a porous boiling surface. Heat fluxes from 3 to 18 kW/m² were achieved with a boiling temperature difference of about 0.2 °C. It was shown that the high heat transfer coefficient was not significantly altered either by mass flux (13 to 125 kg/s m²) or by quality (up to 77%). It was concluded that there is no suppression of nucleate boiling due to forced convection when boiling from a porous surface in a vertical tube, i.e. the factor S in section 7.4.1 is always unity. However, with refrigerants boiling inside horizontal tubes, there was a considerable decrease in the heat transfer coefficient as the quality approached unity or when the mass flux fell below 100 kg/s m². This was attributed to the change from annular to stratified flow, rather than to suppression of nucleate boiling (see section 9.2 for information on flow regions). Thus tubes with internal porous surfaces may be used for enhancing nucleate boiling inside vertical tubes.

Enhanced boiling inside horizontal tubes may be achieved by installing swirling devices, which increase the convective component of the boiling heat transfer coefficient. They may also be used in the vertical tubes of once-through vaporisers. Experiments by Cumo *et al.* (1974) on the swirl flow of Refrigerant-12 in a vertical tube show that the heat transfer may be increased by a factor of up to two by the installation of a twisted tape in a tube. However, the pressure drop is increased by a considerably greater factor. In some applications it may be necessary to take care to avoid erosion or enhanced corrosion due to contact between the tape and the tube. Twisted tapes are also used to increase the critical heat flux, as explained in section 8.4.

CHAPTER 8

Estimation of the critical heat flux

This chapter relates to the difficult topic of predicting the critical heat flux in an industrial vaporiser. It starts with the correlation derived from laboratory experiments on pool boiling (see section 2.1) because this is used as the basis for studying data on the critical heat flux in natural circulation shellside boiling, as described in section 8.3. This correlation is not relevant to the critical heat flux in forced circulation tubeside boiling, which is discussed in section 8.2.

8.1 Laboratory experiments on critical heat flux in pool boiling

Laboratory experiments on pool boiling have been described in section 2.1. When the electrical input to a heated small flat surface or wire submerged in a liquid is increased above a certain critical value, its temperature increases dramatically due to the change from wet-wall heat transfer to the situation in which a film of vapour separates the liquid from the heated surface. Accounts of the experiments on burnout in pool boiling have been given by G. F. Hewitt in section 6.4 of Hetsroni (1982) and in Chapter 9 of Bergles *et al.* (1981), and by D. B. R. Kenning in section 7.11 of Butterworth and Hewitt (1977).

Zuber (1958) proposed the following correlation for estimating the critical heat flux in the pool boiling of a single-component liquid

$$\dot{q}_{crp} = C_{cr} \Delta h_v \rho_g^{\frac{1}{2}} [\sigma g (\rho_l - \rho_g)]^{\frac{1}{4}} \qquad [8.1]$$

where C_{cr} is a dimensionless constant (discussed below), g is the acceleration due to gravity (9.81 m/s^2), $\dot{q}_{crp}$ is the critical heat flux in pool boiling (W/m^2), Δh_v is the latent heat of vaporisation (J/kg), ρ_l and ρ_g are the densities of the liquid and vapour respectively (kg/m^3) and σ is the surface tension of the liquid (N/m).

Zuber recommended that the constant (C_{cr}) should be 0.131. However, Collier and Wallis (1967), p. 853, show that there is a considerable scatter of experimental results, and they recommend that for a safe estimate it should be assumed that $C_{cr} = 0.12$. A. E. Bergles in section 7.6 of Bergles *et al.* (1981) and D. B. R. Kenning in section 7.11 of Butterworth and Hewitt (1977) have suggested that $C_{cr} = 0.18$ fits the data better. More recently, Hewitt in section 6.4.3.1 of Hetsroni (1982) recommends the following correlation for horizontal cylinders, based on the work of Sun and Lienhard (1970) and Lienhard and Dhir (1973a,b). The first step is to determine the nondimensional radius R' from:

$$R' = \tfrac{1}{2}d \left[g \frac{(\rho_1 - \rho_g)}{\sigma} \right]^{\frac{1}{2}} \qquad [8.2]$$

where d is the diameter of the cylinder (m).

The constant C_{cr}, for substitution in equation [8.1] is then given by

if $\quad R' > 1.17 \qquad C_{cr} = 0.118$ \hfill [8.3]

if $1.17 > R' > 0.12$, $C_{cr} = 0.123/(R')^{\frac{1}{4}}$ \hfill [8.4]

This correlation applies to horizontal cylinders at least 20 diameters long, for the boiling of saturated liquids; liquid subcooling increases the heat flux at burnout. It does not apply to viscous liquids, if the viscosity number, Vi given by equation [8.5], is less than 400:

$$Vi = \frac{\rho_1 \sigma^{\frac{3}{4}}}{\eta_1 g^{\frac{1}{4}}(\rho_1 - \rho_g)^{\frac{3}{4}}} \qquad [8.5]$$

where η_1 is the viscosity of the liquid ($N\,s/m^2$), the other symbols being as described above. A correlation for viscous liquids is given by Dhir and Lienhard (1974).

The liquids listed in Table 7.6 all have values of Vi, given by equation [8.5], greater than 400; thus it is likely that equations [8.1] to [8.4] may be applied to the majority of industrial fluids. Also for these fluids, the value of R', given by equation [8.2] is greater than 1.17 for cylinders of diameter greater than 6 mm. Hence the assumption that $C_{cr} = 0.12$, recommended by Collier and Wallis (1967), will cover most industrial vaporisers. Some values of the critical heat flux are given in Table 8.1.

The critical heat flux in a multi-tube shellside vaporiser is less than that for a single tube under otherwise identical conditions, by a factor that depends on the number and the pattern of the tubes, as explained in section 8.3.

The correlations for the critical heat flux in flow through a channel are quite different from those described for pool boiling and shellside boiling, as discussed by Hewitt in section 6.4.3.2 of Hetsroni (1982). The application to industrial in-tube vaporisers is considered in section 8.2.

The work described so far relates to the boiling of a single-component liquid. The state of knowledge on the critical heat flux in the boiling of multi-component fluids is described by R. A. W. Shock in Chapter 3, section 6, of Hewitt *et al.* (1982). The trends reported are bewildering and often conflicting. His Fig. 38 shows values of the constant C_{cr} in equation [8.1] ranging from 0.08 to 0.14 for a mixture of ethanol and benzene.

8.1 Critical heat flux in pool boiling – from equations [8.1] and [8.3] (kW/m^2)

	Temperature (°C)						
	−30	**0**	**50**	**100**	**150**	**200**	**300**
Ammonia	629	989	1475	—	—	—	—
Methanol	—	—	386	766	—	—	—
Toluene	—	—	116	229	346	—	—
Water	—	—	411	997	1819	2710	3498
R-11	—	—	282	380	—	—	—
R-12	192	272	346	—	—	—	—

8.2   Critical heat flux – intube flow

There are two different cases where it is necessary to consider the possibility of critical heat flux with flow inside tubes: dryout or burnout in the heated tubes of steam generators, where water is circulating and boiling at moderate to high pressures with quality at outlet up to 20%; loss of performance in once-through vaporisers, which convert a slightly subcooled liquid into a slightly superheated vapour.

A vast amount of research has been carried out to measure the critical heat flux with high-pressure water flowing up uniformly heated vertical tubes. Section 8.2.1 describes the procedure recommended for estimating the critical heat flux under these conditions. A limited amount of work has been done on the critical heat flux with high-presssure water flowing inside horizontal tubes, up the annular space between two concentric vertical tubes, in inclined tubes and in coiled tubes. Tentative correlations are available for predicting the critical heat flux under these conditions, and some typical results are discussed in section 8.2.2 to 8.2.5. No information is available on liquids other than water.

The second problem is to estimate the quality at which a liquid ceases to wet the wall of a vaporiser in which complete vaporisation is required; this has been seriously neglected by heat transfer researchers. A method is suggested in section 8.2.6.

8.2.1   Burnout in water-tube boilers with flow up vertical tubes

This has been studied extensively, using electrically heated tubes, in connection with the generation of high-pressure steam for the electrical power industry. A comprehensive survey of published data was made by Macbeth (1968). His correlation has been improved and brought up to date by Bowring (1972). The critical heat flux is given by:

$$\dot{q}_{crv} = \frac{A + \frac{1}{4}d\dot{m}\,\Delta h_{sub}}{C + L} \tag{8.6}$$

where A and C are functions of the pressure of the water (see below), d is the internal diameter of the tube (m) and L its length (m), $\dot{m}$ is the mass flux (kg/s m^2), $\dot{q}_{crv}$ is the critical heat flux for flow inside vertical tubes (W/m^2) and Δh_{sub} is the inlet subcooling, i.e. the enthalpy of water at its saturation temperature less its enthalpy at the inlet temperature (J/kg). As the heat flux is assumed to be constant over the whole length, a heat balance gives:

$$\dot{q}_{crv}(\pi dL) = \dot{m}(\tfrac{1}{4}\pi d^2)(\Delta h_{sub} + x_{g\,out}\,\Delta h_v) \tag{8.7}$$

where $x_{g\,out}$ is the quality at outlet and Δh_v is the latent heat of vaporisation (J/kg). Eliminating Δh_{sub} from equations [8.6] and [8.7] gives:

$$\dot{q}_{crv} = B_1 - B_2 x_{g\,out} \tag{8.8}$$

$$\text{where } B_1 = A/C \tag{8.9}$$

$$\text{and} \quad B_2 = \tfrac{1}{4}d\dot{m}\,\Delta h_v/C \tag{8.10}$$

According to Bowring's correlations for A and C, B_1 and B_2 are given by:

$$B_1 = \frac{7.523F_1}{F_3}\left[\frac{1.0 + 0.347F_4(\dot{m}/1356)^n}{1.0 + 0.0143F_2\sqrt{(d)}\,\dot{m}}\right]\Delta h_v \tag{8.11}$$

$$B_2 = [1.0 + 0.347F_4(\dot{m}/1356)^n]\frac{\Delta h_v}{0.308F_3} \tag{8.12}$$

The parameters F_1, F_2, F_3 and F_4, and the mass-flux exponent n, have been correlated with the pressure ratio p_R, which is the ratio of the saturation pressure to the reference pressure of 68.95 bar (1000 lb/in^2 or 6895 kPa). Thus

$$p_R = (p \times 10^{-5}) \times 0.0145 \tag{8.13}$$

where p is the saturation pressure (Pa), $p \times 10^{-5}$ being the saturation pressure in bars. The exponent n is given by

$$n = 2.0 - p_R/2 \tag{8.14}$$

The functions F_1 to F_4 are all equal to one if p_R is equal to one.

Otherwise they are related to p_R as follows:

if $p_R < 1$

$$F_1 = \frac{p_R^{18.942} \exp[20.89(1 - p_R)] + 0.917}{1.917}$$ [8.15]

$$F_2 = \frac{1.309 F_1}{p_R^{1.316} \exp[2.444(1 - p_R)] + 0.309}$$ [8.16]

$$F_3 = \frac{p_R^{17.023} \exp[16.658(1 - p_R)] + 0.667}{1.667}$$ [8.17]

$$F_4 = F_3 p_R^{1.649}$$ [8.18]

if $p_R > 1$

$$F_1 = p_R^{-0.368} \exp[0.648(1 - p_R)]$$ [8.19]

$$F_2 = F_1 p_R^{0.448} / \exp[0.245(1 - p_R)]$$ [8.20]

$$F_3 = p_R^{0.219}$$ [8.21]

$$F_4 = F_3 p_R^{1.649}$$ [8.22]

This correlation may be applied to tubes of internal diameter 3 to 11 mm over the pressure range 4 to 60 bar and to tubes of internal diameter 2 to 40 mm over the pressure range 60 to 120 bar. As this is only an empirical correlation, it might give serious errors if extrapolated to higher diameters.

Some typical values of B_1 and B_2 are given in Table 8.2. For convenience they have been divided by 1000, and so when substituted in equation [8.8] they will give the critical heat flux in kW/m^2.

The experimental results obtained by Robertson (1973) for burnout in a 13-mm bore vertical tube with water flowing at a mass flux of 1360 kg/m^2 and a pressure of 69 bar all lie within $\pm 30\%$ of the almost identical lines predicted by the Macbeth and the Bowring correlations.

In a fired boiler, the heat flux is considerably greater down the side of each tube facing the flame than down the side hidden from the flame. This has been studied by Lee (1966), Butterworth (1971) and Chojnowski and Wilson (1974). Hewitt surveys the available data in section 6.4.4.2 of Hetsroni (1982) and recommends the following procedure to allow for the 'flux tilt' (where the flux varies smoothly from one side of the tube to the other, the flux tilt ratio is the ratio of the maximum to the minimum flux). For flux tilt ratios up to 1.3, the average flux for burnout may be assumed to equal that for uniform heating. For greater flux tilt ratios, the average flux for burnout is less than that predicted by the Bowring correlation. For the extreme case of side-heated tubes, 30% should be deducted to give the mean flux at burnout.

8.2 Parameters for equation [8.8]

Pressure (bar)	Bore (mm)	B_1 and B_2 (kW/m^2) when the mass flux (kg/s m^2) is:					
		$\dot{m} = 500$	750	1000	1250	1500	2000
60	10	$B_1 = 8747$	7602	6842	6315	5942	5476
	10	$B_2 = 6046$	6308	6625	6991	7400	8338
80	10	$B_1 = 5928$	5315	4929	4677	4510	4329
	20	$B_1 = 5078$	4397	3980	3710	3530	3323
	30	$B_1 = 4575$	3882	3468	3202	3025	2820
	40	$B_1 = 4222$	3533	3128	2871	2670	2501
	10–40	$B_2 = 5029$	5420	5871	6373	6918	8126
100	10	$B_1 = 4430$	4134	3954	3843	3774	3712
	20	$B_1 = 3827$	3449	3219	3072	2975	2868
	30	$B_1 = 3465$	3060	2817	2662	2559	2441
	40	$B_1 = 3209$	2794	2549	2393	2290	2170
	10–40	$B_2 = 4712$	5234	5806	6420	7068	8454
120	10	$B_1 = 3397$	3286	3219	3179	3155	3135
	20	$B_1 = 2960$	2765	2643	2562	2506	2439
	30	$B_1 = 2694$	2466	2324	2230	2165	2084
	10–30	$B_2 = 4505$	5135	5793	6472	7170	8610

8.2.2 Dryout in water-tube boilers with horizontal tubes

With horizontal boiler tubes the critical heat flux is considerably lower than with vertical tubes under otherwise identical conditions. Increasing the heat flux with flow up a vertical tube leads to burnout, as described in section 2.5. With horizontal tubes, the temperature of the top of the tube begins to fluctuate upwards, due to an inadequate transfer of water up to the top of the tube to replace water that has vaporised; this is called 'dryout'. There is less information on dryout in horizontal tubes than on burnout in vertical tubes.

The first step is to determine the flow region, as described in section 9.2.1. The importance of the flow region is well illustrated by an analysis of the results of Rounthwaite (1968), who measured the heat transfer coefficients around the periphery of two straight heated tubes, each 6.25 m long and 41 mm bore. The two tubes were connected by a 180° unheated bend of radius 63.5 mm, the flow through the bend being vertically upwards. The two tubes were heated electrically at a constant heat flux. A series of experiments was carried out on water boiling in the tubes at various pressures from 14 to 66 bar, at various heat fluxes up to 80 kW/m^2 and at mass fluxes from 94 to 282 kg/s m^2.

The results of one test, at a pressure of 29 bar, a mass flux of 193 kg/s m^2 and a heat flux of 66 kW/m^2, plotted in Rounthwaite's Fig. 13, showed that at qualities from 87% to 93% the heat transfer coefficients were independent of length, being about 10% higher at the top than at the sides and bottom of the tube. As the quality increased from 93 to 96%, the coefficient at the top fell by a factor of three, the coefficient remaining constant at the other levels. The coefficient at all levels then fell to a constant value within the superheated region. This is explained by the supposition that flow is annular, with a thinner film at the top than at the sides and bottom of the tube. Dryout begins when the liquid film no longer extends to the top of the tube, i.e. when the quality has reached 93% in this case. When the quality has been calculated to be 100%, there is still some enhancement of the coefficient at the bottom of the tube, indicating that superheating of the vapour has begun before all the liquid has been evaporated. Over the range of qualities used in this test, the predicted flow region was always annular.

Rounthwaite also carried out experiments at a pressure of 14.8 bar, a heat flux of 63 kW/m^2 and at five different mass fluxes from 94 to 282 kg/s m^2 (see his Figs. 5 and 6). In some results the predicted flow region was annular; in others it was stratified (liquid flowing mostly along the bottom and vapour along the top of the pipe). High heat transfer coefficients were measured at all points along the bottom of the tube, up to an estimated quality of more than 100%. Along the top of the tube, when there was stratified flow the coefficients were much lower than those at the bottom, but at about the quality estimated for the change to annular flow the coefficient at the top increased sharply to equal that at the bottom.

Using the same facility, Lis and Strickland (1970) studied the effects of the upstream 180° bend on dryout in one 3-m long heated tube of 41 mm bore. The pressure ranged from 15 to 67 bar and the mass flux from 480 to 960 kg/s m^2. The heat flux ranged from 25 to 195 kW/m^2, most of the results having been obtained with heat fluxes below 100 kW/m^2. Initially the boiling water entered the heated length through an unheated 180° bend of radius 124 mm, the direction of flow being upwards. Alternative entry arrangements were tested at the end of the programme, as described later. The increase in the quality of the steam along the heated section was never more than 2%. The flow rates were always sufficient to give annular flow.

Whereas Rounthwaite had found that with stratified flow the heat transfer coefficients were slightly improved by the bend, Lis and Strickland found that dry patches occurred at the top of the tube over some 20 to 50 diameters beyond the bend. There were regions of oscillating temperature before and after the dry patch. The critical heat fluxes for the initiation of fluctuating temperatures at the top of the tube are plotted against quality in Fig. 8.1, Lis and Strickland's curve being lettered PP for the initiation of dryout and KK to NN for the termination of dryout, with increasing quality at a fixed heat flux.

The disturbances were appreciably reduced when an unheated calming section was inserted between the bend and the test length. They were

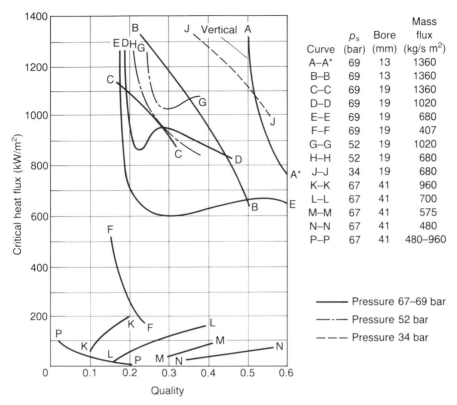

Curve	p_s (bar)	Bore (mm)	Mass flux (kg/s m^2)
A–A*	69	13	1360
B–B	69	13	1360
C–C	69	19	1360
D–D	69	19	1020
E–E	69	19	680
F–F	69	19	407
G–G	52	19	1020
H–H	52	19	680
J–J	34	19	680
K–K	67	41	960
L–L	67	41	700
M–M	67	41	575
N–N	67	41	480
P–P	67	41	480–960

——— Pressure 67–69 bar

—·— Pressure 52 bar

– – – Pressure 34 bar

8.1 Critical heat flux for water boiling in electrically heated horizontal tubes. Curves for 13 and 19 mm bore pipes are from Robertson (1973); curves for 41 mm bore tube are from Lis and Strickland (1970). In all cases flow was annular according to the correlation of Taitel and Dukler (1976) given in Section 9.2.1. In all cases the water entered the heated test section flowing upwards through a 180° bend in a vertical plane

* *a vertical pipe*

eliminated almost completely when the bend was placed in the horizontal plane (without the calming section). With a 57-mm radius bend in the vertical plane immediately before the heated length, there was only a very slight reduction in the mass flux required to avoid dryout.

Robertson (1973) measured dryout in horizontal tubes of 13 and 19 mm bore preceded by a 180° bend placed in a vertical plane, with upward flow. Over the range of qualities used in these tests, flow was always annular, according to the correlation in section 9.2.1. The temperatures along the bottom of the tube were always very steady. Dryout was deemed to have started at the point where fluctuations were first observed in the temperature of the top of the tube. These fluctuations extended over all of the heated length, showing that the presence of a 180° bend with upward flow through it did not have any significant effect on the results. This difference in comparison with the results of Lis and Strickland (1970) is thought to be due to the use by Robertson of tubes of smaller diameter. His results plotted in Fig. 8.1, curves AA to JJ, show that the critical heat fluxes are less with horizontal than with vertical tubes and that there is a reduction in critical heat flux with increasing diameter. No method has been suggested for correlating these results.

In conclusion, the critical heat flux in annular flow (see section 9.2.1) is two to three times smaller with horizontal than with vertical tubes of diameter 12 to 20 mm; with tubes of about 40 mm it is an order of magnitude smaller. Stratified flow should be avoided because this inevitably leads to dryout at the top of the tube. In the low-quality region, the vapour is present as bubbles, but these soon coalesce to form slugs, giving unsteady flow. Annular flow is established, according to section 9.2.1, when the parameter X_{tt} (defined in section 7.2.1) falls to 1.6. With water this occurs when the quality reaches 9% with a pressure of 69 bar, or 5% with a pressure of 34 bar. The upper curves in Fig. 8.1 show that very high values will be obtained for the critical heat flux in small-bore tubes, possibly approaching those calculated from equation [8.1] for pool boiling or from equation [8.6] for vertical tubes, at the low qualities before annular flow is established. This does not apply to the 41-mm bore tubes of Lis and Strickland.

8.2.3   Burnout in flow up a vertical annulus

Information on this topic is required for the design of bayonet-tube waste heat boilers (see section 3.1.1 and Fig. 3.4). Becker and Hernborg (1963) give measurements of burnout conditions for the flow of boiling water up a vertical annulus of inner diameter 9.92 mm and outer diameter 17.42 mm; the pressure ranged from 8 to 37 bar and the heat flux up to 3700 kW/m^2. When heat was supplied to the external surface only, it was found that the data were somewhat lower than those for round tubes under the same conditions. More recent information has been published by Jensen and Mannov (1974) and by Becker and Letzer (1975).

8.2.4   Inclined tubes

Measurements of the critical heat flux in flow up inclined tubes have been made by Chojnowski et al. (1974), Watson et al. (1974) and Morris (1976), using water at high pressure. The critical heat flux was found to lie between the values for vertical and for horizontal tubes. Figure 6.4.18 of Hetsroni (1982) shows some of the results of Watson et al. (1974) with water flowing at 952 kg/s m^2 at a pressure of 184 bar. With a vertical tube, the burnout heat flux was 640 kW/m^2 at zero quality, falling to 420 kW/m^2 at 20% quality; with a tube inclined at 40° from the vertical, the corresponding figures were 480 and 260 kW/m². With a tube inclined at 60° from the vertical, the burnout heat flux fell to 280 kW/m² at zero quality.

8.2.5   Coiled tubes

Data on two-phase flow and boiling in coiled tubes have been reviewed by Hopwood (1972). The burnout flux in a coiled tube tends to be even higher than that in a vertical tube under otherwise identical conditions. Figure 6.4.19 of Hetsroni (1982) shows some of the results of Hopwood (1972) with water flowing at 1360 kg/s m² at a pressure of 180 bar, the average heat flux being 440 kW/m². The burnout quality for a straight vertical tube was 24%. With coiled tubes, burnout occurred first at the point on the circumference of the tube nearest to the centre of the coil. With a helix diameter 300 times the tube diameter, the burnout quality was increased to 40%; with a helix diameter 76 times the tube diameter, the burnout quality was 64%.

Ruffell (1974) has produced a correlation for the dryout quality of boiling water at pressures from 60 to 180 bar in helically coiled tubes. This is valid if the mass flux ($\dot{m}$) is 300 to 1800 kg/s m², the heat flux is $60\dot{m}$ to $340\dot{m}$ in W/m², the tube internal diameter (d) is 0.0107 to 0.0186 m and the coil diameter is $6d$ to $185d$ m.

8.2.6   High-quality dryout in once-through vaporisers

In most vaporisers there is recirculation, and so the quality does not rise to more than about 20%. With once-through vaporisers (see Fig. 5.3), the aim is to produce a liquid-free vapour, so dryout is inevitable. As the heat flux is comparatively low, dryout may not occur until the quality is over 90%. However, this is followed by a dramatic fall in the heat transfer coefficient to roughly the value calculated on the assumption of flow as all vapour, i.e. for a quality of 100% (see section 7.5). There are no well-established correlations for predicting the critical heat flux of any liquid flowing in either a horizontal or a vertical tube. This lack of data is not as serious as it appears at first sight, because only a small fraction of the total area required is for the transfer of heat in the uncertain range of quality between about 90 and 100%.

It is possible to make a safe estimate of high-quality dryout in horizontal tubes by studying published data on wet-wall boiling heat transfer. Shah (1976) recommends that his method, described in section 7.2.2, may be used for qualities up to 100% without introducing serious underdesign. In his Fig. 7 he shows very good agreement between his predictions and the measured values of the boiling heat transfer coefficient by Chawla (1967) with the refrigerant R-11 flowing through a tube of 25 mm bore at a mass flux of 59 kg/s m², a pressure of 0.88 bar abs and a heat flux of 1.75 kW/m². At a quality of 90% the measured heat transfer coefficient slightly exceeded that predicted, the values of the parameters being $1/X_{tt} = 85$ and $Co = 0.0102$. At the highest quality of 95%, the measured heat transfer coefficient was only 85% of the predicted, the values of the parameters being $1/X_{tt} = 166$ and $Co = 0.0056$. This confirms the limiting value for $1/X_{tt}$ of 100, or a minimum value of 0.01 for Co. However, Whiteway (1977) shows some fall-off at $Co = 0.01$, so a safer criterion, in the absence of a systematic study would be to assume that dry-wall heat transfer begins when $Co < 0.02$, as recommended in section 7.2.2.

Palen *et al.* (1982) have studied the transition to mist flow in a vertical thermosyphon reboiler as part of a research programme. The test fluid was one of several organic liquids, or water, or a mixture. Experiments were carried out with three sizes of tube – 19, 32 and 44.5 mm outside diameter. The heat flux ranged from 16 to 246 kW/m² and the mass flux from 68 to 542 kg/s m². The source of the heat was condensing steam. The transition to mist flow was detected by a marked rise in the temperature of the wall. It was shown that the transition to mist flow could be predicted by the empirical correlation developed by Fair (1960). Given $\dot{m}$, the total mass flux of liquid and vapour (kg/s m²), mist flow occurs when the value of X_{tt}, defined in section 7.2.1, falls below the critical value, X_{ttcr}, calculated from:

$$X_{ttcr} = \frac{\dot{m}}{2441} \qquad [8.23]$$

This takes no account of the effect of heat flux. However, it may be used for an approximate estimate of the quality for transition to mist flow with a fixed overall temperature difference, if the heat flux is initially less than 250 kW/m², the mass flux is in the range 60 to 600 kg/s m² and the tube diameter is in the range 19 to 50 mm.

8.3 Critical heat flux – shellside boiling

It is necessary to estimate the critical heat flux in shellside boiling in: (1) a fire-tube waste heat boiler, as illustrated in Fig. 3.16, to avoid corrosion or overheating of the tubes; (2) a horizontal thermosyphon reboiler, as illustrated in Fig. 3.18, to avoid loss in performance; (3) a kettle reboiler, as illustrated in Fig. 3.19; and (4) an internal reboiler, to avoid loss in performance.

The procedure described below may be applied to waste heat boilers and to reboilers. It gives a fairly safe estimate of the highest heat flux that may be used in design, as shown in the example in section 8.3.2, but a further factor of safety may be applied to a waste heat boiler, in view of the seriousness of tube failures, when it cannot be ensured that water treatment

will be of the high standard required on all occasions. Further information
on corrosion in waste heat boilers is given below in section 12.2.

8.3.1   Procedure for estimating critical shellside heat flux

The recommended procedure is that described by J. W. Palen in section
3.6.2 A(f) of Schlünder (1983). The first step is to estimate the critical heat
flux in pool boiling, $\dot{q}_{crp}$, preferably by using equation [8.1] above, taking
$C_{cr} = 0.12$. If lack of physical data prevents this, an approximate estimate
may be made, knowing only the operating pressure and the critical pressure
of the boiling fluid. Equation [8.24] gives the critical heat flux from the
correlation of Mostinski (1963), which has been repeated by Starczewski
(1965).

$$\dot{q}_{crp} = 0.367 p_{cr} p_r^{0.35}(1 - p_r)^{0.9} \qquad [8.24]$$

where p_{cr} is the critical pressure (Pa), p_r is the reduced pressure, defined in
equation [7.30], and $\dot{q}_{crp}$ is the critical heat flux for a single tube (W/m²),
which may be assumed to equal that for pool boiling.
   In a multitubular boiler, the critical heat flux is less than that with a
single tube, due to the flow of the escaping vapour past the upper tubes.
Thus

$$\dot{q}_{crs} = F_{crs}\dot{q}_{cr1} \qquad [8.25]$$

where F_{crs} is the bundle correction factor, calculated as described below, $\dot{q}_{cr1}$
is the critical heat flux for a single tube, equal to $\dot{q}_{crp}$ calculated from
equations [8.1] or [8.24] and $\dot{q}_{crs}$ is the critical heat flux for shellside
boiling (W/m² K). The value of F_{crs} should be calculated from equation
[8.26] which is derived from the work of Palen and Small (1964).

$$F_{crs} = 6.9 D_b L / A \qquad [8.26]$$

where A is the heat transfer surface area (m²), D_b is the diameter of the
outer tube limit of the bundle (m), F_{crs} is the correction factor for equation
[8.25] and L is the heated length of the tubes (m). If F_{crs} calculated from
equation [8.26] is found to be more than 1.0, set $F_{crs} = 1.0$.
   More recent work on the boiling of R-113 in a slice of a reboiler has
been devoted to the study of dryout – Schüller and Cornwell (1984). It is
shown in the example in the next section that the critical heat flux for their
experimental rig, estimated by the method recommended in this section, is
40 kW/m² (this happens to be similar to the value of 38 kW/m²
recommended by Kern (1950) p. 475 for inorganic liquids). Schüller and
Cornwell found that at heat fluxes of 50 kW/m² and more, horizontal
bands a few tubes in height were formed where the heat transfer coefficient
ceased to increase with increasing heat flux but fell back to the value at
only 10 kW/m². The position of these bands did not change with heat flux

in the lower half of the bundle. At the top, the variations in coefficient were large and less stable, that of a particular tube changing from high to low and vice versa as the flux was increased. Even at $100\,\mathrm{kW/m^2}$ there were no cases of excessive overheating (burnout).

8.3.2 Example: critical shellside heat flux

Estimate the critical heat flux for R-113 boiling on the outside of the experimental kettle reboiler described in section 7.4.3. The required physical properties of R-113 are given in the example in section 7.4.6. The diameter of the outer tube limit is 453 mm.

Substituting in equation [8.1] the physical properties from section 7.4.6 gives:

$$\dot{q}_{\mathrm{crp}} = 0.12 \times 145\,715 \times 7.38^{\frac{1}{2}}[14.9 \times 10^{-3} \times 9.81(1512 - 7.38)]^{\frac{1}{4}}$$

$$= 183 \times 10^3\ \mathrm{W/m^2}$$

For the approximate formula given in equation [8.24] it is necessary to determine the reduced pressure from equation [7.30]. The critical pressure has been given as 34.14 bar, so at an operating pressure of 1.0 bar, equation [7.30] gives:

$$p_r = 1.0/34.14 = 0.029\,29$$

Substituting in equation [8.24]:

$$\dot{q}_{\mathrm{crp}} = 0.367 \times 34.14 \times 10^5 \times 0.029\,29^{0.35}(1 - 0.029\,29)^{0.9}$$

$$= 355 \times 10^3\ \mathrm{W/m^2}$$

Thus the preferred method gives a critical heat flux of $183\,\mathrm{kW/m^2}$, whereas the approximate method gives nearly twice this ($355\,\mathrm{kW/m^2}$).

The correction factor required for equation [8.25] must be estimated from equation [8.26], which requires a knowledge of D_b, given in this example as 453 mm, L, given in section 7.4.3 as 25.4 mm and A, calculated in section 7.4.6 as $0.3663\,\mathrm{m^2}$. Hence:

$$F_{\mathrm{crs}} = 6.9 \times 0.453 \times 0.0254/0.3663 = 0.217$$

From equation [8.25]

$$\dot{q}_{\mathrm{crs}} = 183 \times 0.217 = 40\ \mathrm{kW/m^2}$$

8.4 Enhancement of critical heat flux

Various devices are described in section 7.8 for increasing the heat transfer coefficient in boiling where the heating surface is wetted by the liquid. The devices described in section 7.8.1 for enhancing shellside boiling have been developed for increasing the heat transfer coefficient at low temperature differences, and so are operating well below the critical heat flux.

This section is concerned with ways of enhancing the critical heat flux by inducing spiralling flow.

The critical heat flux in in-tube boiling can be increased in one of three ways:

A.   *Twisted tapes* are used to increase the critical heat flux, especially with horizontal tubes, where they perform the function of taking water spirally up to the top of the tube to replace water that has evaporated. Also the induced rotation centrifuges the water to the wall and increases the velocity. They give a considerable improvement in critical heat flux under conditions where the critical heat flux is very much less for horizontal tubes than for vertical tubes, but give no improvement under conditions where relatively high critical heat fluxes are achieved with horizontal tubes. Cumo *et al.* (1974) in experiments on the flow of R-12 up vertical tubes show an improvement in the critical heat flux by the use of twisted tapes.

B.   *Rifled tubes* also increase the critical heat flux by inducing spiral flow. They are not as effective as twisted tapes, but have the advantages of not introducing a large increase in pressure drop and of avoiding the enhancement of corrosion likely due to contact between a twisted tape and the tube.

C.   *Coiled tubes* have a much higher critical heat flux than straight tubes under otherwise identical conditions. The reason for this is that flow along a curved tube induces a secondary flow that keeps the wall well wetted at all locations around the circumference. Consequently the steam quality at dryout is increased and the post-dryout fluctuations in temperature are much lower in a coiled than in a straight tube, as explained by Owhadi *et al.* (1968). This is the basis of the success of the coiled tube boiler described in section 3.3.1.

CHAPTER 9

Calculations in fluid flow

A variety of calculations in fluid flow may be required in the design, rating or testing of a vaporiser. The first six sections of this chapter deal with the flow of the fluid to be vaporised; the last six sections deal with flow in ancillary equipment.

9.1 General equations for steady flow

The pressure drop in the flow of a fluid through a duct is the sum of three terms, due to: (1) wall friction and form drag caused by obstructions in the stream (frictional), (2) increase in elevation (gravitational), and (3) increase in momentum (accelerational). The first is an irreversible loss in pressure, the pressure energy lost being converted into heat; the second and third terms are reversible, i.e. the pressure can be regained by reversing the process. The frictional term, referred to as 'pressure loss', is conventionally expressed in terms of the 'number of velocity heads lost'; this is defined as the frictional loss of head divided by the velocity head, $\bar{u}^2/2g$ (m), where $\bar{u}$ is the mean velocity of flow (m/s) and g is the acceleration due to gravity (9.81 m/s^2). For calculations in boiling flow it is more convenient to relate the loss of pressure to the velocity pressure, which equals $\frac{1}{2}\rho\bar{u}^2$, where ρ is the density of the fluid (kg/m^3). In flow along a duct the density of the vapour decreases as a result of pressure drop, so the velocity increases. It is therefore convenient to replace velocity by mass flux, denoted by $\dot{m}$; this is the mass flow rate ($\dot{M}$) divided by the cross-sectional area for flow. Mass flux is also equal to $\rho\bar{u}$, and it has the dimensions kg/s m^2. It is constant along a duct of uniform cross-section.

   In flow through a duct, the negative pressure gradient, i.e. the pressure drop per unit length, has the dimensions Pa/m or N/m^3; it is given by:

$$-\frac{dp}{dz} = \frac{4f}{d}\frac{\dot{m}^2}{2\rho} + \rho g \sin\theta + \dot{m}^2\frac{dv}{dz} \qquad [9.1]$$

where d is the internal diameter of the pipe (m), f is the Fanning friction factor (see section 9.1.1), p is the local static pressure (Pa), v ($= 1/\rho$) is the specific volume of the fluid (m^3/kg) at the local static pressure and temperature, z is distance along the duct (m) from the inlet, and θ is the upward angle of inclination of the duct to the horizontal. For a duct of non-circular section, d must be replaced by d_e, the equivalent diameter, defined in equation [6.15]. Alternatively $d_e = 4S/P$ where S is the cross-sectional area for flow (m^2) and P is the perimeter (m) of the duct.

When estimating the pressure drop of the liquid to be vaporised in a non-boiling zone, the density may be assumed to be constant, so integrating the pressure gradient in equation [9.1] gives the pressure drop as:

$$p_u - p_d = K\left(\frac{\dot{m}^2}{2\rho}\right) + \rho g\,\Delta H \qquad [9.2]$$

where K is the pressure loss coefficient, p_u and p_d are the upstream and downstream pressures (Pa) and ΔH is the increase in the height of the duct above datum level (m). The pressure loss coefficient is given by

$$K = \frac{4fL}{d} + \sum K_f \qquad [9.3]$$

where L is the length of the duct (m) and $\sum K_f$ is the sum of the pressure loss coefficients for any pipe fittings in the zone (see section 9.1.2). The estimation of pressure loss in flow across banks of tubes is introduced in section 9.1.3.

The estimation of pressure drop in a region of two-phase flow is much more complicated. It is first necessary to calculate the frictional pressure gradient at zero quality from the Fanning friction factor and the pressure loss coefficients of any pipe fittings. Section 9.3 deals with the estimation of the enhancement in frictional pressure drop due to two-phase flow. Section 9.4 deals with the estimation of two-phase density. Section 9.5 describes the calculation of the total pressure drop in the flow of a boiling liquid.

9.1.1   Fanning friction factor

Standard textbooks on fluid flow deal with the estimation of pressure drop in straight pipes. Different authors use different definitions of the friction factor. The convention adopted here is to use the Fanning friction factor, which is defined as the pressure drop divided by the velocity pressure and by the ratio of the length to hydraulic mean depth of the duct; the hydraulic mean depth equals the ratio of cross-sectional area for flow to the perimeter of the duct (S/P), which equals $\frac{1}{4}d$ for a circular pipe. The friction factor depends on the Reynolds number (Re) defined in equation [6.8] and the surface roughess (ε). It can be deduced theoretically that in laminar flow (Re < 2000) the Fanning friction factor (f) is independent of roughness and is related to the Reynolds number by:

$$f = \frac{16}{\text{Re}} \tag{9.4}$$

For fully developed turbulent flow (Re > 4000) ESDU (1966) and Miller (1978) recommend the use of the correlation proposed by Colebrook and White (1937), which gives:

$$\frac{1}{\sqrt{f}} = -4 \log_{10} \left[\frac{1.25}{\text{Re}\sqrt{f}} + \frac{\varepsilon}{3.7d} \right] \tag{9.5}$$

This is difficult to use because it contains the required friction factor on both sides of the equation. A graphical solution is given in Fig. 9.1. The dotted lines may be used to estimate the approximate value of f in the transitional region (2000 < Re < 4000) for which no one has recommended a suitable correlation, due to the considerable scatter of experimental data.

Table 9.1 gives recommended values of the surface roughness, based on the published data of ESDU (1966) and Miller (1978).

When estimating the pressure drop in tubes where scaling is likely to occur, it is first necessary to make an allowance for the increased roughness of the scaled surface. More importantly an estimate must be made of the likely thickness of the deposit, so that the mass flux can be calculated from the true internal diameter of the passage available for flow.

9.1.2 Pressure loss coefficients for single-phase flow through pipe fittings

Based on the work of Idel'chik (1966), ESDU (1972), ESDU (1973a), ESDU (1973b), ESDU (1983a), ESDU (1977a), ESDU (1977b) and Miller (1978), the empirical values of K_f in Table 9.2 are recommended for estimating the pressure loss coefficients for the fittings shown in Fig. 9.2, provided that Re > 10 000. These values, required for equations [9.2] and [9.3], are very approximate; for more accurate estimates, the references quoted should be consulted. The recommended values are more likely to overestimate, rather than underestimate, the pressure loss. For further information on the pressure drop in bends, see ESDU (1983a).

It must be remembered that the pressure loss coefficient gives the irreversible pressure loss in the fitting. For calculations in two-phase flow, it is necessary to calculate the static pressure at various points to determine the local values of the quality. In bends and elbows, there is no change in the mass flux, so the pressure drop is K multiplied by the velocity pressure

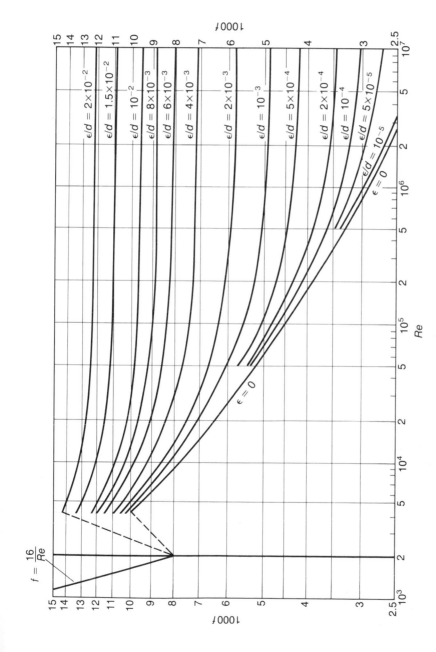

9.1 Friction factor (from equation [9.5])

9.1   Approximate values of surface roughness

Surface	Roughness, ε (m)
Drawn brass, copper or aluminium	2×10^{-6}
Stainless steel or new mild steel	3×10^{-5}
Mild steel, slightly corroded	5×10^{-5}
Rusted or scaled surface	3×10^{-4}

9.2   Approximate pressure loss coefficients for various fittings

Fitting		Pressure-loss coefficient (K_f)				
Bends*	with r/d =	1	1.5	2	3	5
90° circular arc		0.35	0.25	0.2	0.2	0.2
90° 2-cut mitre		—	0.6	0.5	0.4	0.5
90° 3-cut mitre		—	0.6	0.5	0.3	0.2
180° circular arc		0.35	0.25	0.25	0.3	0.35
90° elbow (plain)		1.2				
90° elbow with guide vanes		0.3				
Abrupt contractions[†]		$0.5(1 - \sigma^2)^{[‡]}$				
Abrupt enlargements[†]		$(1 - \sigma)^{2\,[‡]}$				
Junction with blanked manifold $\begin{cases}\text{flow into branch}^{[†]}\\ \text{flow out of branch}^{[†]}\end{cases}$		$0.5 + \sigma^{2\,[‡]}$ $(1 - 0.4\sigma)(1 + \sigma^2)^{[‡]}$				

* To these values for bends must be added the pressure loss coefficient for a straight pipe of length equal to that of the bent pipe
[†] Based on the velocity pressure in the smaller duct
[‡] Where $\sigma = \dfrac{\text{cross-sectional area of smaller duct}}{\text{cross-sectional area of larger duct}}$

($\dot{m}^2/2\rho$). In the other fittings, there is a change in the mass flux, so it is necessary to include the change in pressure due to the change in velocity pressure, according to Bernoulli's equation. The following equations give the drop or rise in static pressure, suffix u referring to upstream and d to downstream conditions. Where there is a change in the cross-sectional area for flow, σ denotes the ratio of the smaller to the larger area.

The pressure drop in a bend or elbow is:

$$p_u - p_d = (K + 4fL/d)(\dot{m}^2/2\rho) \qquad [9.6]$$

where L is the length of the centre-line of the bend or elbow (m).

The pressure drop in an abrupt contraction is:

$$p_u - p_d = 1.5(1 - \sigma^2)(\dot{m}_d^2/2\rho) \qquad [9.7]$$

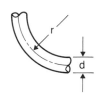

90° circular arc bend

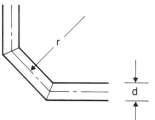

90° 2-cut mitre bend

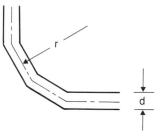

90° 3-cut mitre bend

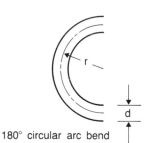

180° circular arc bend

90° elbow

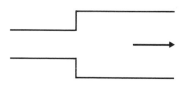

90° elbow with guide vanes

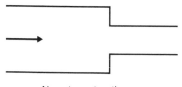

Abrupt contraction

Abrupt enlargement

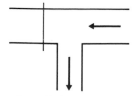

Junction with blanked manifold,
flow into branch

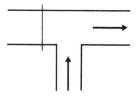

Junction with blanked manifold,
flow out of branch

9.2 Pipe fittings

The pressure rise in an abrupt enlargement is:

$$p_d - p_u = 2\sigma(1 - \sigma)(\dot{m}_u^2/2\rho) \tag{9.8}$$

The pressure drop in flow from a blanked manifold to a branch is:

$$p_u - p_d = 1.5(\dot{m}_d^2/2\rho) \tag{9.9}$$

The pressure drop in flow from a branch to a blanked manifold is:

$$p_u - p_d = (-0.4\sigma + 2\sigma^2 - 0.4\sigma^3)(\dot{m}_u^2/2\rho) \tag{9.10}$$

9.1.3 Pressure loss coefficients for single-phase flow across tube banks

The present state of knowledge on the estimation of pressure drop in flow across banks of plain straight tubes has been summarised by Saunders (1987) Chapter 12, ESDU (1974), ESDU (1979) and by J. Taborek in section 3.3.7 of Schlünder (1983). One of these documents should be consulted if a fairly accurate estimate is required of single-phase pressure drop. They present graphs of friction factor against Reynolds number for different configurations, over a wide range of Reynolds numbers.

When estimating the pressure drop in boiling flow across a tube bank, calculations are inevitably very approximate and the Reynolds number is sufficiently high to have only a minor influence on the friction factor. For a rough calculation, accurate to within a factor of two, the pressure drop due to single-phase friction may be determined from:

$$\Delta p_{fr} = \tfrac{1}{2}n_v(\dot{m}_{max}^2/2\rho) \tag{9.11}$$

where n_v is the number of rows of tubes crossed and $\dot{m}_{max}$ is the maximum mass flux, i.e. the mass flux through the region of minimum cross-sectional area (kg/s m^2), provided that $Re_{lo} > 1000$.

When estimating the shellside pressure drop, it is necessary to include the pressure drop in the nozzles, as described in section 9.7.4.

9.2 Two-phase flow patterns and unsteady flow

When designing a vaporiser, it is important to check that flow will be steady, because fluctuations in quality, velocity and pressure in the regions of two-phase flow may lead to dryout (described in Ch. 8), may upset the steady operation of the process, or may lead to troublesome vibrations of heat exchanger tubes or external pipework (see section 11.3). The various patterns of two-phase fluid flow have been described extensively in the literature, e.g. by Baker (1958), Wallis (1969), Hewitt in section 2 of

Butterworth and Hewitt (1977), and Delhaye in Chapter 1 of Bergles *et al.* (1981).

Some of the patterns are relatively steady, whereas with others the pressure and density at any location vary considerably with time. As a saturated liquid flows through a heated pipe, the vapour formed in the initial stages of boiling is present in the form of small bubbles and flow is steady. As the volume rate of flow of vapour increases and approaches that of the liquid, the bubbles tend to coalesce and the flow pattern may change to some form of unsteady flow. However, at low mass fluxes in horizontal pipes, flow is likely to become stratified, the liquid flowing along the bottom of the pipe and the vapour along the top. If the volume flow rate of vapour greatly exceeds that of the liquid, flow becomes annular, when most of the liquid flows as an annular film on the wall, and in the centre of the pipe there is a faster flowing core of vapour containing some entrained liquid. As this type of flow is steady, it is recommended that flow in heated tubes should become annular before the outlet and that flow should be annular in the pipework transporting a two-phase mixture from the heater to the separator of a vaporiser. The criteria for annular flow are given in sections 9.2.1 and 9.2.2 for flow in horizontal and vertical pipes respectively, and the flow patterns in flow across tube bundles are discussed in section 9.2.3.

Moreover, thermal-hydrodynamic instabilities in flow may occur in boilers, as described by Bailey in section 16 of Butterworth and Hewitt (1977) and by Bergles in Chapter 13 of Bergles *et al.* (1981). The important types of two-phase instability in flow through a heater are described in sections 9.2.4 to 7. The heater consists of many identical parallel paths. The pressure drop must be the same through each path, but the flow rate and quality at outlet may vary from path to path, and the split of the total flow may change from time to time. It will be assumed that the heat flux at any location does not change with time. The special problems with liquid metal systems are not mentioned, because they are considered to be outside the scope of this book.

9.2.1 Annular flow in horizontal pipes

The earliest and most widely used method of predicting the flow pattern in two-phase flow in a horizontal pipe has been to use the flow map published by Baker (1958). Wallis (1969) points out, in section 11.2, that the ordinate in Baker's plot is not dimensionless and cannot be interpreted physically. He suggested that it should be modified to the parameter defined below in equation [9.12]. This is the dimensionless volume flux of the vapour and represents the balance between the inertia force of the gas and the gravity force of the liquid.

$$\dot{v}_g^+ = \frac{\dot{m}x_g}{\sqrt{[gd\rho_g(\rho_1 - \rho_g)]}}$$ [9.12]

where d is the internal diameter of the pipe (m), g is the acceleration due to gravity (9.81 m/s²), $\dot{m}$ is the total mass flux (kg/s m²), x_g is the quality (ratio of mass flow rate of gas to total mass flow rate), and ρ_g and ρ_1 are the densities of the gas and liquid respectively (kg/m³).

Taital and Dukler (1976) show that transition from stratified to annular flow depends on the dimensionless volume flux, defined above, and the Lockhart–Martinelli parameter (X) defined in section 7.2.1. Flow will be annular if $X \leqslant 1.6$ and if $\dot{v}_g^+$ is greater than the value given below.

X	10^{-3}	10^{-2}	3×10^{-2}	0.1	0.3	1.0	1.6
Minimum value of $\dot{v}_g^+$ for annular flow	2.0	2.0	1.8	1.6	1.1	0.45	0.35

The transition from stratified to unsteady flow occurs when $X > 1.6$ and $\dot{v}_g^+$ is greater than the value given below:

X	1.6	4	10	20	50
Minimum value of $\dot{v}_g^+$ for unsteady flow	0.35	0.10	0.02	5×10^{-3}	10^{-3}

At very low vapour rates, bubble flow occurs; this is a steady flow pattern. Afgan and Schlünder (1974), p. 266, say that bubble flow exists if the volume flow rate of vapour is less than 30% of the total volume flow rate, i.e. the quality is less than $0.4\rho_g/\rho_1$.

9.2.2 Annular flow up vertical pipes

The flow patterns in upward vertical flow range from bubble flow to 'plug' or 'churn' flow and ultimately to annular flow. The recommended flow pattern map is that of Hewitt and Roberts (1969), in which the coordinates are the momentum flux of liquid flowing in the absence of vapour, $\dot{m}^2(1 - x_g)^2/\rho_1$, and the momentum flux of vapour flowing without liquid, $\dot{m}^2x_g^2/\rho_g$. Over the normal range of liquid velocities (from about 0.1 to 1.0 m/s), annular flow occurs when the superficial momentum flux of the vapour exceeds 100 kg/m s². Thus flow becomes annular when:

$$\dot{m}x_g > 10\sqrt{\rho_g}$$ [9.13]

where $\dot{m}$ is the mass flux (kg/m² s), x_g the quality (dimensionless) and ρ_g the density of the gas (kg/m³).

Further information is given by Chisholm (1983).

9.2.3 Shellside flow patterns

Chisholm (1983) in his Chapter 16 describes the limited work that has been done on observing the flow patterns in two-phase shellside flow. They propose a map for vertical flow similar to that recommended in the previous section for vertical pipe flow. However, in shellside vaporisers, where the tubes are usually closely spaced, it is unlikely that large slugs of vapour will form and make the overall flow unstable. It is therefore recommended that there is no need to check the flow pattern when designing a shellside vaporiser.

9.2.4 Static instability

Under certain circumstances the curve of pressure drop against rate may be basically unstable and the flow through all or some of the parallel heated paths may be considerably lower than was intended. This problem was first studied by Ledinegg (1938) and is often referred to as 'Ledinegg Instability'. Figure 9.3 compares a stable characteristic (a) with an unstable characteristic (b). In each case, the curve OA shows the relationship between pressure drop and rate. The pressure drop may be across the heated surface, from the inlet header to the outlet header, or it may be from the inlet to the heater to the separator. The curve BC is the pressure rise

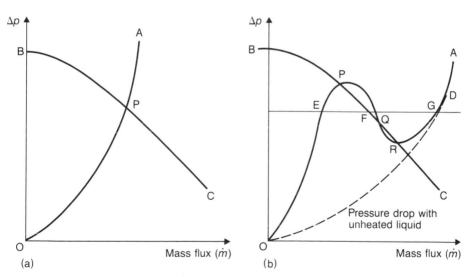

9.3 Stable and unstable pressure drop characteristics for constant heat input
(curve OA is the pressure drop and curve BC the driving force characteristic)
(a) stable
(b) unstable

between the same two locations. With natural circulation, this is the pressure rise due to gravity in the downcomer less the friction in other parts of the loop. With assisted or forced circulation, the characteristics of the pump must be added to the driving force due to gravity. It is seen in Fig. 9.3 (a) that the two curves intersect at only one point (P), which is the operating point. This is a stable situation, because any slight increase in rate leads to an increase in the pressure drop and a decrease in the driving force, and the resulting imbalance decelerates the fluid until the equilibrium rate is restored. The converse occurs if there is a slight decrease in rate. However, in Fig. 9.3 (b) there are three points of intersection (P, Q and R). Points P and R are stable operating points, as described above, but it would not be possible to operate at point Q, because any small increase in rate would result in the driving force exceeding the pressure drop and so accelerating flow until point R is reached; conversely a small decrease in rate would lead to a switch from Q to operating at point P. There is no way of knowing whether operation would be at point P or R. Considering now the similar curves that could be drawn for the pressure difference across the parallel heated paths, some tubes could operate at a point such as E and some at the equal pressure drop of point G, provided that the mean rate was that corresponding to point F. Under these circumstances, the flow through some tubes may be at the lower rate and through the other tubes at the higher rate. This unsatisfactory situation can be avoided by increasing or decreasing the driving force so that it intersects the curve OA at only one point. Alternatively the insertion of a restriction in the region of single-phase flow would raise the curve OA by an amount equal to the pressure drop through the restriction, which is proportional to $\dot{m}^2$, thus avoiding the intersections at Q and R. Obviously if the pressure drop always increases with increasing flow, as is usual, there is no danger of static instability.

If it is suspected that static instability might occur, either through the heated passages or over the recirculation loop, the coordinates of the curve OA should be calculated, using the methods given in sections 9.3 to 5. If a maximum and minimum are found in this curve, as in Fig. 9.3 (b), the driving force curve (BC) must also be determined. If the two curves have three points of intersection, appropriate action must be taken to rectify this trouble. The exercise must cover the full range of possible heat transfer rates. As this is a lengthy task, it is desirable to consider the circumstances that tend to produce an unstable pressure drop characteristic curve. Two geometries are discussed below, for both saturated and subcooled conditions at the inlet to the heater.

With *horizontal flow*, if the liquid is initially saturated, the amount of vapour generated is independent of the amount of liquid recirculated, being equal to the heat transfer rate divided by the latent heat of vaporisation. Thus any increase in the amount of liquid recirculated must lead to an increase in pressure drop, and so a stable curve of pressure drop is obtained, as shown in Fig. 9.3 (a). If the liquid is initially subcooled, then it is possible that, at high recirculation ratios, boiling is suppressed and only sensible heat is transferred. The pressure drop is then as shown in section DA of the pressure drop curve in Fig. 9.3 (b), the dotted curve from O to D being the pressure drop with no heat input. If the mass flux is less than that at point G, then some vapour is generated before the liquid leaves the heated length, and this vapour might lead to a substantial increase in the pressure drop compared with that for no heat input, as at point P. Thus with horizontal flow, static instability may occur if the liquid is initially subcooled, but flow is sure to be statically stable if the liquid is initially saturated.

With *vertical upflow*, the pressure drop due to increase in elevation must be added to frictional pressure drop, estimated as described for horizontal flow. When there is no boiling, this extra gravitational term is equal to $\rho_l g \, \Delta H$, where ρ_l is the liquid density, g the acceleration due to gravity and ΔH the increase in elevation. When boiling occurs, this is reduced by a factor $(1 - \bar{\varepsilon}_g)$ where $\bar{\varepsilon}_g$ is the mean void fraction. Thus reducing the mass flux increases the outlet quality and reduces the gravity term in the calculation of pressure drop. At normal flow rates, the gravity term is always much greater than the frictional term for the flow of liquids, but at low rates the gravity term is small. It follows that the sum of the two terms together gives a stable curve, even when there is subcooling at inlet and the frictional term is as shown in curve OEPQRGDA in Fig. 9.3 (b). Thus the use of vertical upflow avoids static instability. Conversely vertical downflow is likely to be unstable.

9.2.5 Chugging instability

There is a danger that instability may occur if the wall is not sufficiently superheated for the onset of nucleate boiling, as described in section 7.1.1. As explained in section 2.4, there is seldom a significant amount of liquid superheating in industrial vaporisers. However, with very clean liquids and low temperature differences such that there is no nucleate boiling, a significant amount of superheating of the liquid may occur. When operating at low pressures, there is then a danger that when vapour eventually forms it will lead to a sudden expulsion of liquid from the heated tube and a period of liquid deficiency until more liquid can flow into the tube. This form of instability is known as 'chugging flow'. However, superheating of the liquid does not always lead to unsteady flow, as shown by Robertson and Clarke (1981), who described experiments where there was an abrupt change from superheated single-phase liquid flow to saturated two-phase flow, but the location of the plane where boiling began did not fluctuate. There is no way of predicting how much liquid superheat will occur or whether it will lead to chugging instability.

9.2.6 Acoustic effects

In any flow system, pulsations in flow at acoustic frequencies may arise through the resonance of disturbances transmitted through pipelines and reflected at their ends. In two-phase systems, the velocity of sound through the mixture may be even less than that through the vapour alone, but the amplitudes of the resulting pulsations in pressure are considerably less than those encountered in 'water hammer' in liquids. Acoustic effects are potentially dangerous when resonance augments regular disturbances arising from pumps or the regular collapse of bubbles.

Wallis (1969) has deduced his equation [2.46] for the velocity of a compressibility wave in a two-phase homogeneous mixture. If the liquid is assumed to be incompressible compared with the gas, this equation simplifies to:

$$c_{tp} = \frac{c_g}{\sqrt{x_g}} \frac{\rho_g}{\rho_{tp}} \qquad [9.14]$$

where c is the velocity of sound (m/s), x_g the quality and ρ the density (kg/m^3), suffix g referring to the vapour and suffix tp to the two-phase homogeneous mixture. Equation [7.22] gives ρ_{tp}.

Wallis shows from the above that the velocity of sound in a homogeneous mixture is a minimum when the volume rates of flow of the two phases are approximately equal.

It is thus possible to estimate the resonant frequencies in a pipeline conveying a two-phase mixture from the laws of acoustics, if it is assumed that flow is homogeneous and if equation [9.14] is used to determine c_{tp}, the sonic velocity in the two-phase mixture.

9.2.7 Oscillatory instability

This is the most important form of instability and the most difficult to predict. A temporary reduction in the flow rate of a boiling mixture in a heated channel leads to an increased rate of boiling and so reduces the density of the mixture. The consequent perturbation is transmitted to the outlet and reflected to the inlet. If the inlet flow rate is oscillated at a fixed frequency, the components of pressure drop in the two-phase region are additive at low frequencies, but as the frequency is increased there are lags in the components due to gravity, friction and momentum. However, there are no lags in the pressure drop in the single-phase region. At certain frequencies the components tend to cancel each other; hence these are the frequencies that are liable to lead to resonance and the build-up of serious fluctuations in the density. The natural frequency of these density waves is much less than that of the acoustic waves described in the previous section. Further information is given by Bergles in sections 13.2.4 to 13.4 of Bergles *et al.* (1981) and by Potter in section 17 of Butterworth and Hewitt (1977). Proprietary computer programs are available for the solution of this complicated problem. A graphical method of predicting density-wave instability is suggested by Akinjiola and Friedly (1979). Hands (1979) described an experimental study of the subject. Friedly *et al.* (1979) summarise the available experimental data and show good agreement with predictions.

Increasing the pressure drop in the single-phase region by the introduction of a restriction in the inlet will eliminate oscillatory instability if the single-phase pressure drop exceeds the two-phase pressure drop. Conversely, the outlet pipework should be of adequate size and restrictions in it should be avoided if possible. Subcooling at the inlet increases stability by increasing the length of the region of single-phase pressure drop.

It is necessary to consider the danger of density wave (oscillatory) instability over the complete flow path and also between parallel heated paths forming part of the complete path. Thus in the waste heat boiler shown in Fig. 3.13, instability might occur in flow through the complete loop, from the steam drum, through the heated tubes and back to the drum, or it might occur between the parallel paths of the heated tubes from the inlet manifold to the outlet manifold. Similarly in a vertical thermosyphon reboiler (Fig. 3.10) instability might occur in flow through the loop from the distillation column, through the heater and back to the column, or it might occur in flow between the parallel paths of the heated tubes, from the inlet channel to the outlet channel.

If the single-phase pressure drop exceeeds the two-phase pressure drop, instability will not occur.

9.3   Estimation of frictional two-phase pressure gradient

This section is complementary to section 7.2, which deals with the estimation of two-phase convective heat transfer coefficients. The steps in the estimation of local pressure gradient due to friction are as follows:

1. Calculate the liquid pressure gradient $(-dp/dz)_{lo}$ at zero quality, i.e. when the liquid has been heated to its saturation temperature but before any vapour has been established in the bulk of the fluid, using the appropriate correlation given above in section 9.1.
2. Calculate the two-phase pressure drop enhancement factor, ϕ_{lo}^2, from section 9.3.1 for flow inside uniform ducts, or from section 9.3.2 for flow across bundles of tubes.
3. Calculate the two-phase pressure gradient due to friction from

$$(-dp/dz)_{tpfr} = \phi_{lo}^2(-dp/dz)_{lo} \qquad\qquad [9.15]$$

This frictional pressure gradient must be combined with the pressure gradients due to acceleration and increase in elevation, and integrated over the boiling length, as explained in section 9.5, to determine the total pressure drop in the region of two-phase flow.

9.3.1 Two-phase pressure loss in uniform ducts

One of the most commonly used correlations for the calculation of the frictional pressure drop of a two-phase mixture of constant quality is that of Martinelli – see Martinelli *et al.* (1944) and Lockhart and Martinelli (1949); it was based on measurements of pressure drop in the flow of a mixture of liquid and gas in a horizontal pipe under adiabatic conditions (no transfer of heat to or from the fluids). It was modified by Martinelli and Nelson (1948) to apply to the flow of steam and water at pressures approaching the critical pressure. The Martinelli correlations make no allowance for the effects of mass flux on the enhancement factor. Hewitt in Chapter 2 of Hetsroni (1982) recommends that these correlations be used only when the viscosity of the liquid is more than 1000 times that of the gas and when the mass flux is less than $100\,\text{kg/s}\,\text{m}^2$. As it is very unlikely that the first of these conditions will apply in a vaporiser, the Martinelli correlations will not be reproduced in this book.

When the ratio of the viscosity of the liquid to the viscosity of the vapour is less than 1000, Hewitt recommends the use of the Friedel (1979) correlation, which is described below, and illustrated in the example in section 9.5.6.

The value of ϕ_{lo}^2, for substitution in equation [9.15], can be determined by the method of Friedel from the mass flux, $\dot{m}$ $(\text{kg/s}\,\text{m}^2)$, the internal diameter of the duct, d (m) and the physical properties of the fluids as follows:

1. Determine the Reynolds number (defined in equation [6.8]) for all liquid flowing at the saturation temperature (zero quality) and for all vapour (quality of unity), using respectively:

$$\text{Re}_{lo} = \frac{\dot{m}d}{\eta_l} \quad \text{and} \quad \text{Re}_{go} = \frac{\dot{m}d}{\eta_g} \tag{9.16}$$

2. Using the procedure given in section 9.1.1, determine the Fanning friction factors, f_{lo} from Re_{lo} for all-liquid flow, and f_{go} from Re_{go} for all-vapour flow.

3. Determine the parameters E, F and H from

$$E = (1 - x_g)^2 + x_g^2 \frac{\rho_1 f_{go}}{\rho_g f_{lo}}$$

$$F = x_g^{0.78}(1 - x_g)^{0.24} \tag{9.17}$$

$$H = \left(\frac{\rho_1}{\rho_g}\right)^{0.91} \left(\frac{\eta_g}{\eta_1}\right)^{0.19} \left(1 - \frac{\eta_g}{\eta_1}\right)^{0.7}$$

where x_g is the quality, ρ the fluid density (kg/m^3) and η the fluid viscosity ($N\,s/m^2$), suffix l denoting the liquid and suffix g the vapour.

4. Determine the two-phase dimensionless Froude number from

$$Fr_{tp} = \dot{m}^2/gd\rho_{tph}^2 \tag{9.18}$$

where g is the acceleration due to gravity (m/s^2) and ρ_{tph} is the two-phase density (kg/m^3) calculated on the assumption of homogeneous flow from equation [7.22]. (Equation [9.18] simplifies to equation [7.16] for all-liquid Froude number when $x_g = 0$).

5. Determine the dimensionless two-phase Weber number (We_{tp}) from

$$We_{tp} = \dot{m}^2 d/\rho_{tph}\sigma \tag{9.19}$$

where σ is the surface tension (N/m) at the liquid vapour interface. (When $x_g = 0$, equation [9.19] simplifies to equation [9.28], the usual form of the Weber number for liquid flow).

6. Determine ϕ_{lo}^2 from the following correlation which Friedel (1979) gives for horizontal flow or upward vertical flow:

$$\phi_{lo}^2 = E + \frac{3.24FH}{Fr_{tp}^{0.045}We_{tp}^{0.035}} \tag{9.20}$$

A more complicated correlation is given for downward vertical flow; it is not quoted here because this arrangement is seldom found in vaporiser systems.

7. Determine the two-phase pressure gradient from equation [9.15] and multiply this by the length of the duct, to give the total two-phase pressure loss due to friction in the duct.

8. Check that the total pressure drop in the duct is small (say less than 5%) compared with the static pressure in the duct; it may be necessary to include the loss in any fittings, calculated as described in section 9.3.3, and the gravitational and accelerational components, calculated as described in section 9.5. If the total pressure drop is significant compared with the static pressure, the above calculations must be repeated based on the mean value of ρ_g, the gas density, calculated at the arithmetic mean of the inlet and outlet pressures.

This procedure is developed in section 9.5.5 and illustrated in the example in section 9.5.6.

9.3.2 Two-phase pressure loss in flow across tube banks

For the accurate estimation of the loss of pressure in boiling flow across a bank of tubes, it is first necessary to estimate the pressure loss for all-liquid flow, i.e. for zero quality, as described [in section 9.1.3. The procedure of Chisholm (1983), Chapter 16, should be adopted for estimating the enhancement factor, ϕ_{lo}^2.

For approximate calculations of the kind suggested in section 9.1.3 for single-phase pressure drop, it is sufficient to follow the considerably simpler procedure of assuming that flow is homogeneous and fully turbulent. Then the enhancement in pressure drop equals the expansion in volume, and so

$$\phi_{lo}^2 = 1 + (\rho_l/\rho_g - 1)x_g \tag{9.21}$$

9.3.3 Two-phase pressure drop in pipe fittings

This section gives the procedure for estimating the pressure drop (or rise) in two-phase flow through pipe fittings, such as are needed in the pipes conveying a mixture of liquid and vapour from a vaporiser.

1. Calculate the single-phase pressure drop for the flow of liquid at zero quality, using equations [9.6] to [9.10], as appropriate. These equations give the friction loss plus the pressure drop due to increase in momentum (in an enlargement there is a rise in pressure). For the purpose of these calculations, $\dot{m}$ must be taken as the total mass flux and ρ as ρ_l, the density of the liquid.

2. Determine the appropriate value of the empirical constant (C^*) required for estimating the two-phase enhancement factor, as recommended below, based on the work of Ferrell and McGee (1966), Fitzsimmons (1964) and Geiger and Rohrer (1966).

    For *bends* with r/d in the range from 1 to 5,

    $$C^* = 5 - \tfrac{1}{2}r/d \tag{9.22}$$

    For an *elbow* or *junction* with blanked manifold, $C^* = 1.8$.
    For a *contraction*, $C^* = 1.0$
    For an *enlargement*, $C^* = 0.5$.

3. Calculate the two-phase pressure drop enhancement factor from

    $$\phi_{lo}^2 = 1 + x_g[C^*(\rho_l/\rho_g + 1) - 2] - x_g^2[(C^* - 1)(\rho_l/\rho_g + 1)] \tag{9.23}$$

When $C^* = 1$, flow is homogeneous and equation [9.23] simplifies to equation [9.21].

4. Multiply the single-phase pressure drop by the enhancement factor, ϕ_{lo}^2, to obtain the two-phase pressure drop.

9.4 Two-phase density

To be able to estimate the pressure drop in boiling flow due to increase in momentum and due to increase in elevation, it is necessary to be able to estimate local values of the density of the two-phase mixture; this is discussed below. Section 9.5 describes the estimation of the total two-phase pressure drop due to friction (from section 9.3) plus that due to increases in momentum and elevation, for various arrangements.

The two-phase density is given by:

$$\rho_{tp} = \rho_l(1 - \varepsilon_g) + \rho_g \varepsilon_g \qquad [9.24]$$

where ρ is density (kg/m^3) suffixes g denoting that of the vapour, l that of the liquid and tp that of the two-phase mixture, and ε_g is the void fraction, i.e. the fraction of the volume of a short length of channel that is occupied by the vapour. The void fraction depends on $\dot{V}_r$ the local ratio of the volumetric flow rate of vapour to that of liquid, and the mean slip ratio $\bar{u}_r$; these are given by:

$$\dot{V}_r = \frac{x_g}{1 - x_g} \frac{\rho_l}{\rho_g} \qquad [9.25]$$

$$\bar{u}_r = \bar{u}_g / \bar{u}_l \qquad [9.26]$$

where x_g is the quality and $\bar{u}_g$ and $\bar{u}_l$ are the mean velocities of the vapour and liquid respectively (m/s). The local void fraction is given by

$$\varepsilon_g = \frac{\dot{V}_r}{\dot{V}_r + \bar{u}_r} \qquad [9.27]$$

If flow is assumed to be homogeneous, $\bar{u}_r = 1$ and the above equations simplify to equation [7.22].

In most applications, it is sufficiently accurate to estimate the two-phase density on the assumption of homogeneous flow, using equation [7.22]. For upward flow in a vertical pipe, it is advisable to estimate the slip ratio, $\bar{u}_r$, by the method described in section 9.4.1 and then use equations [9.27] and [9.24] to estimate first the void fraction and then the density.

9.4.1 Estimation of slip ratio

Hewitt in Chapter 2 of Hetsroni (1982) recommends the correlation of Premoli et al. (1970), which they derived from data for air and water and for steam and water flowing vertically upward. It is not certain that this is applicable to boiling flow in heated tubes.

The steps in the estimation of the slip ratio by this method are described below and illustrated in the example in section 9.4.2.

1. Calculate the Reynolds and Weber numbers for all liquid flow (at zero quality); the former is given by equation [9.16] and the latter by

$$We_{lo} = \dot{m}^2 d/\rho_l \sigma \qquad [9.28]$$

   where σ is the surface tension at the liquid/vapour interface (N/m).

2. Calculate the parameters C_1 and C_2 from

$$C_1 = 0.0273 \, We_{lo} \, Re_{lo}^{-0.51} \, (\rho_l/\rho_g)^{-0.08} \qquad [9.29]$$

$$C_2 = \frac{\dot{V}_r}{1 + \dot{V}_r C_1} - \dot{V}_r C_1 \qquad [9.30]$$

   where $\dot{V}_r$ is the volumetric flow ratio, determined from equation [9.25].

3. If $C_2 \leqslant 0$, put $\bar{u}_r = 1$ (homogeneous flow); otherwise calculate the parameter C_3 from

$$C_3 = 1.578 \, Re_{lo}^{-0.19}(\rho_l/\rho_g)^{0.22} \qquad [9.31]$$

4. The slip ratio is then given by:

$$\bar{u}_r = 1 + C_3 C_2^{\frac{1}{2}} \qquad [9.32]$$

9.4.2 Example: two-phase density

Water at a saturation temperature of 80 °C is flowing up a vertical pipe at a mass flux of 210 kg/s m². The internal diameter of the pipe is 104 mm. What is the two-phase density if the quality is 5%? The properties of water at this saturation temperature are:

$p_s = 473\,72$ Pa; $\qquad \rho_l = 970$ kg/m³; $\qquad \rho_g = 0.293$ kg/m³;

$\eta_l = 352 \times 10^{-6}$ Ns/m²; $\qquad \sigma = 62.7 \times 10^{-3}$ N/m.

1. From equation [9.16] the liquid Reynolds number is:

$$Re_{lo} = \frac{210 \times 0.104}{352 \times 10^{-6}} = 6.20 \times 10^4$$

from equation [9.28] the liquid Weber number is:

$$We_{lo} = \frac{210^2 \times 0.104}{970 \times 62.7 \times 10^{-3}} = 75.4$$

The density ratio is required for subsequent calculations; this is:

$$\rho_l/\rho_g = 970/0.293 = 3310$$

2. The parameters C_1 and C_2 are obtained from equations [9.29] and [9.30]

$$C_1 = 0.0273 \times 75.4 \times (6.20 \times 10^4)^{-0.51} \times 3310^{-0.08} = 3.87 \times 10^{-3}$$

The volumetric flow ratio is required for determining C_2 and can be determined from equation [9.25], which gives

$$\dot{V}_r = \frac{0.05}{0.95} \times 3310 = 174$$

From equation [9.30]

$$C_2 = \frac{174}{1 + 174 \times 3.87 \times 10^{-3}} - 174 \times 3.87 \times 10^{-3} = 103$$

3. As $C_2 > 0$, the parameter C_3 is required; from equation [9.31]

$$C_3 = 1.578 \times (6.20 \times 10^4)^{-0.19} \times 3310^{0.22} = 1.15$$

4. The slip ratio is determined from equation [9.32]

$$\bar{u}_r = 1 + 1.15 \times 103^{\frac{1}{2}} = 12.7$$

The void fraction can be found from equation [9.27]

$$\varepsilon_g = \frac{174}{174 + 12.7} = 0.932$$

The two-phase density can now be found from equation [9.24]

$$\rho_{tp} = 970(1 - 0.932) + 0.293 \times 0.932 = 66 \text{ kg/m}^3$$

It is of interest to compare this with the homogeneous two-phase density, given by equation [7.22] as

$$\rho_{tph} = \left[\frac{0.05}{0.293} + \frac{0.95}{970} \right]^{-1} = 5.83 \text{ kg/m}^3$$

This shows how, at low operating pressures, the high slip ratio results in the true two-phase density being considerably greater than that estimated on the assumption of homogeneous flow.

Referring to section 9.2.2, it is found that flow is annular, because $\dot{m}x_g = 210 \times 0.05 = 10.5$ is greater than $10\sqrt{\rho_g} = 10\sqrt{0.293} = 5.4$.

9.5 Estimation of total two-phase pressure drop

Equation [9.1] gives the negative pressure gradient for flow through a duct; this equation was integrated in section 9.1 for the special case of incompressible flow, i.e. for the flow of a liquid or of a gas whose density may be assumed to be constant. The object of this section is to recommend methods of integrating equation [9.1] for the more complex case of two-phase flow, where the density of the mixture is dependent on the amount of heat transferred and to some extent on the local pressure. When integrating equation [9.1] for flow through a straight uniform duct, the equation may be split into three components giving the following expression for Δp_{tp} the total two-phase pressure drop:

$$\Delta p_{tp} = \Delta p_{fr} + \Delta p_{gr} + \Delta p_{acc} \qquad [9.33]$$

where Δp_{fr} is the frictional component of the pressure drop, given by:

$$\Delta p_{fr} = \frac{2\dot{m}^2}{d} \int \frac{f_{tp}}{\rho_{tp}} \, dz \qquad [9.34]$$

Δp_{gr} is the gravitational component of the pressure drop, given by:

$$\Delta p_{gr} = g \sin \theta \int \rho_{tp} \, dz \qquad [9.35]$$

and Δp_{acc} is the pressure drop due to acceleration, given by:

$$\Delta p_{acc} = \dot{m}^2 \left[\frac{1}{\rho_{tp\,out}} - \frac{1}{\rho_{tp\,in}} \right] \qquad [9.36]$$

It is seen that the last term in equation [9.33] is the only term on the right-hand side of the equation that has been integrated analytically. Also pressure drop due to acceleration depends only on the initial and final two-phase densities, $\rho_{tp\,in}$ and $\rho_{tp\,out}$ respectively. To solve the integrals in equations [9.34] and [9.35] requires a knowledge of how ρ_{tp} and f_{tp} vary with z.

The determination of the three components of two-phase pressure drop is discussed in the three following subsections. An exact estimation of pressure drop in boiling flow can not be attempted until heat transfer calculations have been completed, giving the quality, and hence the two-phase density, at several chosen locations along the duct. The two-phase density is to some extent reduced by the fall in the density of the vapour due to pressure drop, and it may be necessary to repeat some of the calculations after allowing for this. The integrals must then be determined by stepwise calculations. Such hard work can seldom be justified because of the great uncertainty in estimating two-phase density. Simplified methods of calculation are therefore given below; these should always be carried out as a preliminary step.

9.5.1   Two-phase frictional pressure drop in boiling zone

For an accurate estimation of two-phase frictional pressure drop, equation [9.15] must be integrated. Thus:

$$\Delta p_{fr} = \int \phi_{lo}^2 (-dp/dz)_{lo} \, dz \qquad [9.37]$$

A stepwise procedure must be used to evaluate this integral. The values of ϕ_{lo}^2 and $(-dp/dz)_{lo}$ must be determined at several locations in the boiling zone defined by different qualities; the distance between these locations must be determined from heat transfer calculations. The method given in section 9.3.1 should be used to calculate ϕ_{lo}^2.

The integral in equation [9.34] can be solved analytically if the following simplifying assumptions are made:

1. The friction factor, f_{tp}, is constant and equal to f_l, the value for all liquid flow.
2. Flow is homogeneous.
3. The quality increases linearly with distance.

Under these circumstances it follows from equation [9.34] that if the two-phase density is taken from equation [7.22]

$$\Delta p_{fr} = \frac{2 f_l \dot{m}^2}{d} \int_0^{z_b} \left[\frac{1}{\rho_l} + \left(\frac{1}{\rho_g} - \frac{1}{\rho_l} \right) \frac{x_{g\,out} z}{z_b} \right] dz \qquad [9.38]$$

where z_b is the total length of the boiling zone, the quality at the inlet to the boiling zone is zero and the quality at the end of the boiling zone is $x_{g\,out}$. Integration gives:

$$\Delta p_{fr} = \frac{2 f_l z_b \, \dot{m}^2}{d \quad \rho_l} \, \phi_{lo\,m}^2 \qquad [9.39]$$

where $\phi_{lo\,m}^2 = 1 + \tfrac{1}{2}(\rho_l/\rho_g - 1) x_{g\,out}$ \qquad [9.40]

Although this treatment has been derived for flow inside uniform ducts, it may also be applied to flow across banks of tubes by calculating the pressure drop for single-phase flow of liquid and multiplying this by the enhancement factor given by equation [9.40].

9.5.2   Two-phase gravitational pressure drop in boiling zone

For horizontal flow, Δp_{gr} is zero. For vertical flow equation [9.35] gives, for the second and third assumptions listed in the previous section, and again taking the two-phase density from equation [7.22]

$$\Delta p_{gr} = \rho_l g \int_0^{z_b} \frac{dz}{1 + \left(\dfrac{\rho_l}{\rho_g} - 1 \right) \dfrac{x_{g\,out}}{z_b} z} \qquad [9.41]$$

Integration gives:

$$\Delta p_{gr} = \frac{g\rho_l z_b}{(\rho_l/\rho_g - 1)x_{g\,out}} \ln\left[1 + \left(\frac{\rho_l}{\rho_g} - 1\right)x_{g\,out}\right] \qquad [9.42]$$

This equation may be applied to boiling in upward flow in any type of vaporiser, provided that the second and third simplifying assumptions made in the previous section still apply.

The assumption of homogeneous flow is not likely to introduce any serious error in the case of flow across tubes with a staggered arrangement, because the tubes help to keep the liquid and vapour well mixed. However, with long-tube vertical evaporators or reboilers, there is a sufficient length for slip to be established. As the amount of slip increases with increasing quality, it is likely that at all points the slip is less than that appropriate to the quality at the points, due to delay in accelerating the vapour to the appropriate velocity. It is difficult to allow for this, due to lack of precise data.

9.5.3   Two-phase accelerational pressure drop in boiling zone

Accelerational pressure drop is given by equation [9.36]. Applying this equation to the whole of the boiling zone, taking the two-phase density from equation [7.22] and putting the inlet quality equal to zero, gives:

$$\Delta p_{acc} = \frac{\dot{m}^2}{\rho_l} x_{g\,out}(\rho_l/\rho_g - 1) \qquad [9.43]$$

In the derivation of this equation it is assumed that flow is homogeneous. This is normal practice because this component is usually the smallest of the three and because the value of the density used for calculating the gravity term is not necessarily the same as the density required for estimating the momentum flux at the end of the boiling zone, due to the flow pattern being more complicated than is assumed in the estimation of the mean velocities of the phases. No accurate measurements are available of the momentum flux in two-phase flow out of a pipe. Equation [9.43] is valid whatever the relationship between quality and distance.

9.5.4   Increase in quality in flashing flow

During the adiabatic flow of a saturated liquid and its vapour along a pipe, the pressure drop causes a reduction in the saturation temperature, with the consequent evaporation of some liquid to form more vapour; this is described as 'flashing flow'. Similarly in boiling flow, the final quality is greater than that calculated on the assumption of no pressure drop. From an energy balance, it is possible to calculate the quality at outlet, $x_{g\,out}$, from $x_{g\,in}$, the quality corresponding to the inlet saturation temperature, $T_{s\,in}$, the outlet saturation temperature, $T_{s\,out}$, the specific heat capacity of the liquid, c_{pl} (J/kg K) and the latent heat of vaporisation, Δh_v (J/kg). Thus:

$$x_{g\,out} = x_{g\,in} + \frac{c_{pl}(T_{s\,in} - T_{s\,out})}{\Delta h_v} \qquad [9.44]$$

9.5.5 Two-phase pressure drop in adiabatic flow through a pipeline

In systems where recirculation takes place through an external pipeline, whether by natural or forced circulation, it is necessary to be able to calculate the pressure drop of the two-phase mixture as it flows through the pipeline. Flow is adiabatic (no heat added) so the quality is constant, apart from any increase due to flashing, as described in section 9.5.4. Flashing may be significant in operation under vacuum. The pipeline may contain fittings such as those described in section 9.1.2. The following procedure should be followed for calculating the pressure drop, given the total mass flow rate, $\dot{M}$ (kg/s), the quality at inlet or outlet (x_g) and the physical properties of the liquid and vapour at the given inlet or outlet saturation pressure. As a first approximation, the quality should be assumed to be constant along the whole length of the pipeline.

1. Calculate the mass flux, $\dot{m}$ (kg/s m^2) and the velocity pressure of the total mass flow rate as liquid, $(\dot{m}^2/2\rho_l)$. If there is a change in cross-section, these calculations must be carried out for each size of pipe.

2. Calculate the pressure drop through each fitting of the total mass flow rate as liquid, as described in section 9.1.2.

3. Determine the Fanning friction factor for all liquid flow, f_l, as described in section 9.1.1 based on the Reynolds number for all liquid flow, Re_{lo}, from equation [9.16].

4. Calculate the pressure drop in single-phase flow of the total mass flow rate as liquid for the total length of straight pipe (L) from:

$$\Delta p_{fr\,st} = 4f_l(L/d)(\dot{m}^2/2\rho_l) \qquad [9.45]$$

5. Calculate the two-phase pressure drop in the straight lengths of uniform ducts, as described in section 9.3.1 and add to this the two-phase pressure drop in each fitting, calculated as described in section 9.3.3, to give the total frictional pressure drop, Δp_{fr}.

6. Calculate ρ_{tp}, the two-phase density, as described in section 9.4.

7. Calculate the total two-phase pressure drop from:

$$\Delta p_{tp} = \Delta p_{fr} + \rho_{tp}gH + \Delta p_{acc} \qquad [9.46]$$

where H is the height of the outlet of the pipeline above its inlet (m). At this stage it may be assumed that the density remains constant, so $\Delta p_{acc} = 0$.

8. Check from section 9.5.4 whether there has been any significant change in quality. If there has, repeat the calculations based on the mean quality and taking

$$\Delta p_{acc} = \dot{m}^2 \left(1/\rho_{tp\,out} - 1/\rho_{tp\,in} \right) \qquad\qquad [9.47]$$

9.5.6   Example: pressure drop in water/steam pipeline

The pipeline described in the example in section 9.4.2 is conveying steam and water from an evaporator to a separator. It is 2.3 m long in the vertical upward direction, followed by a 90° circular arc bend and a horizontal section 1.4 m long. The radius-to-diameter ratio of the bend (r/d) is 1.5. If the conditions described in the previous example apply to the inlet to the pipeline, what are the pressures and qualities in the outlet of the evaporator and in the separator? The roughness of the pipe is 3×10^{-5} m. In addition to the physical properties given in the previous example, the following are required: $\eta_g = 11.62 \times 10^{-6}$ N s/m^2; $c_{pl} = 4202$ J/kg K; $\Delta h_v = 2.31 \times 10^6$ J/kg.

This is a problem in the estimation of pressure drop in two-phase flow through a pipeline, so the general procedure is that given above in section 9.5.5. For this problem the stages are:

(A)   Calculate the single-phase pressure drops for each component of the pipeline separately – steps (1) to (4) of section 9.5.5.

(B)   Calculate the two-phase pressure drop enhancement factor by the method of Friedel (1979) – steps (1) to (6) of section 9.3.1.

(C)   Calculate the two-phase pressure drop enhancement factors for each fitting, as described in section 9.3.3.

(D)   Calculate the total two-phase pressure drop from equation [9.46], using the two-phase density calculated in the example in section 9.4.2, which is given as the density at the inlet to the pipeline. Thence calculate the pressure in the evaporator and the separator, neglecting changes in density.

(E)   As this pipeline is operating under vacuum, it is advisable to check whether flashing has any significant effect, as described in section 9.5.4 and make corrections to the previous calculations if necessary.

(A)

(1)   The mass flux is given in section 9.4.2 as 210 kg/s m^2. There are no changes in diameter, so the velocity pressure of the total flow as liquid at

all points in the pipeline is:

$$\frac{\dot{m}^2}{2\rho_1} = \frac{210^2}{2 \times 970} = 22.7 \text{ Pa}$$

(2)   There are three fittings: the inlet contraction, the bend and the enlargement into the separator. For the entrance from the top of the evaporator to the pipe, assuming it to be an abrupt contraction of zero area ratio, the single-phase pressure drop is given by equation [9.7] as:

$$p_u - p_d = 1.5 \times 22.7 = 34.1 \text{ Pa}$$

For a 90° circular bend, with $r/d = 1.5$, Table 9.2 gives $K = 0.25$. The single-phase pressure drop in the bend is given by equation [9.6]

$$p_u - p_d = 0.25 \times 22.7 = 5.7 \text{ Pa.}$$

There is no change in static pressure at the end of the pipeline (equation [9.8]).

(3)   The Reynolds number for the flow as all-liquid has been calculated in the previous example as 6.20×10^4. From an internal diameter of 104 mm, the relative roughness is $3 \times 10^{-5}/0.104 = 2.9 \times 10^{-4}$. Figure 9.1 gives $1000f_{lo} = 5.4$.

(4)   The single-phase pressure drop in the two straight lengths is obtained from equation [9.45] as:

$$\Delta p_{\text{fr st}} = \frac{4 \times 5.4 \times 10^{-3}(2.3 + 1.4)}{0.104} \times 22.7 = 17.4 \text{ Pa}$$

(Note that whereas in single-phase pressure drop calculations all the pressure loss coefficients for an element of pipeline can be added together and then multiplied by the velocity pressure, to give the total pressure drop, in two-phase pressure drop the single-phase pressure loss coefficients must be kept separate, because the enhancement factors are different for different types of element).

(B)

The method of Friedel (1979) is used to calculate the enhancement factor for the straight lengths, as described in section 9.3.1.

(1) Re_{lo} has already been estimated to be 6.20×10^4. For all-vapour flow,

$$Re_{go} = \frac{210 \times 0.104}{11.62 \times 10^{-6}} = 1.88 \times 10^6$$

(2)   For this Reynolds number and a relative roughness of 2.9×10^{-4}, Fig. 9.1 gives $1000f_{go} = 3.7$. $1000f_{lo}$ has already been estimated to be 5.4.

(3) The parameters defined by equation [9.17] are calculated below. It is convenient first to calculate the density ratio, $\rho_l/\rho_g = 970/0.293 = 3310$.

$$E = (1 - 0.05)^2 + 0.05^2 \times 3310 \times 3.7/5.4 = 6.57$$

$$F = 0.05^{0.78}(1 - 0.05)^{0.24} = 0.0955$$

$$H = (3310)^{0.91}(11.62/352)^{0.19}(1 - 11.62/352)^{0.7} = 815$$

(4) To determine the Froude number, it is first necessary to calculate the two-phase density with homogeneous flow. This was calculated at the end of the previous example to be $5.83 \, \text{kg/m}^3$. Hence, from equation [9.18]

$$\text{Fr}_{tp} = 210^2/9.81 \times 0.104 \times 5.83^2 = 1272$$

(5) The surface tension has been given as $62.7 \times 10^{-3} \, \text{N/m}$. Hence the Weber number is given by equation [9.19] as:

$$\text{We}_{tp} = 210^2 \times 0.104/5.83 \times 62.7 \times 10^{-3} = 12\,550$$

(6) The two-phase pressure drop enhancement factor is given by equation [9.20] for both the vertical and the horizontal lengths of pipe. Thus:

$$\phi_{lo}^2 = 6.57 + \frac{3.24 \times 0.0955 \times 815}{1272^{0.045} \times 12\,550^{0.035}} = 138$$

It is of interest to compare this with the value obtained on the assumption of homogeneous flow, which is given by equation [9.21]

$$\phi_{loh}^2 = 1 + (3310 - 1) \times 0.05 = 166.$$

This leads to only a small increase (20%) in the estimated pressure drop by this simple method rather than by Friedel's method.

(C)

Section 9.3.3 gives the enhancement factors for the pipe fittings. For the bend, equation [9.22] gives:

$$C^* = 5 - \tfrac{1}{2} \times 1.5 = 4.25$$

For the contraction at inlet, $C^* = 1.0$.

The two-phase pressure drop enhancement factor for the bend can now be calculated from equation [9.23].

$$\phi_{lob}^2 = 1 + 0.05[4.25(3310 + 1) - 2] - 0.05^2[(4.25 - 1)(3310 + 1)]$$

$$= 678$$

For the contraction, $C^* = 1$, so flow is homogeneous; the enhancement for homogeneous flow has been calculated in paragraph (B)(6) as 166.

(D)

The two-phase pressure drop across the several elements can now be determined by multiplying the single-phase values by the appropriate enhancement factors.

At entry, $\Delta p_{tp} = 34.1 \times 166 = 5660$ Pa

At the bend, $\Delta p_{tp} = 5.7 \times 678 = 3865$ Pa

In the straight lengths, $\Delta p_{tp} = 17.4 \times 138 = 2401$ Pa, due to friction.

The pressure drop due to gravity in the vertical length, for which $H = 2.3$ m, is obtained approximately by assuming that the two-phase density does not change significantly from the initial value of 66 kg/m³, calculated in the previous example in section 9.4.2. Hence equation [9.35] simplifies to

$$\Delta p_{gr} = g\rho_{tp}H = 9.81 \times 66 \times 2.3 = 1489 \text{ Pa}$$

If flashing is negligible, $\Delta p_{acc} = 0$. Thus equation [9.33] gives, for the pipeline (excluding entry)

$$\Delta p_{tp} = (3865 + 2401) + 1489 + 0 = 7755 \text{ Pa}$$

The pressure at the bottom of the line is given as 47 372 Pa.

Hence the pressure in the evaporator is $47\,372 + 5660 = 53\,032$ Pa. The pressure at the end of the pipeline is $47\,372 - 7755 = 39\,617$ Pa.

(E)

From the steam tables, the corresponding saturation temperatures are 82.5 °C in the evaporator, 80.0 °C at entry to the pipeline and 75.3 °C at the end of the pipeline.

Equation [9.44] can be used to determine x_{ge} (the quality at the top of the evaporator) and x_{gs} (the quality in the separator) from the saturation temperatures and the known quality of 0.05 at the bottom of the pipeline. Thus:

$$0.05 = x_{ge} + \frac{4202(82.5 - 80.0)}{2.31 \times 10^6}; \qquad \text{whence } x_{ge} = 0.045$$

$$x_{gs} = 0.05 + \frac{4202(80.0 - 75.3)}{2.31 \times 10^6} = 0.0585$$

The fact that the quality increases slightly from 0.05 at entry to the pipeline to 0.0585 at the outlet means that the frictional pressure drop will be somewhat greater than that calculated on the assumption that the quality is constant at 0.05; also there will be a small pressure drop due to acceleration. However, the greater quality means that the two-phase density, and hence the gravitational pressure drop, will be less than that calculated.

9.6 Flow around circulation loop

In any vaporiser employing recirculation, there is a location where the vapour generated separates from the remaining liquid which is recirculated around the circulation loop. This section is concerned with the calculations relating to flow around the loop. With forced or assisted circulation, the objective is to determine the required head to be provided by the pump, for the specified recirculation rate. With natural circulation, the objective is to determine the unspecified recirculation ratio; this is defined as the ratio of the flow rate of liquid entering the heater to the rate of generation of vapour. The recirculation ratio is equal to the reciprocal of the quality at the outlet from the heater. In a boiler, separation takes place in a steam drum. In an evaporator, separation may occur above the heater, as in Figs 3.6 and 3.7, or in an external separator, as in Figs 3.8 and 3.9. With a reboiler, separation occurs in the distillation column, except in the case of a kettle reboiler, where separation occurs in the shell, the pressure of vapour in the shell being equal to that in the base of the column plus the pressure drop in the vapour return main. In all except the kettle reboiler, the pressure in the location where separation takes place will be given in the specification.

In all cases, the pressure rise in the downcomer must equal the pressure drop in the riser. Figure 9.4 shows a generalised arrangement.

With forced or assisted circulation the recirculation ratio will have been specified, but with natural circulation this must be determined by trial and error, as described in section 9.6.4 and illustrated in the example in section 9.6.5.

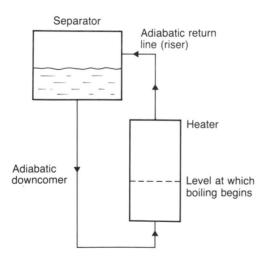

9.4 Generalised diagram of a circulation loop for a vaporiser with recirculation. A pump must be added in the downcomer with forced circulation or between the downcomer and the heat source with assisted circulation

9.6.1 Pressure rise in a downcomer

The downcomer is taken as the region between the separator and the entry to the heater. This ought to be a region of single-phase flow, but some vapour may be present due to faulty operation of the separator; this vapour is called 'carry-under' (see section 9.8). With some designs of water-tube boiler the downcomer tubes may be heated, so here there will also be small amounts of vapour present. The pressure rise in the downcomer is normally calculated on the assumption of single-phase flow, following the procedure in section 9.1. With forced and assisted circulation, the pressure rise in the pump must be added to this to give the total pressure rise. The amount of head to be supplied by the pump must be sufficient to make the total pressure rise in the downcomer equal to the pressure drop in the riser. It may be necessary to estimate the static pressure at entry to the pump to check that it is sufficient to avoid cavitation (the formation of bubbles in the pump, leading to damage when the bubbles collapse).

9.6.2 Pressure drop in a riser

There are three regions where this must be calculated:

1. Single-phase heated flow in the heater.
2. Two-phase heated flow in the heater.
3. Two-phase adiabatic flow in the return pipeline.

The pressure drop in the region of single-phase heated flow can be calculated by following the procedure in section 9.1, having first determined the length of this region, as described below in section 9.6.3. Section 9.5 gives the procedures for determining the pressure drop in the region of two-phase flow, sections 9.5.1 to 9.5.3 dealing with the heated region and section 9.5.5 with adiabatic flow. Allowances must be made for the effects of hydrostatic pressure on quality, as described in section 9.6.3, but this may be neglected with high-pressure steam and water.

With high-pressure steam and water, and when the separator is not very high above the point of entry to the heater, it will also be found that the amount of subcooling at entry to the heater is negligible, and so region (1) may be neglected. There is no region (3) in the vaporisers shown in Figs 3.6, 3.7, 3.11, 3.12, 3.17, 3.19, or with an internal reboiler.

With a flash system, a restriction is installed in the return line immediately before entry to the separator, in order to increase the pressure in the heater and prevent the formation of any vapour there. Thus the wall temperature at the top of the heater must not exceed the saturation temperature there, the calculation of the latter being described in the next section.

9.6.3   Estimation of level at which boiling begins

It is necessary to be able to calculate the boiling temperature of the liquid at any point of the recirculation loop, to ensure that flashing has not occurred where single-phase flow is required, or to determine the level in the heater at which boiling begins. As the liquid flows down the downcomer its boiling temperature increases due to the greater hydrostatic pressure. There is also a fall in boiling temperature due to friction. The frictional term is negligible in normal arrangements of internal downcomers. The gravitational term is negligible in flow through a restriction. The change in boiling temperature is given by the Clapeyron equation, which is given below and can be found in most textbooks on thermodynamics or physical chemistry.

$$\frac{\mathrm{d}p_v}{\mathrm{d}T_s} = \frac{\Delta h_v}{T_s(1/\rho_g - 1/\rho_l)}$$ [9.48]

where p_v is the local vapour pressure (Pa), T_s is the local boiling temperature (K), Δh_v is the latent heat of vaporisation (J/kg) and ρ_g and ρ_l are the densities of vapour and liquid respectively (kg/m^3). This equation can be used to determine the drop in saturation temperature as a result of pressure drop through a restriction, using stepwise integration if the physical properties and T_s change significantly as a result of the pressure drop. Alternatively values of T_s can be read from tables of physical properties.

If z (m) is the height of the location above a datum level, equation [9.1] gives, in the absence of friction:

$$\frac{\mathrm{d}p}{\mathrm{d}z} = -\rho_l g$$ [9.49]

Combining equations [9.48] and [9.49] gives:

$$\frac{\mathrm{d}T_s}{\mathrm{d}z} = -\frac{T_s g}{\Delta h_v}\left(\frac{\rho_l}{\rho_g} - 1\right)$$ [9.50]

Values of $\mathrm{d}T_s/\mathrm{d}z$ (K/m) have been calculated for water at four pressures from equation [9.50] using the physical properties given by ESDU (1980) and have been tabulated in Table 3.1.

Once the temperature and pressure at the base of a heater with upward flow have been established, it is possible to determine the extent of single-phase heating, as illustrated in Fig. 9.5. Here the ordinate is z (m), the height above the point of entry into the bottom of the heater. The abscissa is temperature. The point O gives the temperature of the liquid entering the heater. The point A gives the corresponding saturation temperature, the distance OA being the amount of subcooling of the liquid due to the hydrostatic pressure at the base of the heater being greater than that in the

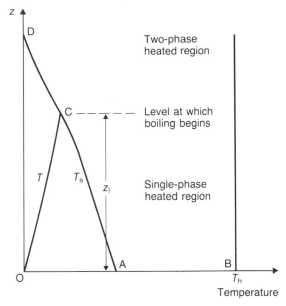

9.5 Temperature/height diagram for determining temperature profile in heated vertical upward flow

separator. The line at B shows a typical relationship between the temperature of the heating fluid and height, represented by T_h. The line OC gives the relationship between T_l, the temperature of the liquid to be vaporised, and height of the single-phase heating region.

A heat balance shows that the temperature rise of the liquid in the single-phase region, ΔT_l, can be calculated from:

$$\Delta T_l = \frac{\dot{q}_{mean} A z_l}{\dot{M} c_p L} \qquad [9.51]$$

where A is the total heat transfer surface area (m²), c_p is the specific heat capacity of the liquid to be vaporised (J/kg K), $\dot{M}$ is the mass flow rate of liquid up the tubes (kg/s). $\dot{q}_{mean}$ is the mean heat flux in the liquid region (W/m²), z_l is the length of the liquid zone. The use of this equation is illustrated in the Example in section 10.6.3. If there is nucleate boiling in this zone (see section 7.1.2) the heat transfer coefficient for this should be added to the single-phase convective coefficient, as described in section 7.4. Any small bubbles formed as a result of nucleation will condense in the subcooled liquid, so this does not constitute the boiling region. The effects of such bubbles on pressure drop may be neglected.

The line AC gives the relationship between T_s, the saturation temperature, and height. The slope of this line, (dT_s/dz) is the sum of two terms, that due to gravity, given by equation [9.50], and that due to friction. The latter is usually the smaller, and is often negligible. If the physical properties are assumed to be constant over the single-phase region, the drop in saturation temperature in this region can be determined from:

$$-\Delta T_s = [\Delta p_f + \rho_l g z_l] \left/ \frac{dp_v}{dT_s} \right. \tag{9.52}$$

where Δp_f is the pressure drop due to friction in the liquid region plus the pressure drop at entry, and dp_v/dT_s has been obtained from equation [9.48].

The rise in liquid temperature between points O and C in Fig. 9.5, given by equation [9.51], plus the fall in saturation temperature from A to C, given by equation [9.52], must equal the amount of subcooling at inlet, represented by OA. From this equality it is possible to determine the length of the single-phase region, z_l. The length of the boiling region is obtained by difference

$$z_b = Z - z_l \tag{9.53}$$

The justification for the assumption that there is no superheating of the liquid and therefore that boiling starts at the point C is discussed in section 2.4. Thereafter flashing occurs due to two-phase pressure drop and further vapour is formed as a result of the further addition of heat in the boiling zone. The point D in Fig. 9.5 represents the top of the heater, where the liquid has been cooled to the saturation temperature corresponding to the pressure there. With an internal downcomer, this will be the same as the temperature at the base. With an external downcomer, there will be a further drop in temperature and pressure due to the pressure drop in the adiabatic return line (see Fig. 9.4).

9.6.4 Estimation of recirculation ratio in natural circulation

In a natural circulation vaporiser, the mass flow rate of the vapour to be generated, $\dot{M}_g$ (kg/s) will have been given. Once the geometry has been decided upon, $\dot{M}_g$ can be divided by the cross-sectional area for flow to give $\dot{m}_{g\,out}$, the mass flux of the vapour at the outlet from the heater (kg/s m^2). The total mass flux of liquid and vapour, $\dot{m}$, is obtained by multiplying $\dot{m}_g$ by the recirculation ratio, which is the reciprocal of $x_{g\,out}$ the quality at outlet. Hence

$$\dot{m} = \dot{m}_g / x_{g\,out} \tag{9.54}$$

It is now possible to guess a value of $x_{g\,out}$ for substituting in equation [9.54] to give $\dot{m}$, from which can be calculated the pressure drop in the single-phase and two-phase regions, as described above, using the procedure given in section 9.6.3 to determine the level at which boiling begins.

In shellside boiling, it is necessary to make certain assumptions about the

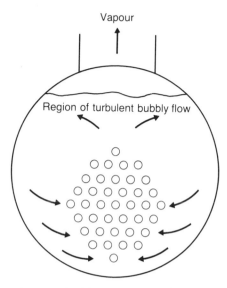

9.6 Typical bundle layout with double vortex flow pattern (with 39 tubes as in Fig. 9.7)

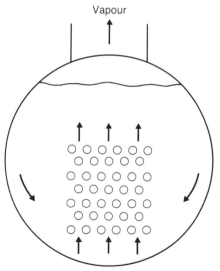

9.7 Simplified rectangular layout with one-dimensional flow pattern assumed for calculation (with 39 tubes as in Fig. 9.6)

flow pattern and the arrangement of tubes. Figure 9.6 shows a typical kettle reboiler with flow in the form of two vortices, as shown by Cornwell *et al.* (1980). To be able to perform the required calculations, a simpler, one-dimensional flow pattern through tubes on a rectangular layout must be assumed, as shown in Fig. 9.7. The imaginary layout is devised to give the same pressure drop as the real one, as devised by Brisbane *et al.* (1980).

9.6.5   Example: recirculation ratio in kettle reboiler

Determine the recirculation ratio in the experimental kettle reboiler described in the example in section 7.4.6, at a heat flux of 20 kW/m², which gives a vapour generation rate of 50.3 g/s.

There are 241 short tubes on a square pattern, each occupying 0.0254^2 m² of tubesheet area arranged in-line in an octagonal space. Hence the total tubesheet area occupied is $241 \times 0.0254^2 = 0.15548$ m². It will be supposed that the same number of tubes could be arranged on the same pitch in a square space. Thus the side of the square would be $\sqrt{0.15548} = 0.394$ m, and so the vertical length up the bundle (z) is 0.394 m. The cross-sectional area for flow up the bundle is obtained from the plan area. The length of the tubes is 28 mm, giving a plan area without any tubes of $0.028 \times 0.394 = 0.011032$ m². But this is obstructed by tubes of 19.05 mm diameter on a pitch of 25.4 mm. The area ratio is:

$$\sigma = \frac{25.4 - 19.05}{25.4} = 0.25$$

Thus the minimum cross-sectional area for flow through the space between tubes is:

$$S = 0.011032 \times 0.25 = 0.002758 \text{ m}^2$$

The mass flow rate is $\dot{M} = 50.3 \times 10^{-3}/x_{g\,out}$. Hence the mass flux between the tubes of the bundle is:

$$\dot{m} = \frac{\dot{M}}{S} = \frac{50.3 \times 10^{-3}}{0.002758 x_{g\,out}} = 18.23/x_{g\,out} \text{ kg/s m}^3$$

Equation [9.50] can be used to determine whether it is necessary to allow for the effects of the subcooling of the liquid at entry to the bundle. In the region where vapour-free liquid is recirculating the increase in its saturation temperature with increasing depth is given by equation [9.50], as:

$$-\frac{dT_s}{dz} = \frac{(47 + 273) \times 9.81}{145715}(1512/7.38 - 1) = 4.4 \text{ K/m}$$

This shows that there will be a significant amount of subcooling. It may be assumed that the space above the bundle is a region of highly turbulent bubbly flow, as indicated in Fig. 9.6, but that the space at the side of the bundle is free of vapour. This approximation gives the effective height of liquid in recirculation as equal to the heated length, $Z = 0.394$. Neglecting the decrease in dT_s/dz with increasing pressure gives the amount of subcooling at entry to the bundle

$$T_{so} - T_0 = 4.4 \times 0.394 = 1.73 \text{ K}$$

where the extra suffix 0 denotes the values at entry, where $z=0$.

It is now possible to follow the procedure in section 9.6.3 to determine z_1, the length of the single-phase region. From equation [9.51] the temperature rise of the liquid as it flows from point O to point C in Fig. 9.5, with a constant heat flux of $20 \, \text{kW/m}^2$ is

$$\Delta T_1 = \frac{20\,000 \times (\pi \times 0.019\,05 \times 0.0254 \times 241)z_1}{(0.0503/x_{g\,\text{out}}) \times 955 \times 0.394} = 387 x_{g\,\text{out}} z_1$$

From equation [9.48]

$$\frac{dp_v}{dT_s} = \frac{145\,715}{(47 + 273)(1/7.38 - 1/1512)} = 3377 \, \text{Pa/K}$$

The velocity pressure is:

$$\frac{\dot{m}^2}{2 \times 1512} = \frac{18.23^2}{x_{g\,\text{out}}^2 \times 2 \times 1512} = \frac{0.1099}{x_{g\,\text{out}}^2}$$

No information is available on the pressure drop at inlet to a bundle; assuming it to be the same as for an abrupt inlet to a tube, equation [9.7] gives the drop in static pressure at inlet, for an area ratio $\sigma = 0.25$

$$\Delta p_{\text{in}} = 1.5(1 - 0.25^2) \times 0.1099/x_{g\,\text{out}}^2 = 0.1545/x_{g\,\text{out}}^2$$

The number of tubes crossed in the liquid zone is:

$$n_v = z_1/0.0254$$

Substituting this in equation [9.11] gives the frictional loss in cross flow in the liquid zone:

$$\Delta p_{\text{fr}} = \tfrac{1}{2} n_v \times 0.1099/x_{go}^2 = 2.16 z_1/x_{go}^2$$

The total frictional pressure drop in the single-phase (liquid) zone is:

$$\Delta p_{\text{fr}} = 0.1545/x_{g\,\text{out}}^2 + 2.16 z_1/x_{g\,\text{out}}^2$$

Substituting in equation [9.52] gives $-\Delta T_s$, the drop in saturation temperature in the liquid zone (between the points A and C in Fig. 9.5). Thus

$$-\Delta T_s = \frac{0.1545/x_{g\,\text{out}}^2 + 2.16 z_1/x_{g\,\text{out}}^2 + 1512 \times 9.81 z_1}{3377}$$

Expressions are now available for ΔT_1 and $-\Delta T_s$ in terms of $x_{g\,\text{out}}$ and z_1. But it is known that the sum of these is equal to the amount of subcooling, already estimated at 1.73 K. Hence:

$$387 x_{g\,\text{out}} z_1 + \frac{0.1545/x_{g\,\text{out}}^2 + 2.16 z_1/x_{g\,\text{out}}^2 + 14\,833 z_1}{3377} = 1.73$$

This leads to:

$$z_1 = \frac{0.394 - 1.04 \times 10^{-5}/x_{g\,\text{out}}^2}{1 + 1.455 \times 10^{-4}/x_{g\,\text{out}}^2 + 88 x_{g\,\text{out}}} \qquad [9.55]$$

The next step is to derive an expression for the pressure drop in the boiling zone. The pressure drop for liquid alone has already been established as:

$$\Delta p_{lo} = 2.16z/x_{g\,out}^2$$

Then equation [9.39] becomes:

$$\Delta p_{fr} = (2.16z_b/x_{g\,out}^2)\phi_{lo\,m}^2.$$

Combining with equation [9.40] gives

$$\Delta p_{fr\,b} = (2.16z_b/x_{g\,out}^2)[1 + \tfrac{1}{2}(1512/7.38 - 1)x_{g\,out}]$$

where z_b is the boiling length

$$z_b = Z - z_1 = 0.394 - z_1$$

Hence

$$\Delta p_{fr\,b} = \frac{2.16}{x_{g\,out}^2}(1 + 102x_{g\,out})z_b$$

Adding together the pressure drop at inlet and the single-phase and two-phase friction losses gives the pressure drop through the bundle, excluding gravity, as

$$\Delta p_{frT} = 0.1545/x_{g\,out}^2 + 2.16z_1/x_{g\,out}^2 + (2.16/x_{g\,out}^2)(1 + 102x_{g\,out})(0.394 - z_1)$$

$$= (1.006 + 220z_b x_{g\,out})/x_{g\,out}^2 \qquad [9.56]$$

The driving force is the difference between the pressure rise due to gravity in the downcomer over the boiling length less the pressure drop due to gravity in the two-phase (boiling) region. The former is $\rho_l g z_b$; the latter is given by equation [9.42]. Hence the driving force, Δp_{dr}, is given by:

$$\Delta p_{dr} = 1512 \times 9.81z_b\left[1 - \frac{\ln(1 + 204x_{g\,out})}{204x_{g\,out}}\right] \qquad [9.57]$$

The acceleration loss in the boiling region is given by equation [9.43]:

$$\Delta p_{acc} = 2(0.1099/x_{g\,out}^2)204x_{g\,out} = 44.8/x_{g\,out} \qquad [9.58]$$

Values of $x_{g\,out}$ are chosen until $\Delta p_{dr} - \Delta p_{acc} - \Delta p_{frT} = 0$

$x_{g\,out}$ (assumed)		0.03	0.04	0.041	0.0409	0.04083	—
z_1 (equation [9.55])	=	0.100	0.084	0.083	0.083	0.083	m
Δp_{frT} (equation [9.56])	=	3269	2334	2269	2276	2280	Pa
Δp_{acc} (equation [9.58])	=	1493	1120	1093	1095	1097	Pa
Δp_{dr} (equation [9.57])	=	2956	3350	3384	3380	3377	Pa
$\Delta p_{dr} - \Delta p_{acc} - \Delta p_{frT}$	=	−1806	−104	+22	+9	0	Pa

Hence $x_{g\,out} = 0.040\,83$ and the recirculation ratio is $1/0.040\,83 = 24.4$.

9.7 Design of distributors

This section deals with the problems in fluid flow involved in the design of the distributors that convey fluid from a single pipe to many parallel paths through a vaporiser and then collect the fluid from the paths and return it to a single pipe. The arrangements considered are:

1. Liquid distributor for conveying a liquid from a pipe to many points inside a vessel, as when conveying a liquor to be concentrated to the top tubesheet of a falling-film evaporator.
2. Manifolds taking fluid to and from parallel heat exchanger tubes.
3. Flow into and out of spaces, such as the channels and shells of shell-and-tube heat exchangers.

Inevitably the fluid will not be distributed evenly to all of the parallel paths; some will take more than the intended amount, others less, depending on the changes in pressure in the distributors. The amount of this maldistribution may be expressed as the ratio of the greatest deviation of the flow rate through any path to the mean flow rate. Thus the maximum maldistribution, denoted by $\Delta \dot{M}_{max}$, is:

$$\Delta \dot{M}_{max} = (\dot{M}_m - \dot{M}_{min})/\dot{M}_m \qquad\qquad [9.59]$$

or $\qquad (\dot{M}_{max} - \dot{M}_m)/\dot{M}_m \qquad$ whichever is larger

where $\dot{M}_m$ is the mean flow rate per path, i.e. the total rate divided by the number of parallel paths, $\dot{M}_{max}$ is the maximum rate through any path and $\dot{M}_{min}$ the minimum rate through any path. The tolerable amount of the maximum maldistribution may be 5–10% in some cases, whereas in other cases up to 20% may be acceptable.

The aim in the design of distributors is to choose suitable standard sizes for the pipes to give a sufficiently uniform distribution, without excessive capital cost. Restriction plates may be inserted in the pipes, but this increases the cost of pumping.

If maldistribution looks like being a problem, the effect on thermal performance may be estimated by making the simplifying assumption that the heat exchanger consists of two parallel exchangers, one operating at a rate of $\dot{M}_{max}$ per path, and the other at $\dot{M}_{min}$ per path, the number of paths per exchanger being such as to give the actual total flow rate. This approach has been developed by Mueller (1977).

It is unwise to split two-phase flow because this leads not only to a large amount of maldistribution in mass flow rates but also to big differences in quality from one path to another. It is therefore desirable to split the flow of the liquid to be vaporised into its parallel paths before any vaporisation has taken place. In a 'once-through' vaporiser the collecting system conveys a single-phase vapour. When recirculation is employed, the collecting system conveys a two-phase mixture of liquid and vapour, but the quality is the same in all paths, apart from the effects of maldistribution. Thus, if the mixture is regarded as homogeneous, there should be no great error in applying correlations developed for single-phase flow to two-phase collecting manifolds. However, in any vaporising system there is no way of allowing for the big reduction in density. A similar problem arises if the heating fluid is a gas, whose density is likely to be much increased. In such cases the collecting and distributing manifolds must be designed separately. Advice is given below on the best overall arrangement of manifolds. It is important to realise that a given design of manifold will give much more maldistribution if it is used as a collecting manifold than if it is used as a distributing manifold. The reason for this is that in a distributing manifold, the momentum of the incoming fluid makes it leave preferentially from the furthest offtakes, but this is partly offset by friction; however in a collecting manifold the effects of momentum and friction are additive in giving a bigger rate through the near branches. Thus it may be advisable to use lower velocities in collecting than in distributing manifolds.

The methods given below for estimating the maximum maldistribution are those of Idel'chik and Shteinberg (1974). They apply only to single-phase flow of a fluid whose density is the same in the outlet as in the inlet channels. Thus they are applicable to the arrangement in paragraph (1) above, but arrangements (2) and (3) are often complicated by two-phase flow.

In order to be able to use the correlations of Idel'chik and Shteinberg to predict the amount of maldistribution in a manifold, it is necessary to calculate L/d, the length-to-diameter ratio of the manifold, and S_r, the effective ratio of flow areas in the branches and the manifold, which is defined by:

$$S_r = \frac{n d_b^2}{d^2 \sqrt{K_b}} \qquad [9.60]$$

where d is the internal diameter of the manifold and d_b is the internal diameter of each branch (m), n is the number of branches and K_b is the pressure loss coefficient for the branch. Section 9.1 defines K_b and explains its significance. The amount of maldistribution can then be read from the appropriate curve in Fig. 9.8. Advice on the use of this method follows.

9.7.1 Liquid distributor

Figure 3.21 shows a small falling-film evaporator with a distributor consisting of a single manifold with seven downcomers taking liquid to the top of the upper tubesheet. With a larger evaporator of this type it may be necessary to have a more complicated system consisting of a primary distributor taking the liquor to several secondary distributors, each with its own downcomers taking the liquor to the tubesheet. The curve on the right

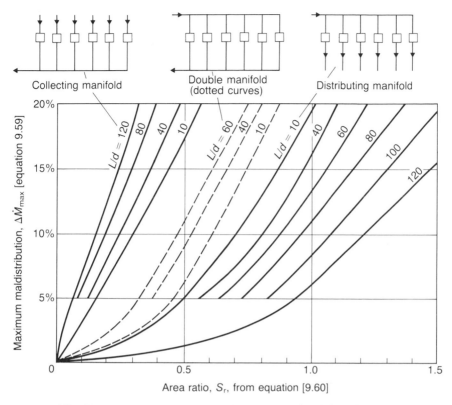

9.8 Maldistribution in three simple manifolds (from Idel'chik and Shteinberg (1974))

of Fig. 9.8 may be used for sizing the distributors. If the distributor consists of a perforated pipe discharging the liquor into a downcomer open to atmosphere, the value of K_b for substitution in equation [9.60] may be taken as 2.5. In this case d_b is the diameter of the perforations. If the distributor consists of short branches, of length greater than about $5d_b$, K_b may be taken as 1.5. In either case it is necessary to check that the pressure in the distributor is greater than atmospheric, otherwise it will not run full, and this will lead to very poor distribution. In particular, if downcomers are allowed to run full and are positioned vertically, they will produce a vacuum at their inlet because the pressure rise due to decrease in level will greatly exceed the pressure drop due to friction. It is therefore better that downcomers should be slightly downward inclined. A hole drilled near to the top of a downcomer will act as a vacuum breaker, ensuring that the pressure in the downcomer is atmospheric, as illustrated in Fig. 9.9.

The simplest form of liquid distributor is the 'sparge pipe', which is a perforated pipe. A simple design procedure is to choose a standard size of pipe such that the cross-sectional area of flow through it is more than twice

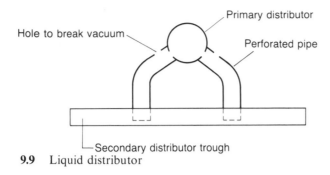

9.9   Liquid distributor

the total area through all the holes. It then follows from equation [9.60] that $S_r < 0.3$. Under these circumstances, the solid curves for the distributing manifold on the right-hand side of Fig. 9.8 will be less than 3%.

9.7.2   Design of manifolds

When a heat exchanger consists of many tubes connected to inlet and outlet manifolds, as in Fig. 3.13, flow in the two manifolds should be in opposite directions, as shown in the sketch for a double manifold in Fig. 9.8. In this way the pressure drop in the outlet manifold may be partly balanced against the pressure rise in the inlet manifold. The curves shown in the middle of Fig. 9.8 may be used for sizing the manifolds when there is no significant change in the density of the fluid as it flows through the heat exchanger and when the two manifolds are of the same diameter.

   When boiling takes place in the tubes the velocity is much higher at outlet than at inlet, so it may be assumed that the outlet manifold determines the amount of maldistribution. Under these circumstances the amount of maldistribution may be estimated by assuming that the system acts as a collecting manifold and using the curves on the left of Fig. 9.8. The value of K_b to be substituted in equation [9.60] is the total pressure drop in the tubes divided by the velocity pressure at the outlet. For an approximate calculation it may be assumed that flow is homogeneous and that quality increases linearly with distance; then the pressure drop may be determined from equations [9.39] and [9.40] and K_b is given by the following approximate equation:

$$K_b = \frac{4f_1 z_b}{d}\left(\frac{1 + \frac{1}{2}(\rho_1/\rho_g - 1)x_{g\,out}}{1 + (\rho_1/\rho_g - 1)x_{g\,out}}\right) \tag{9.61}$$

where f_1 is the Fanning friction factor for the non-boiling region and z_b is the length of the boiling region (m).

9.7.3 Inlet channel

Shell-and-tube heat exchangers are normally installed in a horizontal
position with the inlet and outlet nozzles for the tubeside fluid on the top
or the bottom of the channel, as shown by TEMA (1978). This
arrangement introduces no problems with regard to obtaining good
distribution of fluid between the tubes. However, many vaporisers are
installed vertically with the inlet nozzle at the bottom, as in Figs 3.4, 3.5,
3.8, 3.10, or at the top. There is then a tendency for more fluid to enter the
tubes in the centre than those at the outside. A ring vortex is set up, as
illustrated in Fig. 9.10 (a). The installation of an impingement plate merely
reverses the direction of rotation of the ring vortex, as illustrated in Fig.
9.10 (b). This would not be a satisfactory way of introducing gas into a
falling film evaporator because the high radial velocity would upset the flow
of the liquor over the top tubesheet. Flow straighteners may be installed, as
shown in Fig. 9.11; (a) is a perforated spherical plate, known as a 'pepper
pot'; (b) consists of four plates in the form of concentric truncated cones.
Similar devices (inverted) may be used for introducing the heating fluid, as
in Figs 3.4 and 3.5. When introducing a gas into the side of the bottom
channel of a falling-film evaporator, it is necessary for the gas to flow at a
low velocity through a wide duct, as shown in Fig. 3.21, to give good
distribution through the tubes and also to avoid entraining the falling drops
of liquid. It is necessary to install a duct in the form of a gradual
enlargement between the pipe conveying the gas and the wide duct, to
reduce the velocity of the gas to well below that in the pipe. (It must be

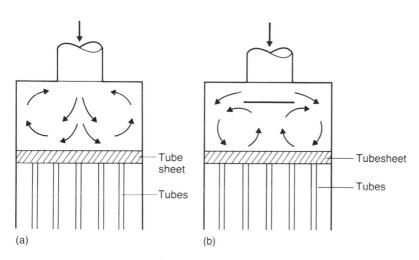

9.10 Flow patterns in inlet channels
    (a) without impingement plate
    (b) with impingement plate

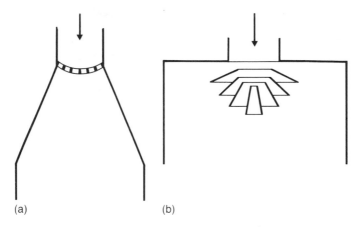

(a)   (b)

9.11 Flow straighteners for inlet channels
     (a) pepper pot
     (b) concentric conical plates

remembered that it is more difficult to reduce the velocity of a fluid than to increase it).

9.7.4   Shell nozzles

The pressure-loss coefficient for the shellside nozzles can be calculated from the following approximate formula:

$$K_n = 1 + \frac{1}{(R + 0.6S)^2} \qquad [9.62]$$

where K_n is the pressure-loss coefficient based on the velocity in the nozzle, R is the ratio of escape area to nozzle area and S is the ratio of slot area for flow between tubes to nozzle area, calculated as described below. The escape area is the circumference of the inside of the nozzle multiplied by the average depth of the outer tubes below the end of the nozzle (H). Thus

$$R = \frac{4H}{d_n} \qquad [9.63a]$$

where d_n is the internal diameter of the nozzle.

    When there is an impingement plate, H is the average depth of the plate below the end of the nozzle, and $S = 0$. When there is no impingement plate

$$S = 1 - \frac{d_o}{P} \qquad [9.63b]$$

where d_o is the external diameter of the tubes and P is their pitch.

    It is important to ensure that there is sufficient escape area around the nozzles to avoid excessive pressure drop. It is good practice for R, from equation [9.63], to be in the region of unity.

With shellside boilers, illustrated in Figs 3.16 and 3.17, the positioning of the inlet and outlet nozzles is important, as discussed in section 3.4.1. Also the velocity through the segmental area between the upper tubes and the top of the shell must be similar in magnitude to the velocity between the tubes. Much higher velocities are used in the outlet nozzles.

9.8   Carry-over and carry-under

In a vaporiser with recirculation there is a separator where the vapour that has been generated is separated from the liquid that is recirculated. This may take place in the space at the' top of the heater, as in Figs 3.6, 3.7, 3.19 and 5.1, or in an external separator, as sketched in Fig. 9.4. Separation is never perfect: drops of liquid are carried up by the vapour (carry-over); bubbles of vapour are carried down by the liquid (carry-under). The amount of carry-over increases rapidly with increasing mass flux of vapour.

Advice on the amount of disengagement space that should be provided above the liquid level to permit separation is given by R. A. Smith in section 3.5.4 of Schlünder (1983). The main source of the information is the work of Davis (1940), who studied carry-over from marine water-tube boilers. From this, the minimum interfacial area between liquid and vapour is:

$$S_{i\,min} = \frac{0.8\dot{M}}{\sqrt{\rho_g}} \qquad\qquad [9.64]$$

where $S_{i\,min}$ is the minimum interfacial area (m^2), $\dot{M}$ is the mass flow rate of vapour (kg/s) and ρ_g is the density of the vapour (kg/m^3). This equation is not valid for foaming liquids – see section 13.3.

In a kettle reboiler (Fig. 3.19) or a vapour generator using a K-shell, the disengagement space above the static liquid level should be at least 200 mm high.

In large evaporators such as those shown in Figs 3.6 and 3.7, there is a considerable space above the level of boiling liquid, to allow for the surges in level that may result from unsteady flow.

Equation [9.64] may also be used as a rough check on the size of external separators, including steam drums.

9.9   Design of spray eliminators

Even when the amount of disengagement space recommended in the previous section has been provided, the vapour leaving the separator will contain some droplets of entrained liquid. This presents no problem in the design of reboilers, but small amounts of carry-over of liquid containing dissolved solids may be troublesome in the design of evaporators or if the saturated steam generated in a boiler has to pass into a superheater. It may then be necessary to install one of the liquid/vapour separators described in section 4.1. The four types described there and illustrated in Figs. 4.1 to 4.6, in order of increasing efficiency, are: (1) impingement separator, (2) cyclone separator, (3) knitted wire mesh separator, and (4) superfine glasswool filter. The important data needed for design are the loading, expressed as grams of liquid per cubic metre of vapour, and the approximate cut, which is the size of the smallest drop that will be removed. For a given value of $\dot{M}$, the mass flow rate (kg/s) of vapour, it is required to estimate the value of S, the cross-sectional area for flow that will give the optimum mass flux, $\dot{m}_g = \dot{M}_g/S$ (kg/s m²). Too low a mass flux leads to inefficient removal of spray; too high a mass flux gives a high pressure drop and often leads to the re-entrainment of some of the liquid that has been removed. An important property needed for the design of spray eliminators is ρ_g, the density of the vapour (kg/m³).

The design procedures given in the following subsections have been taken from Stearman and Williamson (1972).

9.9.1 Impingement separator (Fig. 4.1)

This will handle heavy loadings with a cut of 100 μm. The optimum mass flux through the body of the separator is:

$$\dot{m}_g = 13\sqrt{\rho_g} \qquad [9.65]$$

The pressure drop through such a device is negligible.

9.9.2 Cyclone separator (Figs 4.2 and 4.3)

This will handle loadings up to 100 g/m³ with a cut of 10 μm. The optimum mass flux in the inlet pipe is:

$$\dot{m}_g = 36\sqrt{\rho_g} \qquad [9.66]$$

The pressure drop is likely to be several kPa

9.9.3 Knitted wire mesh separator (Figs 3.7, 4.4 and 4.5)

This will handle loadings up to 20 g/m³ with a cut of 10 μm. The optimum mass flux through the pad is:

$$\dot{m}_g = 3.3\sqrt{\rho_g} \qquad\qquad [9.67]$$

The pressure drop at this mass flux through a pad of the dimensions given in section 4.1.3 is 100 Pa when dry; the presence of droplets will increase this to 200–300 Pa.

9.9.4 Superfine glasswool filter (Fig. 4.6)

This will handle loadings of 1–2 g/m³ with a cut of 1 μm. It should be designed for a velocity ($\bar{u} = \dot{m}_g/\rho_g$) of 0.1 m/s. It gives a high pressure drop, similar to that of a cyclone.

9.10 Limit to counter-current flow in vertical tubes

In the design of a falling-film evaporator in which gas is introduced into the bottom channel, as shown in Fig. 3.21, it is necessary to check that the velocity of the gas at entry to the tubes is sufficiently small to allow the liquid to flow steadily out of the tubes. This requires an estimate of the limit to counter-current flow (sometimes called the 'flooding limit'). It has been the subject of extensive research, mostly with air and water at ambient conditions. The subject has been studied by Wallis (1969) in his section 11.4, where he develops an empirical relationship between $\dot{v}_g^+$, the dimensionless volume flux of the gas, and $\dot{v}_l^+$, the dimensionless volume flux of the liquid. The former has already been defined in equation [9.12]; the latter is given by:

$$\dot{v}_l^+ = \frac{\dot{m}_l}{\rho_l\sqrt{gd}} \qquad\qquad [9.68]$$

where d is the internal diameter of the tube (m), g is the acceleration due to gravity (9.81 m/s²), $\dot{m}_l$ is the mass flux of the liquid leaving the pipe (kg/s m²) and ρ_l its density (kg/m³).

Wallis found that the results could be correlated by an equation of the form:

$$(\dot{v}_g^+)^{\frac{1}{2}} + (\dot{v}_l^+)^{\frac{1}{2}} = C \qquad\qquad [9.69]$$

Wallis showed that the comprehensive studies of flooding in towers filled with random packing could also be correlated by equation [9.69], basing $\dot{m}$ on the free space for flow and taking d as the equivalent diameter, d_e, given by equation [6.15], A being the surface area of the packing and V its free volume. Where forces due to viscosity and surface tension are negligible, he showed that good agreement was obtained with $C = 0.775$.

Flooding velocities for air and water in vertical tubes, of internal diameter ranging from 19 to 32 mm, gave $C = 0.725$ for tubes with plain horizontal ends. With a well-rounded air inlet, values of C from 0.85 to 1.05 were obtained. The use of a double-chamfer cut at the bottom of the tubes, as described at the end of section 3.5.1, also increases the limit to countercurrent flow.

More recently, an extensive study of available data has been made by McQuillan and Whalley (1984). They showed that Wallis underpredicted the limit at high values of $\dot{v}_1^+$. They proposed the following equation, which gave better predictions than any of the published correlations.

$$\dot{v}_g^+ = 0.388(\dot{v}_1^+)^{-0.22}\left[\frac{d^2 g(\rho_1 - \rho_g)}{\sigma}\right]^{-0.155}(1 + 1000\eta_1)^{-0.18} \qquad [9.70]$$

where η_1 is the viscosity of the liquid ($N\,s/m^2$) and σ is the surface tension (N/m). This gave excellent agreement over the range $0.01 < \dot{v}_1^+ < 1.2$ and $3 \times 10^{-4} < \eta_1 < 1.0$.

It is recommended that equation [9.70] should be used when checking that there will not be flooding inside the tubes of a falling-film evaporator with upflow of gas. In view of the doubts about the method of estimation and the serious consequences of flooding, it is suggested that the maximum possible mass flux of vapour should not exceed 50% of the estimated limit for countercurrent flow. It is also important to design the gas entry to avoid swirling around the ends of the tubes, as discussed in section 9.7.3.

9.11   Hydrodynamics of falling-film flow

Falling films are encountered in condensation on vertical tubes and in falling-film evaporators. The problems dealt with here are the estimation of the minimum wetting rate, the thickness of the film and the pressure drop of gas flowing up or down the tube. The relevant parameters are the liquid film Reynolds number, which can be calculated from equation [6.29], and the dimensionless volume flux of the liquid, which is defined by equation [9.68].

9.11.1   Minimum wetting rate

Under adiabatic conditions, there is a minimum rate for the complete wetting of a pipe by a falling film of liquid. Below that rate, dry patches appear on the wall. The minimum wetting rate depends on the Reynolds number, the roughness of the surface and the interfacial surface tension between the liquid and the surface. The minimum wetting rate can be decreased by degreasing the surface or by roughening it.

Lower wetting rates can be tolerated when there is condensation. If the surface is not easily wetted or if a suitable material is applied to the surface or added to the vapour, 'dropwise condensation' may be promoted. Then the condensate forms as discrete drops, which grow until they are big enough to run down the surface under gravity. The different modes of condensation are described in textbooks on heat transfer, e.g. Kern (1950) pp. 252–3 and McAdams (1954) chapter 13 section I. They point out that dropwise condensation, which gives higher heat transfer coefficients than filmwise condensation, is very difficult to achieve and maintain under industrial conditions. Hirschburg and Florschuetz (1981) report results where the heat transfer coefficient in condensation agrees with that predicted from Nusselt's theory for filmwise condensation (given in section 6.3.1), at film Reynolds numbers as low as 10. The estimation of the heat transfer coefficient in condensation is dealt with in section 6.3.4.

When an evaporating film is flowing down a heated surface there is a greater tendency, compared with adiabatic flow, for dry spots to form on the surface. Andros and Florschuetz (1976) reported visual observations of film flow in an evaporator breaking up into rivulets with dry patches in between. ESDU (1981a) have analysed published data on the performance of heat pipes operating by thermosyphon inside vertical pipes. It was concluded that the heat transfer coefficient was considerably less than that predicted from Nusselt's theory when the liquid film Reynolds number was less than 50 in the adiabatic region between the upper (condensing) region and the lower (evaporating) region, presumably due to incomplete wetting of the heated surface. However, the performance was as good as or better than predicted at Reynolds numbers greater than 100. Also Hirschburg and Florschuetz (1981) report results where the heat transfer coefficient in laminar film evaporation agrees with that in laminar film condensation at local Reynolds numbers between 60 and 1000. To be safe, film theory should not be used to estimate the heat transfer coefficient if the film Reynolds number falls to below 100 at the bottom of the tubes of a falling-film evaporator, as described in section 7.6.

9.11.2 Thickness of falling film

Wallis (1969) deals in his section 11.3 with falling-film flow, assuming that the gas velocity is sufficiently low for interfacial shear stress and pressure drop to be negligible. The theoretical film thickness, assuming flow to be laminar, is:

$$\delta = \left[\frac{3\eta_1 \dot{M}}{\pi g d \rho_1 (\rho_1 - \rho_g)} \right]^{\frac{1}{3}} \qquad [9.71]$$

where d is the internal diameter of the pipe (m), g is the acceleration due to to gravity (9.81 m/s^2), $\dot{M}$ is the mass flow rate of liquid (kg/s), δ is the thickness of the film (m), η_l is the viscosity of the liquid (N s/m^2) and ρ_l and ρ_g are the densities of the liquid and gas respectively (kg/m^3).

Experimental results plotted by Wallis in his Fig. 11.8 show that equation [9.71] gives a good prediction of the thickness for values of the film Reynolds number from 200 to 3000. This figure also shows that for Reynolds numbers between 3000 and 30 000, the thickness is given by Wallis' equation [11.76], which simplifies to:

$$\delta = 0.063d(\dot{v}_1^+)^{2/3} \qquad [9.72]$$

where $\dot{v}_1^+$ is the dimensionless volume flux of the liquid given by equation [9.68].

With upward flow of gas, equations [9.71] or [9.72] may be used to predict the thickness of a falling film provided that the gas rate does not exceed 50% of the predicted flooding rate, estimated as recommended in section 9.10. Furthermore, it may be assumed that, if this restriction on gas rate is met, the liquid film heat transfer coefficient may be calculated by the method recommended in section 7.6.

With downward flow of gas, vapour shear slightly reduces the film thickness and increases the liquid film heat transfer coefficient.

9.11.3 Pressure drop in gas flow up or down a vertical tube containing a falling film of liquid

The introduction of a liquid falling as a film on the internal surface of a tube increases the pressure drop of a gas flowing up or down inside the tube for two reasons:

1. It reduces the internal diameter of the tube from d to $(d - 2\delta)$, where δ is the thickness of the film, calculated as described in the previous section;
2. It increases the interfacial roughness to a value which may be estimated as follows.

Wallis (1969) shows in his Fig. 11.3 that the interfacial friction factor, which is the same as the Fanning friction factor (f) defined in section 9.1, increases linearly with increasing thickness of the liquid film, in accordance with:

$$f = 0.005\left(1 + 300\frac{\delta}{d}\right) \qquad [9.73]$$

The results apply to flow through 1-in, 2-in and 3-in vertical pipes with δ/d ranging from 0.001 to 0.022. Substituting in equation [9.72] shows that equation [9.73] is valid if $\dot{v}_1^+ < 0.21$.

Wallis assumed that the surface velocity of the liquid was negligible in comparison with the mean velocity of the gas. Hence with either upflow or downflow, the pressure loss can be estimated by using equation [9.72] to estimate the film thickness and equation [9.73] to estimate the friction factor.

9.12 Design of weirs and drains

This section is concerned with the flow of a liquid with a free surface over some form of weir into a region where flow is controlled by gravity. There are three examples of this that might be of interest to the designer of vaporisers: (1) flow over a rectangular weir cut in the side of a container or at the top of a vertical tube, (2) flow out of a drain in the bottom of a vessel, (3) flow out of a drain in the side of a vessel.

9.12.1 Rectangular weir

Tubes with castellated tops may be used in falling-film evaporators as shown in Fig. 3.21. There are usually six notches at the top of each tube; each notch may be treated as a weir. The distributor used for conveying the liquid to the space above the upper tubesheet may take the form of a trough with weirs cut in its sides at regular intervals along its length.

The problem in the design of a weir is to estimate the height above the base of the weir of the liquid in the upstream region where the velocity is low; this is the height H in Fig. 9.12. This subject is dealt with in textbooks on fluid mechanics, e.g. Francis (1975), Chapter 14. As the liquid accelerates towards the weir the height above the base of the weir falls to $2/3H$. The height H is given by:

$$H = 1.5 \left[\frac{\dot{V}}{b\sqrt{g}} \right]^{2/3} \hspace{3cm} [9.74]$$

where b is the breadth of the weir (m), g is the acceleration due to gravity (9.81 m/s^2) and $\dot{V}$ is the volume flow rate per weir (m^3/s).

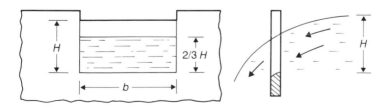

9.12 Rectangular weir

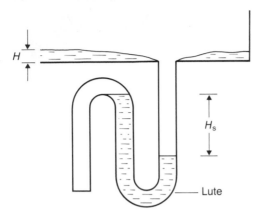

9.13 Bottom drain

9.12.2 Flow out of the bottom of a vessel or duct

The arrangement is as shown in Fig. 9.13. The flow rate of liquid is given, and the problem is to determine the minimum size of pipe to avoid entrainment of gas and vapour with the liquid as it enters the pipe, to calculate H, the height of liquid in the vessel and H_s, the depth of liquid required to seal the lute, shown in Fig. 9.13. McDuffie (1977) shows that for weir flow the depth of liquid in the slow moving zones is:

$$H = 0.56d(\dot{v}_1^+)^{2/3} \qquad\qquad [9.75]$$

where d is the internal diameter of the drainpipe (m), $\dot{v}_1^+$ is the dimensionless volume flux of the liquid, defined in equation [9.68]. McDuffie also shows that entrainment starts when $\dot{v}_1^+$ reaches a value between 0.3 and 0.55.

It follows from equation [9.69] for the limits of countercurrent flow, taking $C = 0.725$, that when there is no gas flow ($\dot{v}_g^+ = 0$), flooding occurs when $\dot{v}_1^+ = 0.725^2 = 0.53$. In other words, McDuffie's detection of entrainment indicates that flooding is occurring, in consequence of which the small amount of gas entrained by the current of descending liquid at its surface is unable to escape upwards in the centre of the pipe.

It is recommended that the minimum size of a drain hole in the bottom of a vessel or duct be estimated so that $\dot{v}_1^+ = 0.3$. For this, equation [9.75] gives $H = \frac{1}{4}d$. Substituting $\dot{v}_1^+ = 0.3$ in equation [9.68] shows that the smallest standard size of drainpipe should be chosen with an internal diameter greater than $d_{\min}$ (m), given by:

$$d_{\min} = 1.78\left(\frac{\dot{M}_1}{\rho_1 g^{\frac{1}{2}}}\right)^{0.4} \qquad\qquad [9.76]$$

where $\dot{M}_1$ is the flow rate of liquid (kg/s), ρ_1 is its density (kg/m^3) and g is the acceleration due to gravity (9.81 m/s^2).

The height H_s must be sufficient to balance the maximum possible pressure difference that might occur between the vessel and the receiver of the liquid. The minimum value of H_s is thus:

$$H_s = \Delta p / \rho_l g \qquad [9.77]$$

where Δp is the maximum pressure difference across the lute, the other symbols being as defined above.

If the procedure described above gives an inconveniently great value of H_s or of the internal diameter of the pipe, it is necessary to include a level-control valve in the pipe. McDuffie (1977) shows that to make sure that the level of liquid in the vessel is sufficient to flood the pipe and to avoid any entrainment of gas, the level control must be set so that:

$$H > 0.9 \frac{\sqrt{\dot{V}}}{(gd)^{\frac{1}{4}}} \qquad [9.78]$$

where $\dot{V}$ is the volume flow rate (m³/s). When delivering to a pump, it is advisable to place a vortex breaker above the top of the pipe. This may take the form of a plate of diameter about twice the internal diameter of the pipe supported horizontally above the hole by radial support plates, with a clearance of 0.5d between the plate and the hole.

9.12.3 Flow out of the side of a vessel

The arrangement is as shown in Fig. 9.14. The overflow pipe must have a downward slope sufficient for it to remove the liquid rapidly from the point of entry, making the flow similar to that over a weir. It can be shown by simple experiment that entrainment will not occur if at entry the overflow pipe is running half full or less. From the data of Ackers (1969) it can be shown that the condition for this is that $\dot{v}_l^+ < 0.3$. Hence equation [9.76] also gives the minimum diameter for an overflow pipe.

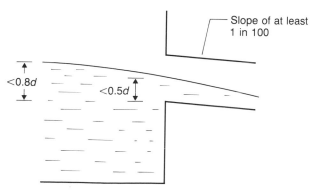

9.14 Side overflow pipe

The minimum slope of 1 in 100, shown in Fig. 9.14, for the overflow pipe, is adequate provided that the Fanning friction factor (f) is less than 0.014. For water, or any liquid of a similar viscosity, this arrangement is satisfactory if the internal diameter of the overflow pipe is more than 25 mm, for materials of the normal range of roughness (given in Table 9.1).

CHAPTER 10

Estimation of surface area required for heat transfer

Having estimated the various individual heat transfer coefficients and substituted these in equation [6.3] to determine U, the overall heat transfer coefficient, it is necessary to be able to solve the integral in equation [6.4] in order to calculate the surface area required for the specified rate of heat transfer ($\dot{Q}$). If U does not increase significantly from the cold to the hot end of the exchanger (say by more than 10%) and if the temperatures of both fluids vary linearly with the amount of heat transferred, then analytical solutions are available, as explained in section 10.1. Otherwise one of the approximate methods of integration described in section 10.3 must be used.

In a vaporiser in which a liquid is preheated, vaporised and then superheated, the values of U and the shapes of the temperature profiles are likely to be very different in the three zones. In this case, the surface area required for each zone should be estimated separately, and summed to give the total area required. The method of section 10.1 may be applied to preheating and superheating zones, but it may be necessary to use section 10.3 for the vaporising zone. When flash evaporation is used, boiling is suppressed and all the heat is transferred to liquid in single-phase flow; hence section 10.1 may be used to determine the surface area required, provided that there is no subcooled nucleate boiling (see sections 7.1.2 and 7.4.4).

Sections 10.2 and 10.4 deal with specific problems in the estimation of temperature, section 10.5 deals with design and rating problems and with the analysis of critical performance tests. Section 10.6 gives a design example and section 10.7 discusses the use of proprietary computer programs.

10.1 Constant overall heat transfer coefficient

In a single-phase heat exchanger, the value of U, the overall heat transfer coefficient, is likely to increase slightly from the cold to the hot end of the exchanger, due to changes in physical properties, but the specific heat capacity of each fluid is unlikely to increase significantly with increasing temperature. Under these circumstances, U may be assumed to be constant and equal to the arithmetic mean of the initial and final values.
Alternatively U may be evaluated from the physical properties of the fluid at the mean temperature. This section is concerned with the estimation of the required surface area when U is constant.

Figure 10.1 shows a typical graph of the temperature of the two fluids in a single-phase heat exchanger plotted against the fraction of the length of the exchanger; the slope of each curve is proportional to the temperature difference at every location. Figure 10.2 shows the same temperatures plotted against x_h, the fraction of the heat transfer rate that takes place from the inlet of the liquid to be vaporised to the specified location. Because the specific heat capacities are constant for each fluid, the lines in Fig. 10.2 are straight. Hence the temperature difference at any location is:

$$\Delta T = \Delta T_0 + x_h(\Delta T_1 - \Delta T_0) \tag{10.1}$$

where ΔT is the difference between the temperature of the hot fluid (T_h) and the temperature of the cold fluid (T_c) at the location where the fraction x_h of the heat load has been transferred, ΔT_0 and ΔT_1 being the initial and final values of ΔT, i.e. at the locations where $x_h = 0$ and 1 respectively. Substituting for $(T_h - T_v) = \Delta T$ from equation [10.1] in equation [6.4] and

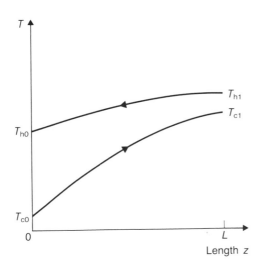

10.1 T/z diagram for single-phase heat exchanger

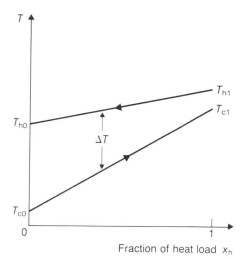

10.2 T/x_h diagram for single-phase heat exchanger

integrating gives, if U is constant:

$$A = \frac{\dot{Q}}{U \, \Delta T_{lm}}$$ [10.2]

where ΔT_{lm} is the log mean temperature difference given by:

$$\Delta T_{lm} = \frac{\Delta T_0 - \Delta T_1}{\ln(\Delta T_0 / \Delta T_1)}$$ [10.3]

This equation holds when the flow of the two fluids is wholly co-current or wholly countercurrent, the latter being shown in Figs 10.1 and 10.2. When there is cross-flow or several passes alternating between co-current and countercurrent flow, more complex solutions are available, as described by Saunders (1987) Chapter 7 and in standard textbooks such as Kern (1950).

In many vaporisers, the heat is supplied by condensing steam. Although the heat transfer coefficient in condensation varies along the length of a heat exchanger, the thermal resistance of the film of condensed steam is usually very small in comparison with the sum of the other resistances, and so U may be assumed to be constant in a preheating or superheating zone. Under these circumstances, the T/x_h diagram is as shown in Fig. 10.3, and the mean temperature difference is equal to the log mean. Under these circumstances

$$\Delta T_{lm} = \frac{T_{c1} - T_{c0}}{\ln\{(T_{hs} - T_{c0})/(T_{hs} - T_{c1})\}}$$ [10.4]

where T_{hs} is the saturation temperature of the steam and T_{c0} and T_{c1} are the temperatures of the vaporising liquid at inlet and outlet respectively.

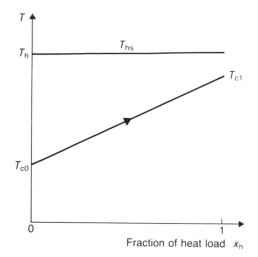

10.3 T/x_h diagram for single-phase heating by isothermal condensation

This equation may also be applied to the condensation of initially superheated steam, taking T_h as the saturated, and not the actual, temperature of the steam, and U as the overall heat transfer coefficient from the steam/condensate interface to the fluid to be vaporised, assuming condensation to occur over the whole length of the tube (see section 6.3). Similarly for isothermal vaporisation and sensible heat transfer from the heating fluid, the T/x_h diagram is as shown in Fig. 10.4 and, provided that U may be regarded as constant, the mean temperature difference is equal to the log mean. Under these circumstances equation [10.3] simplifies to:

$$\Delta T_{lm} = \frac{T_{h0} - T_{h1}}{\ln\{(T_{h0} - T_{vs})/(T_{h1} - T_{vs})\}} \qquad [10.5]$$

where T_{vs} is the saturation temperature of the vaporising fluid, and T_{h0} and T_{h1} are the inlet and outlet temperatures of the heating fluid.

When both fluids are isothermal, equations [10.4] and [10.5] simplify to:

$$\Delta T_{lm} = T_{hs} - T_{vs} \qquad [10.6]$$

When making the assumption of isothermal flow, it is advisable to check that this can be justified. Pressure drop along the heat exchanger may significantly reduce the saturation temperature, particularly at low pressures. When boiling in upward flow, the changes in hydrostatic head usually have a significant effect, as described in section 10.3. In a concentrating evaporator, the increase in concentration leads to an increase in boiling temperature, as described in section 10.4. The drop in saturation temperature due to a pressure drop of Δp is obtained from equation [9.48] as:

$$\Delta T_s = \Delta p T_s (1/\rho_g - 1/\rho_l)/\Delta h_v \qquad [10.7]$$

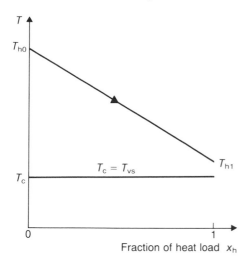

10.4 T/x_h diagram for isothermal vaporisation by a single-phase heating fluid

where T_s is the saturation temperature (K), Δh_v is the latent heat of vaporisation (J/kg) and ρ_g and ρ_l are the densities of the vapour and liquid respectively (kg/m³). When the assumption of isothermal flow is not valid, a stepwise method must be used, as described in section 10.3.

10.2 Effect of hydrostatic head on temperature difference in boiling zone

When a boiling liquid is recirculated from the vapour/liquid separator down to the bottom of the heater, there is an increase in its boiling temperature as a result of the hydrostatic head. Table 3.1 shows that this amounts to 2.6 °C per metre increase in depth for water at atmospheric pressure, being greater under vacuum and negligible at very high pressure. Section 9.6.3 explains how to estimate the level at which boiling begins. Equation [9.51] defines the line OC in Fig. 9.5 and thus gives the temperature difference throughout the single-phase region. If nucleate boiling begins between O and C, the temperature will rise more steeply in the region of subcooled nucleate boiling, due to the greater heat transfer coefficient. The shape of the curve CD may be determined by estimating the pressure drop from the bottom channel to various levels in the boiling zone, the temperature at each point being the saturation temperature corresponding to the local pressure. In the case of a reboiler, the pressure drop in the pipeline returning the vapour to the distillation column leads to the temperature at D being greater than that in the column. Thus it is important to have an adequate return pipeline when the temperature difference is small, to minimise this loss of temperature difference. The point C in Fig. 9.5 is known as a 'pinch point' because the reduced temperature difference might cause a serious loss in performance, and so should not be neglected.

10.3   Estimation of area required for the boiling zone

Throughout the boiling zone there is likely to be a large change in the value of α_v, the local heat transfer coefficient for the fluid to be vaporised. Increasing the quality increases the convective coefficient but slightly reduces the nucleate component, which is increased by an increase in temperature difference. Minor changes result from changes in physical properties with changing temperature. In a concentrating evaporator the increase in concentration leads to a fall in heat transfer coefficient, as discussed in section 10.4. If the local value of U, the overall heat transfer coefficient, found by substituting α_v in equation [6.3], does not increase by more than 10% from the inlet to the outlet of the boiling zone, the area required for this zone may be determined as described above in section 10.1, provided that the temperature difference, calculated as described in the previous section, varies roughly linearly with height.

If the increase in U lies between 10% and 40%, it may be assumed to vary linearly with the fraction of the heat load; if the temperature difference also varies linearly with the fraction of the heat load, the integration of equation [6.4] gives the following equation:

$$A = \dot{Q} \frac{\ln(U_0 \Delta T_1 / U_1 \Delta T_0)}{U_0 \Delta T_1 - U_1 \Delta T_0} \qquad [10.8]$$

where A is the surface area required (m²), $\dot{Q}$ is the heat transfer rate in the zone under consideration (W), U is the overall heat transfer coefficient, referred to area A (W/m² K) and ΔT is the difference between the bulk temperature of the heating and the vaporising fluids (K). Suffix 0 refers to the conditions at the inlet to the zone and suffix 1 to the outlet. (This equation is quoted by McAdams (1954) as his equation [8.17] and by Kern (1950) as his equation [5.23].)

If the variations in U or ΔT are such that equation [10.8] would not give adequate accuracy, U and ΔT should be calculated at one or more values of x_h, the fraction of the heat transfer rate, intermediate between the values at the inlet and outlet of the zone, so that equation [10.8] may be used to determine the surface area required for each subdivision of the zone.

The most accurate way to solve equation [6.4] is to determine the integral graphically. This is achieved by estimating U, T_h and T_v at several values of x_h, calculating the heat flux, $\dot{q} = U(T_h - T_v)$ and plotting $1/\dot{q}$

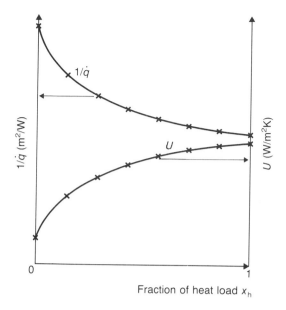

10.5 Typical curves relating U and $1/\dot{q}$ to the fraction of the heat load in a boiling zone

against x_h, as illustrated in Fig. 10.5, which also shows a plot of U against x_h, for interest. The number of intermediate values of x_h chosen must be sufficient to permit the drawing of a curve through the points. The surface area required is given by the following modified form of equation [6.4]:

$$A = \dot{Q} \int_0^1 \left(\frac{1}{\dot{q}}\right) dx_h \qquad\qquad [10.9]$$

where A is the surface area required for the zone under consideration (m^2), $\dot{Q}$ is the rate of heat transfer in the zone (W), and the integral is the area under the curve of $1/\dot{q}$ plotted against x_h for the zone; the dimensions of $1/\dot{q}$ are m^2/W.

The estimation of the critical heat flux is dealt with in Chapter 8. The methods of calculating two-phase heat transfer coefficients given in sections 7.2, 7.3 and 7.4 are for wet-wall boiling, and they apply when the heat flux is less than the critical appropriate to the local quality, or until the critical quality is reached. Dry-wall convection is discussed in section 7.5. Once this condition is reached, the heat transfer coefficient should be calculated for dry gas of quality unity, even if vaporisation is not complete. Separate estimates should be made of the area required for the wet-wall and the dry-wall regions.

10.4 Effect of concentration on physical properties of aqueous solutions

When estimating the thermal performance of a concentrating or crystallising evaporator, it must be remembered that the heat flux will be less than that with pure water under otherwise identical conditions. The reasons for this are twofold: the boiling temperature is elevated as a result of the concentration of dissolved solids, and this leads to a reduction in the temperature difference; the single-phase convective heat transfer coefficient is less than for pure water at the same temperature and the same mass flux, because the solution has a higher viscosity, a lower specific heat capacity, and possibly a lower thermal conductivity. Methods of allowing for these two effects are discussed below.

10.4.1 Elevation of boiling temperature due to dissolved solids

It is shown in textbooks on physical chemistry, e.g. Maron and Lando (1974), pp. 450–4, how to estimate the elevation of the boiling temperature of solutions of non-volatile solutes. The following approximate equation is obtained by combining Raoult's equation for the vapour pressure of a solution with the Clausius–Clapeyron equation, which is obtained by integrating the Clapeyron equation – equation [9.48] – assuming that the density of the vapour can be determined from the ideal gas laws.

$$\Delta T_s = \frac{RT_s^2}{\Delta h_v} N \qquad [10.10]$$

where N is the molar fraction of the solute in the solution, R is the gas constant ($8314/\tilde{M}$ J/kg K), T_s is the boiling temperature (K) of the solvent at the prevailing pressure, Δh_v is the latent heat of vaporisation of the solvent at the temperature T_s (J/kg) and ΔT_s is the elevation of the boiling temperature (K). Thus for one gram mole of a non-volatile solute dissolved in 1 kg of water, the quantity of water ($\tilde{M} = 18$) is $1000/18 = 55.56$ gram mole. Hence $N = 1.0/(1.0 + 55.56) = 0.01768$.

For water at atmospheric pressure, $T_s = 100 + 273 = 373$ K and $\Delta h_v = 2257$ kJ/kg. Substituting in equation [10.10] gives:

$$\Delta T_s = \frac{(8314/18) \times 373^2}{2257 \times 10^3} \times 0.01768 = 0.5 \text{ K}$$

This shows that the atmospheric boiling temperature is increased from 100.0 to 100.5 °C.

10.4.2 Reduction in single-phase convective coefficient due to dissolved solids

Very little information is available on the physical properties of aqueous solutions. ESDU (1977c) gives thermal conductivity, viscosity and specific heat capacity of sea water and its concentrates. These values have been used in Table 6.1, which shows that the convective coefficients for a saturated solution are likely to be about 80% of those for pure water at the same temperatures.

10.5 Application to design, rating or testing of vaporisers

So far this chapter has been concerned with the estimation of the surface area required for a specified rate of heat transfer, given the mass flux, temperatures and properties of both fluids, and the thermal resistance of the wall, including fouling. It is now necessary to consider how this procedure may be applied to the practical problems that arise when it is required to design a vaporiser, estimate what heat transfer rate would be achieved in a given vaporiser for which the conditions of the fluids at inlet have been specified, or analyse the results of tests on a vaporiser. The steps in the solution of these three problems are described below.

10.5.1 Design procedure

Here some preliminary calculations are required to make an approximate estimate of the surface area required. The given resistance of the wall and dirt may be combined with approximate values for the heat transfer coefficients of the heating fluid and the vaporising fluid. Preferably this should be based on an analysis (as described in section 10.5.3) of the results of tests on a similar vaporiser operating on an existing plant or in a semi-technical plant. If no such data are available the following approximate values of heat transfer coefficient may be of assistance.

Heating fluid:	Condensing steam	$10\,000$ W/m^2 K
	Hot water	$4\,000$ W/m^2 K
	Hot dry gas at pressure	
	1 bar (10^5 Pa)	100 W/m^2 K
	10 bar (10^6 Pa)	200 W/m^2 K
Fluid to be vaporised, boiling suppressed:	water	$5\,000$ W/m^2 K
	aqueous solution	$3\,000$ W/m^2 K
	hydrocarbon	$1\,000$ W/m^2 K

Boiling liquid: water   10 000 W/m^2 K

aqueous solution   8 000 W/m^2 K

hydrocarbon   2 000 W/m^2 K

The approximate surface area required can be estimated from equation [10.2]. A preliminary estimate must now be made of the number and length of tubes (of the desired diameter) and of the arrangement on the shellside, so that suitable fluid velocities are obtained and the surface area is as estimated. The fluid velocities should be such that the maximum specified pressure drop is not exceeded and there is an economic arrangement that minimises the sum of the capital charges and the operating costs. The following simplified expressions, based on Smith (1981), may be used to determine the economic velocity in single-phase flow. It is significant that erring from the economic velocity by $\pm 40\%$ increases the total cost by only a few percent.

Approximate economic velocity (m/s) for single-phase flow in shell-and-tube heat exchangers:

gases on tubeside $v_E = 10\rho_g^{-\frac{1}{3}}$

gases on shellside $v_E = 10\rho_g^{-\frac{1}{3}}$

[10.11]

liquids on tubeside $v_E = 1.0 \, (1000\eta_l)^{0.1}$

liquids on shellside $v_E = 0.5 \, (1000\eta_l)^{0.1}$

where ρ_g is the density of the gas (kg/m^3) and η_l is the viscosity of the liquid (N s/m^2).

When heat is supplied by the condensation of steam, the velocity has no effect on the thermal calculations, provided that it is not so high that the resulting pressure drop will produce a significant drop in saturation temperature. For shellside condensation of steam, the spacing of the baffles must be close enough to provide adequate support for the tubes and avoid damage due to vibration of the tubes (see section 11.3).

Having decided upon the approximate geometry of the vaporiser, it is then necessary to follow the procedure described in Chapters 6 and 7 to determine the local values of U, the overall heat transfer coefficient, so that the appropriate method described in the earlier sections of this chapter can be used to make a more accurate estimate of the required surface area. If this differs significantly from the approximate value previously estimated, adjustments must be made to the geometry of the vaporiser and the calculations repeated until the area calculated from the geometry equals that required for transferring the specified heat load. The pressure drop of the fluids may then be estimated. The procedure is illustrated by the example in section 10.6.

10.5.2 Rating calculations

Here the geometry is defined and the flow rates are given, so it is possible to estimate the surface area available. It is then necessary to guess the heat transfer rate and deduce from this the outlet conditions of the two fluids. The procedure described above can now be used to determine the required surface area. If this exceeds the available area, a proportionately smaller heat transfer rate should be chosen, and vice versa. The calculations must be repeated until the required area equals the available area.

10.5.3 Analysing test results

The procedure here is the same as for rating calculations, except that it is the fouling resistance (not the heat transfer rate) that must be determined by trial and error.

10.6 Design example

Design a vertical thermosyphon reboiler, similar to that illustrated in Fig. 3.10, to vaporise 0.83 kg/s of water, using steam condensing at 120 °C as the heating medium. The tubes should be of 25 mm external and 22 mm internal diameter, made of stainless steel, with the space between baffles being 1.8 m. The pressure at the base of the distillation column is 1.12 bar, 112 kPa. A heat transfer coefficient of 10 000 W/m² K may be taken for the dirt on both the inside and outside of the tubes. The thermal conductivity of the stainless steel is 16 W/m K. The physical properties of water may be taken from ESDU (1980).

10.6.1 Preliminary calculations

If there are n_t tubes of length L (m), the surface area in contact with the vaporising fluid is:

$$A = \pi d_i L n_t = \pi(22 \times 10^{-3})L n_t = 0.0691 L n_t \qquad [10.12]$$

The procedure (described in section 10.5.1) will be to determine an approximate value of A, guess a value of n_t, calculate the value of L to give the approximate area, and then determine L more accurately from heat transfer calculations. The calculations will be repeated, if necessary, making different assumptions about the inlet and outlet piping and for different numbers of tubes.

The temperature at the base of the distillation column, i.e. the saturation temperature of water giving a vapour pressure of 1.12 bar (112 000 Pa) is found from the physical data on water/steam to be 102.5 °C. The heat load must be based on the latent heat of vaporisation at this temperature, which is 2250 kJ/kg. Hence the required heat transfer rate is:

$$\dot{Q} = 0.83 \times 2250 \times 10^3 = 1868 \times 10^3 \text{ W}$$

The overall heat transfer coefficient can be calculated from equation [6.3]. This requires a knowledge of α_w, the heat transfer coefficient for the wall, which, for a stainless steel tube 1.5 mm thick, is given by equation [6.49] as:

$$\alpha_w = \frac{16}{1.5 \times 10^{-3}} = 10\,667 \text{ W/m}^2 \text{ K}$$

As there are no fins, the surface area ratios in equation [6.3] may be taken as the diameter ratios. The wall diameter is the log mean of the internal and external diameters, which in this case does not differ significantly from the arithmetic mean of 23.5 mm. Hence equation [6.3] gives:

$$\frac{1}{U} = \frac{1}{\alpha_v} + \frac{1}{10\,000} + \frac{1}{10\,667}\frac{22}{23.5} + \left[\frac{1}{10\,000} + \frac{1}{\alpha_h}\right]\frac{22}{25}$$

where α_v and α_h are the heat transfer coefficients of the water to be vaporised and the heating steam respectively. Hence:

$$U = \left[\frac{1}{\alpha_v} + \frac{1}{3625} + \frac{0.88}{\alpha_h}\right]^{-1} \qquad [10.13]$$

From Table 3.1 it is seen that the increase in the boiling temperature of water with depth at the pressure prevailing is about $2\frac{1}{2}$ °C/m. If boiling begins at a depth of 1 to 2 m, there would be a temperature rise in the non-boiling zone of about 4 °C. Referring to Fig. 9.5, it follows that at points O and D, the temperature difference between the two fluids is the difference between the saturation temperature of the condensing steam and that of the water in the base of the distillation column, namely, $120 - 102.5 = 17.5$ °C. At point C, the temperature difference is $17.5 - 4 = 13.5$ °C. The approximate mean temperature difference over the length of the tubes is the log mean of these two differences; thus from equation [10.3]:

$$\Delta T'_m = \frac{17.5 - 13.5}{\log_e(17.5/13.5)} = 15.4 \text{ °C}.$$

From this, the mean temperature of the water as it passes through the tubes is $120 - 15.4 = 104.6$ °C. From ESDU (1980) the physical properties of water and steam at this temperature are:

Vapour pressure,	$p_s = 1.206 \times 10^5$ Pa
Latent heat,	$\Delta h_v = 2.244 \times 10^6$ J/kg
Surface tension,	$\sigma = 0.058$ N/m
Density of liquid and vapour,	$\rho_l = 954, \rho_g = 0.701$ kg/m^3
Specific heat capacity of liquid,	$c_{pl} = 4228$ J/kg K
Thermal conductivity of liquid,	$\lambda_1 = 0.6804$ W/m K
Dynamic viscosity of liquid and vapour,	$\eta_l = 268 \times 10^{-6}$,
	$\eta_g = 12.42 \times 10^{-6}$ Ns/m^2

From equation [6.7], the Prandtl number of the liquid is

$$Pr_l = 268 \times 10^{-6} \times 4228/0.6804 = 1.665$$

From equation [6.8] the Reynolds number of the liquid is:

$$Re_{lo} = \dot{m} \times 22 \times 10^{-3}/268 \times 10^{-6} = 82\dot{m} \qquad [10.14]$$

where $\dot{m}$ is the mass flux through the tubes, given by equation [9.54] as

$$\dot{m} = \frac{0.83/x_{g\,out}}{(\pi/4)(0.022)^2 n_t} = \frac{2183}{x_{g\,out} n_t} \qquad [10.15]$$

where $x_{g\,out}$ is the quality of the steam/water mixture leaving the tubes.
From equations [6.6] and [6.9], the heat transfer coefficient for the water in the non-boiling zone is:

$$\alpha_{lo} = \frac{0.6804}{0.022} \times 0.022\,46\,Re_{lo}^{0.795} \times 1.665^{(0.495 - 0.0225\ln 1.665)}$$

or $\alpha_{lo} = 0.889\,Re_{lo}^{0.795}$ \qquad [10.16]

10.6.2 Approximate design

From section 10.5.1, the approximate heat transfer coefficients are: for condensing steam, $\alpha_h = 10\,000$ W/m^2 K; for boiling water (assuming that there is no subcooled nucleate boiling in much of the non-boiling zone), $\alpha_v = 5000$ W/m^2 K. Substituting in equation [10.13], the approximate overall heat transfer coefficient is:

$$U' = \left[\frac{1}{5000} + \frac{1}{3625} + \frac{0.88}{10\,000} \right]^{-1} = 1773 \text{ W/m}^2 \text{ K}$$

From equation [10.2], the approximate surface area required is:

$$A' = \frac{1868 \times 10^3}{1773 \times 15.4} = 68.4 \text{ m}^2$$

This gives the mean heat flux, $\dot{q}_m = 1868 \times 10^3/68.4 = 27\,300$ W/m^2.

Substituting in equation [10.12] gives:

$$n_t L' = 68.4/0.0691 = 990 \text{ m} \tag{10.17}$$

where L' is the approximate length of the tubes (m).

The approximate economic velocities given by equation [10.11] show that for water only on the tubeside $v_E = 1.0$ m/s and for steam of density 0.701 on the tubeside $v_E = 20 \times 0.701^{-1/3} = 22.5$ m/s. It follows that the economic tubeside mass flux is $1000 \times 1.0 = 1000$ for water and $0.701 \times 22.5 = 16$ kg/s m² for 100% vapour. Interpolating suggests that for boiling water it should be in the region of 200 to 500 kg/s m². With vertical thermosyphon reboilers, a much more important economic consideration is to avoid the use of very long tubes, which would entail considerable expense due to the need to build the distillation column at a level higher than that originally planned.

As a first approximation, consider using 250 tubes. Substituting $n_t = 250$ in equation [10.17] gives the approximate length:

$$L' = 990/250 = 4.0 \text{ m};$$

from equation [10.12], $A = 0.0691 \times 4.0 \times 250 = 69.1 \text{ m}^2$.

This length of 4.0 m may not be considered to be excessive and it does not call for any revision of the estimate above of the temperature of the water when boiling begins.

The recirculation ratio might be in the region of 40, giving $x_{g\,out} = 1/40 = 0.025$. Substituting in equation [10.15] gives:

$$\dot{m} = \frac{2183}{0.025 \times 250} = 349 \text{ kg/s m}^2$$

This is within the rough range given above.

The next steps are to carry out pressure drop calculations to revalue the recirculation ratio, followed by heat transfer calculations to revalue the length of tube required.

10.6.3 Recirculation ratio

It is now possible to follow the procedure in section 9.6 to estimate the pressure drop around the circulation loop at different values of the recirculation ratio, to determine which value gives zero pressure drop around the loop. To avoid oscillatory instability, the pressure drop in the single-phase region should approximately equal that in the two-phase region (see section 9.2.7). With an arrangement such as that shown in Fig. 3.10, where the return pipe is short and straight, it should be possible to keep the pressure loss coefficient for this pipe to no more than two. Using a pipe of approximately the same cross-sectional area as the tubes will give a pressure drop in the return line not much greater than that in the two-phase region. Thus if 250 tubes of 22 mm bore are adequate, the return pipe should have an internal diameter of about $22\sqrt{250}$, or 350 mm, so that its cross-sectional area for flow is similar to that of the tubes.

Equation [9.48] gives the relationship between changes in saturation pressure (Δp_s) and saturation temperature (ΔT_s) at the mean water temperature as:

$$\Delta p_s = \Delta T_s \frac{2244 \times 10^3}{(273.15 + 104.6)(1/0.701 - 1/954)} = 4167 \, \Delta T_s \quad \text{Pa} \qquad [10.18]$$

Equation [9.49] gives the relationship between static pressure and height in regions of single-phase flow of liquid, neglecting friction.

$$\frac{\mathrm{d}p}{\mathrm{d}z} = -954 \times 9.81 = -9359 \, \text{Pa/m} \qquad [10.19]$$

Combining equations [10.18] and [10.19] gives the relationship between saturation temperature and height in regions of single-phase flow of liquid, neglecting friction.

$$\frac{\mathrm{d}T_s}{\mathrm{d}z} = -9359/4167 = -2.246 \, ^{\circ}C/m \qquad [10.20]$$

It is normal practice to control the level of liquid in the base of the distillation column to coincide with the top of the tubes. This gives the static head in the lower channel of the heater as $L - \Delta H_{in}$, where L is the length of the tubes and ΔH_{in} is the frictional loss of head in the inlet pipe from the base of the column to the lower channel of the heater (m). Combining this with equation [10.19] gives the following equation for Δp, the pressure in the lower channel of the heater less the pressure in the vapour space at the bottom of the column. This must equal the pressure drop in flow through the tubes and through the return pipe back to the still.

$$\Delta p = 9359(L - \Delta H_{in}) \qquad [10.21]$$

The temperature rise of the liquid as it flows through the non-boiling liquid zone is obtained from equation [9.51]. The mass flow rate of liquid ($\dot{M}$) is the vapour generation rate (0.83 kg/s) multiplied by the recirculation ratio, which is $1/x_{g\,out}$ where $x_{g\,out}$ is the quality at the outlet of the tubes. The mean heat flux in the liquid zone ($\dot{q}_{mean}$) cannot be determined at this stage; as an approximation, it may be assumed to equal the mean over the total length of the tube. Under this assumption, $\dot{q}_{mean}A$ is equal to $\dot{Q}$, the heat transfer rate, which has been estimated to be 1868 kW. Hence the temperature rise of the liquid in the length z_1 before boiling begins is:

$$\Delta T_1 = \frac{1868 \times 10^3 z_1}{(0.83/x_{g\,out}) \times 4228L} = 532 z_1 x_{g\,out}/L \qquad [10.22]$$

Equation [9.52] gives the drop in saturation temperature in the non-boiling zone as:

$$-\Delta T_s = (\Delta p_{frl} + \Delta p_{grl})/4167 \qquad [10.23]$$

where Δp_{frl} is the pressure drop due to the acceleration of the liquid at entry into the tubes, plus the frictional losses at entry and in the tubes as far as the point where boiling begins, and Δp_{grl} is the pressure drop due to gravity in this zone of flowing liquid, before boiling begins. From equations [9.2], [9.3] and [9.7], for a negligible area ratio at entry to the tubes, the single-phase pressure drops are:

$$\Delta p_{frl} = \left[\frac{4f_1 z_1}{0.022} + 1.5\right]\frac{\dot{m}^2}{2 \times 954} = \dot{m}^2\left[\frac{f_1 z_1}{10.5} + 0.000\,786\right] \qquad [10.24]$$

$$\Delta p_{grl} = 9359 z_1 \qquad [10.25]$$

where f_1 is the Fanning friction factor, determined as described in section 9.1.1 for flow in the liquid zone.

But the sum of ΔT_1 and $-\Delta T_s$ is the amount of subcooling of the liquid in the bottom channel of the heater, given by equation (10.20) as 2.246 $(L - \Delta H_{in})$. Thus:

$$523 z_1 x_{g\,out}/L + (\Delta p_{frl} + \Delta p_{grl})/4167 = 2.246(L - \Delta H_{in}).$$

Substituting for the frictional and gravitational pressure drops in the liquid zone from equations [10.24] and [10.25] and rearranging gives:

$$z_1 = \frac{2.246(L - \Delta H_{in}) - (\dot{m}/2303)^2}{532 x_{g\,out}/L + 2.246 + f_1\dot{m}^2/43\,750} \qquad [10.26]$$

Turning now to the two-phase (boiling) zone, the length of this is

$$z_b = L - z_1 \qquad [10.27]$$

Combining equations [9.39] and [9.40] gives the frictional pressure drop in the two-phase (boiling) zone as:

$$\Delta p_{fr2} = \frac{2f_1 z_b}{0.022}\frac{\dot{m}^2}{954}(1 + \tfrac{1}{2}E) = \frac{f_1 z_b \dot{m}^2(2 + E)}{21.0} \qquad [10.28]$$

where E is the increment in specific volume (assuming homogeneous flow) during flow through the boiling zone divided by the liquid specific volume. Thus:

$$E = \left[\frac{954}{0.701} - 1\right]x_{g\,out} = 1360 x_{g\,out} \qquad [10.29]$$

Equation [9.42] gives the two-phase gravitational pressure drop as

$$\Delta p_{gr2} = \frac{9359 z_b}{E}\ln(1 + E) \qquad [10.30]$$

Equation [9.43] gives the two-phase accelerational pressure drop:

$$\Delta p_{acc} = \frac{\dot{m}^2 E}{954} \qquad [10.31]$$

Assuming homogeneous flow in the return line, which is designed to have a pressure drop of two velocity heads, based on the tubeside velocity, the pressure drop in the return line is:

$$\Delta p_R = \frac{\dot{m}^2 (1 + E)}{954} \qquad [10.32]$$

The total pressure drop from the bottom channel to the distillation column is the sum of the six individual losses. Thus:

$$\Delta p = \Delta p_{frl} + \Delta p_{grl} + \Delta p_{fr2} + \Delta p_{gr2} + \Delta p_{acc} + \Delta p_R \qquad [10.33]$$

The residual pressure drop is the difference between Δp calculated from equation [10.33] and Δp calculated from equation [10.21], the former giving the pressure rise from the still to the lower channel and the latter the pressure drop in the remainder of the loop. When the correct value of $x_{g\,out}$ has been substituted in the above equations, the residual pressure drop is zero. These equations may be solved with a computer or with a programmable calculator. Table 10.1 gives the steps in pressure drop calculations. For the estimation of the friction factor, it was assumed that the surface roughness was 25 μm, giving a relative roughness of $25 \times 10^{-3}/22 = 1.14 \times 10^{-3}$.

The first column of Table 10.1 gives the symbol which is used to denote the various quantities. The second column gives the source, the figures in parentheses giving the numbers of the equations used. The third and fourth columns give the first and last steps in the estimation of recirculation ratio for the approximate design proposed in section 10.6.2. This gives an outlet quality of 0.0203. Hence the recirculation ratio is $1/0.0203 = 49$. The temperature rise in the single-phase region is obtained from equation [10.22] as $\Delta T_1 = 532 \times 1.574 \times 0.0203/4.0 = 4.25°C$. The temperature difference at inlet to the tube has been found as 17.5 °C. At the point where boiling begins it is $17.5 - 4.25 = 13.25$ (cf. 13.5 in the preliminary calculations). This was based on the assumption that the mean heat flux in the single-phase zone equals the mean over the whole length; if it is only 90% of this, the mass flux is reduced from 430 to 418 kg/s m^2, the length of the non-boiling zone is increased from 1.57 to 1.64 m, ΔT_1 would be increased from 4.25 to 4.52 °C, all of which would lead to a small loss in performance. Hence this must be investigated after completing the thermal calculations.

The pressure drop in the single-phase zone due to friction is the sum of the inlet loss and the loss in the pipe up to the point where boiling begins; thus, for the last step, Δp_f (single-phase) $= 9359 \times 0.5 + 323 = 5002$ Pa. In the two-phase zone, the frictional pressure drop is 4051 in the tubes plus

10.1 Reboiler recirculation

Item (symbol)	Source	First step	Last step	Alternative designs		Units
L	assumed	4.0	4.0	4.0	3.0	m
n_t	assumed	250	250	250	333	—
ΔH_{in}	assumed	0.5	0.5	0.8	0.8	m
$x_{g\,out}$	assumed	0.025	0.0203	0.0221	0.0158	—
$\dot{m}$	[10.15]	349	430	395	415	kg/s m^2
Re_{lo}	[10.14]	28 640	35 270	32 399	34 025	—
$f \times 10^4$	Fig. 9.1	66	64	65	65	—
E	[10.29]	34.0	27.6	30.06	21.49	—
z_1	[10.26]	1.402	1.574	1.375	1.100	m
Δp_{fr1}	[10.24]	203	323	256	253	Pa
Δp_{gr1}	[10.23]	13 124	14 728	12 866	10 299	Pa
Δp_{fr2}	[10.28]	3586	4051	4066	2378	Pa
Δp_{gr2}	[10.30]	2542	2758	2809	2575	Pa
Δp_{acc}	[10.31]	4348	5355	4918	3878	Pa
Δp_R	[10.32]	4476	5548	5082	4059	Pa
Δp	[10.33]	28 279	32 763	29 997	23 441	Pa
Residual pressure drop	[10.33] and [10.21]	4477	8	37	44	Pa
Recirculation ratio		40	49	45	63	—
α_{lo}	[10.16]		3665	3425	3561	W/m^2 K
ΔT_1	[10.22]		4.25	4.04	3.08	°C
z_b/L	$1 - z_1/L$		0.607	0.656	0.633	—
Frictional pressure drop ratio*			1.92	1.18	1.31	—

* This is the ratio $\dfrac{\Delta p_{fr2} + \Delta p_R}{\Delta p_{fr1} + 9359\,\Delta H_{in}}$

5548 in the return pipe, giving a total of 9599 Pa, which is 92% greater than the corresponding figure for the single-phase zone. The application of the method of Akinjiola and Friedly (1979) to this problem, after extrapolation, showed this design to be on the verge of oscillatory instability, described in section 9.2.7. This suggests that the pressure drop in the pipe from the distillation column to the inlet channel of the heater should be increased. This would lead to a reduction in the circulation rate. Further consideration is given in section 10.6.5.

10.6.4 Thermal calculations

For this, the heat flux will be estimated at six locations: (i) at inlet; (ii) $\frac{1}{2}$-way up the single-phase zone; (iii) at the point where boiling begins; (iv) $\frac{1}{3}$-way up the two-phase zone; (v) $\frac{2}{3}$-up the two-phase zone; (vi) at the outlet from the tubes. Prior to this, the condensation coefficient will be discussed and the requirements for nucleate boiling will be established.

With tubes 4 m long, the minimum number of shellside passes is five, because this gives a baffle pitch of $4/5 = 0.8$ m with a maximum unsupported length of twice this for tubes in the window, i.e. 1.6 m, which is just within the maximum of 1.8 m specified. With 250 tubes, condensation will take place on $250 \times 5 = 1250$ baffle spaces. Assuming that each baffle removes all the condensate formed in the baffle space above it, equation [6.32] gives:

$$\Gamma = 0.83/\pi 0.022 \times 0.009\,61 \text{ kg s m}$$

(neglecting the reduction in latent heat with increasing temperature). Equation [6.29] gives:

$$Re_f = 4 \times 0.009\,61/268 \times 10^{-6} = 143$$

(neglecting the reduction in viscosity with increasing temperature). Figure 6.3 gives, for water at 120 °C, $\phi_c = 2150$. Equation [6.34] gives the theoretical heat transfer coefficient for the laminar flow of condensate:

$$\bar{\alpha}_{cl} = 2150 \times 0.009\,61^{-\frac{1}{3}} = 10\,113 \text{ W/m}^2 \text{ K}.$$

As Re_f is between 30 and 1600, the heat transfer coefficient for the film of condensate is given by equation [6.43]:

$$\bar{\alpha}_c = 10\,113 \left[\frac{143^{4/3}}{1.62 \times 143^{1.22} - 7.8} \right] = 11\,080 \text{ W/m}^2 \text{ K}$$

The revised condensing coefficient (11 080) is slightly greater than the value assumed in the approximate design (10 000). Putting $\alpha_h = \bar{\alpha}_c = 11\,080$ in equation [10.13] gives the following equation for U in terms of α_v:

$$U = \left[\frac{1}{\alpha_v} + \frac{1}{3625} + \frac{0.88}{11\,080} \right]^{-1} = \left[\frac{1}{\alpha_v} + \frac{1}{2815} \right]^{-1} \qquad [10.34]$$

Equation [7.1] gives the minimum temperature difference between the wall and the saturation temperature for the onset of nucleate boiling.

$$T_{wONB} - T_s = \left[\frac{8 \times 0.58\dot{q}(273.15 + 104.6)}{0.6804 \times 2244 \times 10^3 \times 0.701} \right]^{\frac{1}{2}} = 0.0128\dot{q}^{\frac{1}{2}} \qquad [10.35]$$

For the liquid region, $\alpha_v = \alpha_{lo}$, which is given by equation [10.16]

$$\alpha_{lo} = 0.889 \times 35\,270^{0.795} = 3665 \text{ W/m}^2 \text{ K}$$

If there is subcooled nucleate boiling, the coefficient for this estimated as described below, must be added to α_{lo}.

The pressure in the lower channel has already been estimated from equation [10.21] to be 32 756 Pa above the pressure in the base of the distillation column. Table 10.2 gives this figure as the pressure at location (i), neglecting the pressure drop at entry to the tubes. Table 10.1 shows the pressure drop in the single-phase zone to be 323 due to friction and 14 728 Pa due to gravity. Subtracting these from the excess pressure at (i) gives the excess pressure at location (iii), where boiling begins. The pressure at (ii) was found by linear interpolation between those at (i) and (iii). The pressure drop from location (iii) to location (iv) may be determined from equations [10.28], [10.30] and [10.31]. As location (iv) is one third of the way along the boiling zone, and as in the derivation of these equations it was assumed that quality increases linearly with length, the values z_b and E substituted must be one third of their final values, namely:

$$z_b = (4.0 - 1.574)/3 = 0.809, \quad \text{and} \quad E = 27.6/3 = 9.2. \text{ Thus:}$$

10.2 Thermal calculations for $L = 4.0$ m, $n_t = 250$, $\Delta H_{in} = 0.5$, $x_{g\,out} = 0.0203$

Location	(i)	(ii)	(iii)	(iv)	(v)	(vi)	Units
x_g	0	0	0	0.0068	0.0135	0.203	—
$p - p_{st}$	32 756	25 230	17 705	13 500	9836	5548	Pa
$T_s - T_{st}$	7.86	6.05	4.25	3.24	2.36	1.33	K
$T_b - T_{st}$	0	2.12	4.25	3.24	2.36	1.33	K
$T_h - T_b$	17.50	15.38	13.25	14.26	15.14	16.17	K
$1/X_{tt}$	0	0	0	0.306	0.570	0.828	—
F	1	1	1	1.450	1.936	2.421	—
α_{cb}	3665	3665	3665	5284	7117	8729	W/m² K
S	0.654	0.654	0.654	0.525	0.417	0.347	—
α_{nb}	0	1303	2395	1854	1389	1088	W/m² K
$T_w - T_b$	7.60	6.22	4.21	4.04	3.77	3.60	K
$\dot{q}$	27 860	25 785	25 460	28 790	32 020	35 370	W/m²
$T_w - T_s$	−0.26	2.29	4.21	4.04	3.77	3.60	K
$T_{wONB} - T_s$		2.06	2.04	2.17	2.29	2.41	K
Δz		0.787	0.787	0.809	0.808	0.809	m
$\Delta \dot{Q} \times 10^{-3}$		365	348	379	425	471	kW

Where p is the local static pressure and p_{st} is that in the vapour space at the base of the still (Pa), T_b is the local bulk temperature of the boiling water, T_h is the saturation temperature of the condensing steam, T_s is the local saturation temperature and T_{st} is the temperature of the vapour at the base of the still. The calculations are described in section 10.6.4

$$P_{(iii)} - P_{(iv)} = \frac{0.0064 \times 0.809 \times 430^2(2 + 9.2)}{21.0} +$$

$$\frac{9359 \times 0.809}{9.2} \ln(1 + 9.2) + \frac{430^2}{954} \, 9.2 = 4206 \text{ Pa}$$

Similarly the pressure drop over two thirds of the boiling length is

$$P_{(iii)} - P_{(v)} = \frac{0.0064 \times 1.617 \times 430^2(2 + 18.4)}{21.0} +$$

$$\frac{9359 \times 1.617}{18.4} \ln(1 + 18.4) + \frac{430^2}{954} \, 18.4 = 7869 \text{ Pa}$$

Subtracting these figures from the excess pressure at (iii) gives the excess pressure at (iv) and (v), as recorded in Table 10.2. The excess pressure at location (vi) is the pressure drop in the return pipe, which has already been estimated from equation [10.32] to be 5548 Pa (Table 10.1).

Using equation [10.18] the excess pressure at each location has been divided by 4167 to give $T_s - T_{st}$, the local saturation temperature (T_s) less the temperature in the distillation column ($T_{st} = 102.5 °C$). The tabulated values of $T_b - T_{st}$, the local bulk temperature less the temperature in the column, are equal to $T_s - T_{st}$ in the two-phase zone, at locations (iii) to (vi) inclusive. At (i) no change in liquid temperature has taken place, so $T_b = T_{st}$. The excess bulk temperature at location (ii) has been determined by linear interpolation between the values at (i) and (iii). The local values of ΔT, the temperature difference between heating and vaporising fluids, has been determined by subtracting $T_b - T_{st}$ from the overall temperature difference between the heating fluid and the vapour in the base of the column, viz. $120 - 102.5 = 17.5 °C$.

The preliminary stage in the estimation of the local values of the heat flux is the estimation of the parameter $1/X_{tt}$ from equation [7.13], which gives:

$$\frac{1}{X_{tt}} = \left[\frac{x_g}{1 - x_g}\right]^{0.9} \left[\frac{954}{0.701}\right]^{0.5} \left[\frac{12.42}{268}\right]^{0.1} = 27.13 \left[\frac{x_g}{1 - x_g}\right]^{0.9} \qquad [10.36]$$

Table 10.2 gives the values of $1/X_{tt}$ from equation [10.36], F from equation [7.14], α_{cb} from equations [7.10] and [7.15], S from equations [7.34] and [7.35], α_{nb} from equation [7.37] and $\dot{q}$ from equations [7.46] and [7.47] or [7.39].

Proceeding from the bottom of the tubes to the top:

At (i) assuming no nucleation, $\alpha_v = \alpha_{lo}$, so equation [10.34] gives:

$$U = (1/3665 + 1/2815)^{-1} = 1592 \text{ W/m}^2 \text{ K. Hence:}$$

$$\dot{q}_{(i)} = 1592 \times 17.5 = 27860 \text{ W/m}^2$$

The temperature drop across the liquid is $T_w - T_b$, denoted by θ_b. This is obtained by dividing the heat flux by the liquid heat transfer coefficient, so:

$\theta_b = 27\,860/3665 = 7.60\,°C$

But at this location $T_s - T_b = 7.86$, so here the wall temperature is below saturation temperature and there can be no nucleate boiling.

At (ii) if there is nucleate boiling here, the heat flux must be calculated from equations [7.46] and [7.47]. It has already been estimated in equation [10.34] that U_h, the overall coefficient, excluding α_v, is 2815. Also $T_h - T_b = 15.38$; the amount of subcooling is $T_s - T_b = 6.05 - 2.12 = 3.93$. Hence from equation (7.46):

$$\dot{q} = 2815(15.38 - \theta_b) \tag{10.37}$$

Also from equation [7.47]:

$$\dot{q} = 3665\theta_b + \alpha_{nb}(\theta_b - 3.93) \tag{10.38}$$

For water at $104.5\,°C$, Table 7.6 gives $\phi_{nbi} = 870$. Hence from equation [7.37]:

$\alpha_{nb} = 870 \times 0.654(\theta_b - 3.93) = 569(\theta_b - 3.93)$

Substituting in equation [10.38]

$$\dot{q} = 3665\theta_b + 569(\theta_b - 3.93)^2 \tag{10.39}$$

The solution to equations [10.37] and [10.39] is $\theta_b = 6.22$; $\dot{q} = 25\,785\,W/m^2$.

For the boiling zone, $\dot{q}$ is given by equations [7.46] and [7.39], so:

At (iii) $\dot{q} = 2815(13.25 - \theta_b) = 3665\theta_b + 569\theta_b^2$; whence $\theta_b = 4.21$, $\dot{q} = 25\,460$

At (iv) $\dot{q} = 2815(14.26 - \theta_b) = 5284\theta_b + 870 \times 0.525\theta_b^2$; whence $\theta_b = 4.04$, $\dot{q} = 28\,790$

At (v) $\dot{q} = 2815(15.14 - \theta_b) = 7117\theta_b + 870 \times 0.417\theta_b^2$; whence $\theta_b = 3.77$, $\dot{q} = 32\,020$

At (vi) $\dot{q} = 2815(16.17 - \theta_b) = 8729\theta_b + 870 \times 0.347\theta_b^2$; whence $\theta_b = 3.60$, $\dot{q} = 35\,370$.

For each value of $\dot{q}$ at all but the first location, $T_{wONB} - T_s$ has been calculated from equation [10.35]. The results are included in Table 10.2 and show that in all cases $T_{wONB} < T_w$, thus justifying the assumption that nucleate boiling occurs at all these locations.

The heater can now be divided into five elements, between the six locations. The lengths of each element (Δz) are given in Table 10.2. The heat fluxes never differ by more than 10% from one location to the next, so the mean heat flux may be taken as the arithmetic mean of the initial and final values. The total surface area is $69.1\,m^2$ with tubes 4 m long. The heat transfer rate in each element is then given by:

$$\Delta \dot{Q} = \dot{q}_{mean}\,\Delta z\,\frac{69.1}{4} \qquad\qquad [10.40]$$

Summing the values of $\Delta \dot{Q}$ gives a total of 1988 kW, which is 6% greater than the required heat transfer rate of 1868 kW. This design is therefore suitable, allowing a margin for a small loss in performance if it is found necessary to insert a restriction in the inlet pipe to avoid pulsations. Possible improvements to the design are considered in the next section.

10.6.5 Improved designs

The last two columns in Table 10.1 show the results of fluid flow calculations on two alternative designs. In the first the inlet pipe has been further restricted to increase the loss of head in it from 0.5 to 0.8 m. This improves the ratio of two-phase to single-phase frictional pressure drops. The overall heat transfer coefficient at inlet is reduced by 3% but the log mean temperature difference is increased by 1%. Also the fraction of the tube in boiling flow is increased by 8%. Thus the safety margin of 6% should not be changed significantly by this increase in single-phase resistance, and the danger of unsteady flow would be avoided.

The last column shows the effect of reducing the length of tubes from 4 to 3 m and increasing the number to retain the same surface area. This leads to quite a small reduction in mass flux. The overall heat transfer coefficient is reduced by 1% at inlet and 4% at outlet, the latter being due to the lower vapour generation rate per unit cross-sectional area of tube. However, the log mean temperature difference is increased by 4% and the fraction of the tube in boiling flow is also increased by 4%. So this modified design should also be thermally adequate. Furthermore, the ratio of the two-phase to single-phase pressure drops has been improved compared with the original design.

Further considerations must be taken into account before finalising the design. The design margin could be reduced if there is no danger of the fouling resistances being greater than the assumed value of $10\,000\ \mathrm{W/m^2\,K}$. However, the design methods are based on correlations that are accurate, at best, to only $\pm 30\%$, so there is always a danger of underdesign if margins are reduced.

10.7 Computer programs for the rating and design of vaporisers

If the reader wants to prepare a computer program for rating a vaporiser, all of the equations given in Chapters 6 to 10 inclusive are suitable for inclusion in rating programs. Alternatively programmable calculators can be used to carry out rating calculations.

During the past 20 years there has been a great advance in the complexity of the methods used for designing heat exchangers, due to the advent of the computer. The facility for quick design permits the preparation of many different designs, so that the most economical may be chosen. The chief gain has been in the solution of the following difficult problems:

1. To estimate the split in the flow of fluid on the shellside of a shell-and-tube heat exchanger between cross-flow, bypass and leakage paths, so that the velocities, and hence the coefficients, may be estimated. The effects of leakage on temperature difference at each baffle pass may also be allowed for.

2. To estimate the amount of recirculation in natural-circulation vaporisers.

3. To integrate when making a stepwise estimation of the surface area required from local values of the heat flux.

4. To make stepwise estimates of pressure drop along a vaporiser, so that local values may be estimated of the pressure, and hence of the saturation temperature.

No computer programs are yet available for the most difficult problem, which is to design a multi-component vaporiser where the source of heat is a hot gas containing a condensable vapour.

Proprietary computer programs are available and these are now used for the majority of design calculations. Commonly used proprietary computer programs are those supplied by Heat Transfer Research Inc. in Alhambra, California, Heat Transfer and Fluid Flow Services at Harwell, England, and B'Jac Int. Ltd at Richmond, Virginia. Their methods of performing rating calculations are based on standard correlations, mostly taken from the published literature, which have been modified and improved so that they accord with available data on tests on large heat exchangers.

It is important to remember that computer designs are no more accurate than the equations in the programs.

The design modes in the proprietary computer programs are intended for use by the non-expert. With more complex problems it is preferable to produce approximate designs, using the computer in the rating mode to check each design, and then to try to produce a more efficient, a more economical, or a more convenient design, as appropriate.

CHAPTER 11

Special problems relating to mechanical design

Successful design of vaporisers involves an appreciation of both thermal and mechanical aspects of design. Details of mechanical design procedures are outside the scope of this book. They are given in the appropriate codes and are discussed by Saunders (1987) and in Volume 4 of Schlünder (1983). The purpose of this chapter is to remind the thermal designer of limitations that may be imposed by problems relating to mechanical design. First the various design codes are introduced; this is followed by a consideration of restrictions in the design of shell-and-tube heat exchangers; finally the very important topic of flow-induced vibrations of tubes is considered.

11.1 Design codes

The mechanical design of vaporisers is covered by pressure-vessel codes. A complete list, covering 65 countries, is given by BSI (1980). As well as giving rules for design and fabrication, most codes are specific to a wide range of acceptable materials. A choice must be made taking account of corrosion resistance, mechanical properties, cost, availability and fabrication techniques. Some general advice on the choice of materials for heat exchangers is given by Saunders (1987) in his Chapter 18 and problems specific to vaporisers are discussed in Chapter 12 of this book. The basic principles of stress analysis required in the mechanical design of vaporisers have been described by C. Ruiz in Chapter 4.1 of Schlünder (1983). Due in part to the use of different safety factors, design stresses for similar materials may be different in the various codes, resulting in different required component thicknesses.

11.2   Mechanical design of shell-and-tube heat exchangers

The pressure parts of a shell-and-tube heat exchanger are designed in accordance with a pressure vessel design code, such as the American Society of Mechanical Engineers' ASME VIII or the British Standards Institution's BS 5500. However, these do not deal with the special features of tubular exchangers; the universally accepted code dealing with these features is the Standards of Tubular Exchanger Manufacturers Association – TEMA (1978). Although TEMA is intended specifically to supplement the ASME Boiler and Pressure Vessel Code, Section VIII, Division 1, most of it may be used to supplement other codes. It is strictly applicable to exchangers provided that the shell diameter does not exceed 1.5 m and the pressure does not exceed 210 bar (21×10^6 Pa). The product of these must not exceed 10.5×10^6 N/m, although in practice many parts of TEMA are used for exchangers outside these limits.

The various restraints imposed by TEMA that affect thermal and hydraulic design are discussed below.

11.2.1   Minimum tube pitch

When striving to achieve a high shellside heat transfer coefficient or a compact exchanger, and when excessive shellside pressure drop is not likely to be a problem, the thermal designer wants to achieve a high shellside velocity, and so he wants to minimise the clearance between tubes. TEMA requires that the tubes shall be spaced with a minimum pitch (centre-to-centre distance) of 1.25 times the outside diameter of the tubes; they also prescribe a minimum pitch according to the tube pattern and the duty. With small tubes, say less than 20 mm diameter, care must be taken that the ligament in the tubesheet is not too small to achieve an adequate tube-to-tubesheet joint. When it is required to be able to clean the outside of the tubes mechanically, there must be an adequate cleaning lane between layers, i.e., a free space of 6 to 7 mm.

11.2.2   Maximum baffle pitch

When it is necessary to avoid a high shellside pressure drop, the thermal designer must minimise the shellside velocity. This may be achieved either by increasing the pitch of the tubes or by increasing the pitch of the baffles. The former leads to an increase in the diameter of the shell; the latter leads to an increased space between the supporting baffles. TEMA sets the minimum standard and recommends the maximum unsupported straight tube length that may be used with tubes of different external diameter, for two classes of materials. In many exchangers it is necessary to restrict the

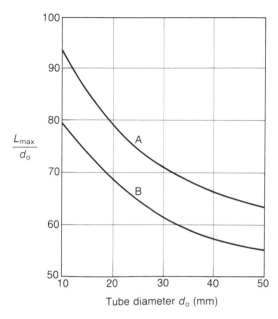

Curve A: Carbon and high-alloy
steel (up to 399°C), low-alloy steel
(up to 454°C), nickel-copper
(up to 316°C), nickel (up to 454°C),
nickel-chromium iron (up to 538°C).

Curve B: Aluminium, aluminium
alloys, copper and copper alloys
at Code Maximum Allowable
Temperature.

11.1 TEMA's maximum unsupported length (L_{max}) divided by external diameter,
for different materials and different diameters

maximum unsupported length to below the TEMA maximum, to avoid
tube vibration (see section 11.3). Figure 11.1 shows the ratio of TEMA's
maximum unsupported length (L_{max}) to the external diameter of the tube
(d_o) for the two classes defined beside the figure. If there are tubes in the
windows (segmented openings) of the baffles, L_{max} is twice the pitch of the
baffles when these are all on the same pitch; if the pitch at one end of the
shell is greater than central pitch, L_{max} is the sum of the end and central
pitches. If it is difficult to keep below the specified maximum unsupported
length, consideration should be given to designing a shell with no tubes in
the window, because with such an arrangement, L_{max} is the maximum pitch
of the baffles. This is a satisfactory arrangement for large shells conveying a
high volume rate of fluid, but it leads to some increase in the diameter
compared with the more usual arrangement in which some tubes are in the
windows of the baffles.

11.2.3 Nozzles

Nozzles are normally of the same diameter as the pipes conveying the fluids
to and from the heat exchanger, and should be large enough to avoid
excessive pressure drop, see sections 9.7.3 and 9.7.4. If the diameter of the
nozzle is more than half that of the shell, this may lead to difficulties with
the code adopted and require special stress analysis.

11.2.4 Impingement plates

It is normal practice to follow the recommendations of TEMA with regard to whether it is necessary to fit an impingement plate opposite the inlet nozzle on the shell, in order to protect the nearby tubes from erosion. It is always necessary with saturated vapours, including steam for heating, with liquid/gas mixtures, and with abrasive or corrosive fluids. For clean liquids, including a liquid at its boiling temperature, an impingement plate is not required if the momentum flux in the nozzle ($\dot{m}^2/\rho$) is less than 2200 N/m^2, where $\dot{m}$ is the mass flux in the nozzle (kg/s m^2) and ρ is the density of the fluid (kg/m^3). The momentum flux between the impingement plate and the shell should be less than 6000 N/m^2.

It is important to remember that the addition of an impingement plate leads to an appreciable increase in the pressure drop across the nozzle, which may be calculated as described in section 9.7.4. This extra pressure drop may be alleviated by perforating the plate, but the perforations must lie between the tubes, so that the fluid flowing through them does not impinge on the first layer of tubes. If the diameter of the nozzle is appreciable in comparison with that of the shell, this may leave a small gap between the edge of the plate and the shell, which would induce a very high velocity through that gap, thus increasing the danger of impingement damage to the outermost tubes. To avoid this, the plate should, under such circumstances, be extended up to the shell, as shown in section 11.3.3.

11.2.5 Tube count

The shell of a shell-and-tube heat exchanger may not be completely filled with tubes for one or more of the following reasons.

A. An unpierced annulus (see Fig. 11.2(a))

This is required when using one of the TEMA types of floating head at the rear. These are used to avoid trouble due to differential thermal expansion, or to facilitate the removal of the bundle for cleaning. They have the disadvantage of increasing the cost and permitting a considerable amount of bypassing of shellside fluid through the space between the bundle and the shell. The latter is not a serious disadvantage if the shellside fluid is a single-component boiling liquid or condensing vapour. An unpierced annulus is also required in the fire-tube waste heat boiler shown in Fig. 3.16, because the internal refractory lining would prevent fluid from entering any tubes near to the shell. With fixed tubesheets (as in Fig. 3.16) the presence of an unpierced annulus reduces the stresses due to differential thermal expansion. A further advantage is that there is some internal recirculation down through the sides of the annular space.

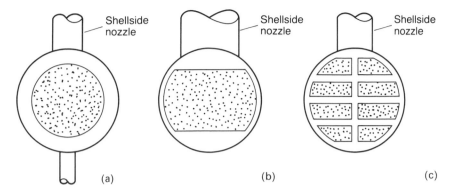

11.2 Three examples of cross-section through a shell to show the parts of the
tubesheet containing tubes (tubes may be fitted in the speckled regions only)
(a) Unpierced annulus
(b) Large spaces top and bottom for clearance under nozzles
(c) Arrangement for eight tubeside passes

B. Nozzle escape area

This must be adequate to avoid excessive nozzle pressure drop, and so it is
necessary to avoid having tubes too close to the nozzles; this means that
there must be unpierced segments near the nozzles, as shown in Fig.
11.2 (b). This is a costly problem with large nozzles; it can be eliminated by
using a vapour belt of the type shown in Fig. 4.10 or 4.11, but these are
also costly.

C. Pass partitions

These are required when there are several tubeside passes; there can be
no tubes in these spaces, illustrated in Fig. 11.2 (c). These pass partition
spaces permit bypassing of the bundle when they lie in the direction of flow,
e.g. the central pass in Fig. 11.2 (c).

D. Tie rods and spacers

These are needed to retain all baffles securely in position, but reduce the
number of spaces for tubes. The diameter and minimum number of tie rods
are specified by TEMA. A decision must be made on the best location of
the tie rods and whether there should be more than the minimum number.

E. Tube-end attachment

This restricts the minimum clearance between the outermost tubes and the
shell.

Typical tube counts are given by Saunders in Table 2 of section 4.2.5 of
Schlünder (1983).

11.2.6 Tubesheet

The thickness of the tubesheet must be calculated according to the rules of
the design code adopted. With one fixed tubesheet and a floating head at
the rear or with U-tubes, the tubesheet must be thick enough to withstand
the pressure of either fluid on one side only. With two fixed tubesheets it
may be possible to design for only the differential pressure, e.g., in a heat
interchanger where it is impossible to introduce either fluid separately. With
two fixed tubesheets, allowance must be made for differential thermal
expansion. Such calculations must be based on the operating conditions
that give the greatest thermal expansion; this requires that the mean
temperature of the tubes should be carefully calculated, as described in
section 6.6. For fixed tubesheets it is usually necessary to limit the
difference between mean tube temperature and mean shell temperature to
about 40 °C. Sophisticated computer programs are available for estimating
tubesheet stresses, and these may be used in difficult cases. They show how
stresses are reduced by having an unpierced annulus.

11.3 Flow-induced vibration of tubes

As the size of shell-and-tube heat exchangers has grown during the last two
decades, there has been an alarming increase in the number of failures due
to vibration of the tubes. Some have failed after only two weeks of
operation; others have operated satisfactorily for twelve years and then
failed. The problem is similar to the collapse of large chimneys due to
wind-induced vibrations, but is complicated by the interactions between the
closely spaced tubes. The vibration of tubes in a bundle may be induced by
turbulent buffeting, in which case the direction and amplitude vary in a
random manner, or the vibration may take on a complicated pattern, some
tubes vibrating at right-angles to the direction of flow, whilst adjacent tubes
vibrate in the direction of flow; the pattern is steady until some minor
change in flow causes a different pattern of motion to be established. The
latter type of vibration is similar to the vibration of chimneys at right-
angles to the wind direction, due to the shedding of vortices. Such
vibrations occur only if the exciting frequency equals or exceeds the
fundamental natural frequency of the chimney or the tubes. Thus there is
seldom trouble in heat exchangers where the velocity is low or where the
tube supports are so close together that their natural frequency is very high,
as quantified below. In nearly all of the failures reported, the unsupported
length exceeded 80% of the TEMA maximum (defined in section 11.2.2).
Flow in a longitudinal direction (inside or outside the tubes) does not
induce vibration except at velocities much higher than those normally used
in vaporisers. Some exchangers with vertical tubes and large unsupported

lengths have not suffered from vibration, either because the shellside flow was axial or because it was in cross flow at a comparatively low momentum flux $(\dot{m}^2/\rho)$. Failure by vibration of tubes in kettle or firetube boilers is uncommon, but a method of dealing with shellside boiling is suggested in section 11.3.5.

In vaporisers, vibrations are most likely to occur when the tubeside fluid is being heated by a gas or condensing vapour on account of the high velocities normally used. Particularly liable to vibration damage are the tubes near to the inlet nozzle of an exchanger in which the shellside fluid is a condensing vapour. Heat exchanger tubes fail either by fatigue, by impact with each other, or by impact with a baffle or tubesheet. If the environment becomes slightly corrosive, corrosion fatigue becomes a danger.

Before checking the design of a heat exchanger to ensure that it will not suffer from tube vibration, it is worthwhile considering all possible modes of operation. For example, in many cooling water systems the coolers operate in parallel between a rising and a falling water main. Incorrect positioning of valves may lead to an unexpectedly high flow of cooling water through one or more coolers, or the operator may improve the performance of one cooler by deliberately increasing the flow of water to it. In vaporisers where the heating fluid is a gas or vapour passing through the shell, operation at a reduced pressure, but at the same mass flow rate may increase the shellside velocities to the extent of inducing tube vibration. The equipment data sheets will seldom include all the possibilities that should be considered.

A more comprehensive treatment of the subject of flow-induced vibration of tubes is given by Saunders (1987) in his Chapter 11.

11.3.1 Prediction of vibration

The first step in predicting whether vibration may cause trouble is to estimate the natural frequency of vibration of the tubes. Normally the tubes are supported by many baffles, and the first and last spaces may be longer than the central space. The tubes should be treated as clamped at the tubesheets and freely supported at the baffles. In practice, if the clearance between tubes and baffles is excessive, the latter assumption may be invalidated, and the natural frequency will be less than that calculated, as discussed at the end of section 11.3.2. The natural frequency may be significantly affected by the mass of the fluids inside and outside the tubes, or by axial loads in the tubes. A thorough treatment has been given by Kissel (1972). For many spans of equal length, McDuff and Felgar (1957) give the following approximate equation for the fundamental natural frequency:

$$f_n = 1.6 \sqrt{\frac{EI}{m_e L^4}} \qquad [11.1]$$

where f_n is the fundamental natural frequency (Hz), $I = (\pi/64)(d_o^4 - d^4)$ is the sectional moment of inertia of the tube (m^4), E is the modulus of elasticity of the material of the tubes (N/m^2), L is the length of each span (m), and m_e is the effective mass per unit length (kg/m); the last is the sum of the mass per unit length of the tube itself, of the fluid inside the tube, and of the amount of shellside fluid that would fill the space occupied by the tube (calculated as illustrated in the example in section 11.3.4). Equation [11.1] is applicable to bundles with more than 10 equal spans; if there are only six, four or two spans, the frequency is greater by 6%, 14%, or 53% respectively.

There are higher modes of vibration, but with more than four spans these include frequencies that are only slightly greater than that of the lowest mode. Thus it is unwise to operate with an exciting frequency between two modes, because no plant operates at the design rate all the time, and the range of permissible rates is small and difficult to estimate accurately.

Chen (1968) measured the frequency of pressure pulsations occurring in a fluid flowing through several arrangements of tubes, with inline and staggered arrangements, over a wide range of pitch-to-diameter ratio. The vortex frequency is given by:

$$f_v = \text{Sr}\,\frac{\bar{u}}{d_o} \qquad\qquad [11.2]$$

where f_v is the frequency of the pulsations (Hz), $\bar{u}$ is the cross-flow velocity at the minimum section (m/s), d_o is the external diameter of the tubes (m) and Sr is the Strouhal number. When calculating $\bar{u}$, allowance should be made for the fact that some of the shellside fluid bypasses the main cross-flow stream, as explained in section 6.2.6. For flow past a single cylinder the Strouhal number is approximately 0.2. Chen found the Strouhal number to vary considerably with the lateral and longitudinal pitches. With normal tube layouts, his critical Strouhal numbers ranged from 0.17 to 0.8 over the normally used range of pitch-to-diameter ratios of 1.25 to 1.5.

In an excellent summary of the present state of knowledge on flow-induced vibrations, J. M. Chenoweth recommends the values in Table 11.1

11.1 Strouhal number for various tube patterns, from Fig. 2 of section 4.6.4 of Schlünder (1983)

Pitch-to-diameter ratio	Tube layout angle			
	30°	45°	60°	90°
$P/d_o = 1.25$	0.18	0.58	0.80	0.42
1.33	0.22	0.59	0.75	0.35
1.50	0.29	0.57	0.61	0.29

for the critical Strouhal number, taken from Fig. 2 of section 4.6.4 of Schlünder (1983), for square and triangular layouts.

Tube vibration can be expected to occur if the value of f_v, calculated from equation [11.2] and using the value of Sr obtained from Table 11.1, exceeds the value of f_n, calculated from equation [11.1].

Connors (1970) studied the 'fluidelastic whirling' in an ideal tube bank. He showed that the momentum flux required to produce whirling is proportional to the damping factor for the tubes. Unfortunately it is difficult at present to use this more rational approach due to the lack of data on damping, particularly in the complex situation of an industrial tube-bundle, where interaction at the baffles may considerably affect the amount of damping. Current research on damping of tubes in bundles may lead to it being possible to use the method of Connors to determine whether heat exchanger tubes will vibrate in service. For further information, see Chapter 11 of Saunders (1987).

11.3.2 Prediction of damage

If the energy input is not high, the tubes of a heat exchanger may vibrate continuously but never rupture. Thorngren (1970) devised, from theoretical considerations, two non-dimensional numbers for the prediction of the velocity that could cause damage to the tubes, one for damage by the baffles and one for damage by collision of tubes at the mid-span with adjacent tubes. From observations of exchangers with water on the shellside he found that damage occurred if the value of the appropriate number exceeded unity. These criteria show that baffle damage is more likely than collision damage, except when the span is in the region of (or exceeds) the TEMA maximum unsupported length.

Erskine and Waddington (1973) studied 20 large heat exchangers, of which 14 had suffered baffle damage and six had experienced no damage. They found that one exchanger had been damaged by water at a damage number of 0.6 and one by steam under vacuum at a damage number of only 0.03. They proposed that the critical damage number should be taken as 0.56 for water and reduced for other fluids according to their density and viscosity according to the empirical formula:

$$\frac{\bar{u}_d}{\bar{u}_{dw}} = (\rho/1000)^{-0.265}(1000\eta)^{-0.118} \qquad [11.3]$$

where $\bar{u}_{dw}$ is the critical velocity for vibration damage by water, calculated as described below, and $\bar{u}_d$ is the critical velocity (m/s) for vibration damage by the actual shellside fluid, whose density is ρ (kg/m^3) and whose viscosity is η (N s/m^2).

No allowance for bypassing was made when analysing the results, so no such allowance should be made when using the above method. Figure 11.3 gives the critical velocity for vibration damage by water flowing past

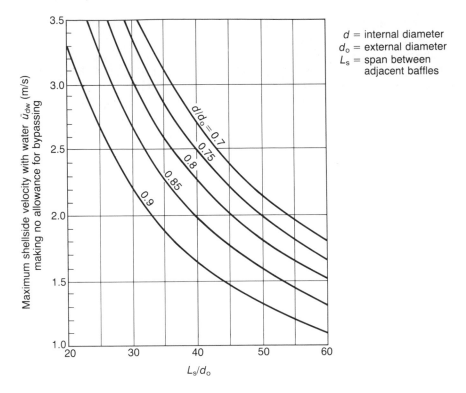

11.3   Maximum water velocity with carbon steel tubes to avoid baffle damage

carbon steel tubes supported by baffles whose thickness is half the external diameter of the tubes, assuming that the critical Thorngren damage number is 0.56 for baffle damage. No collision damage will occur if the unsupported length is less than 60 diameters. The velocity should be multiplied by 0.75 if the tubes are of a copper alloy or by 0.4 if they are of aluminium. The critical velocity is proportional to the square root of the baffle thickness.

Adopting the above procedure reduces the chances of tube rupture resulting from flow-induced vibrations, but it is no guarantee; occasional failures have occurred with lower baffle damage numbers, when the method of Chen has indicated that there would be vibrations. This may be the result of corrosion fatigue. If there is any danger of slight corrosion, the Chen method should be used to prevent vibration.

In the calculations described above it is assumed that there is good contact pressure between tubes and baffles. If this is not so, the effective unsupported length will be greater than was intended, so tube vibration and the danger of damage will be initiated at velocities considerably lower than those estimated. A technique sometimes used in large vacuum condensers to ensure good contact is to put a slight camber on the tubes, which also helps to drain liquid from the tubes at shutdown.

11.3.3   Methods of reducing the maximum unsupported length

Tube vibrations are likely in shell-and-tube heat exchangers where there is
a high volume rate of flow on the shellside; in such cases there is probably
a large shellside nozzle. The impingement plate should extend as far as the
shell, as shown in Fig. 11.4, to protect the outer tubes in the top layer of
tubes from a stream at a very high velocity. A central half-support plate, as
shown in Fig. 11.5 helps to steady the tubes under the inlet nozzle; it costs
little and introduces no extra pressure drop.

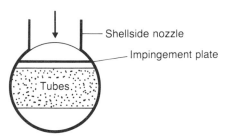

11.4   Full-width impingement plate

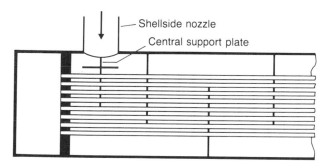

11.5   Additional support under inlet nozzle with tubes in the windows

Various systems of rod baffles have been patented; these support the
tubes and permit longitudinal flow. They are described in detail in Saunders
(1987).

Often the best solution to the problem of avoiding tube vibrations is to
use an exchanger with no tubes in the windows. It is then easy to fit extra
support plates between baffles if further reduction in span is needed. A
central support plate (in the centre of the inlet nozzle) helps to steady tubes
in the region of high turbulence where the shellside fluid first contacts the
tubes. Figure 11.6 shows a shell fitted with a central support plate, with no
tubes in the windows, and with one support plate per baffle space. The
support plates have windows at both ends, so that they support all the
tubes without affecting flow.

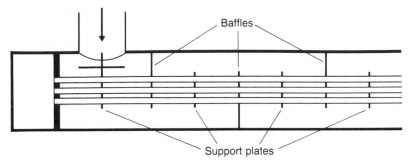

11.6   Additional support plates with no tubes in the windows

11.3.4   Example on tube vibrations

Steam is condensing at $120\,°C$ on the outside of steel tubes of 20 mm external diameter, 2 mm thick on 25 mm triangular pitch at $30°$ tube layout angle. They are supported by baffles 12 mm thick with a span of 1000 mm. What velocity is likely (a) to induce vibration of the tubes, (b) to cause damage? The density and viscosity of saturated steam at $120\,°C$ are $1.119\,\text{kg/m}^3$ and $12.9\,\mu\text{N s/m}^2$ respectively. Steel has a modulus of elasticity of $2 \times 10^{11}\,\text{N/m}^2$ and a density of $7690\,\text{kg/m}^3$. The density of the tubeside fluid is $950\,\text{kg/m}^3$.

1. To determine f_n, the natural frequency of the tubes, it is necessary to calculate I, the sectional moment of inertia of the tube, and m_e, the effective mass per unit length of tube. In this example, the internal diameter, $d = 20 - 2 \times 2 = 16$ mm. Hence:

$$I = \frac{\pi}{64}(d_o^4 - d^4) = \frac{\pi}{64}(0.020^4 - 0.016^4) = 4.637 \times 10^{-9}\,\text{m}^4$$

The effective mass is the sum of three items: that due to the tube, that due to the steam and that due to the tubeside liquid.

$$m_e \text{ (of the tube)} = 7690 \times \frac{\pi}{4}(0.020^2 - 0.016^2) = 0.8697\,\text{kg/m}$$

$$m_e \text{ (of the steam)} = 1.119 \times \frac{\pi}{4} \times 0.020^2 = 0.0004\,\text{kg/m}$$

$$m_e \text{ (of the tubeside) liquid} = 950 \times \frac{\pi}{4} \times 0.016^2 = 0.1910\,\text{kg/m}$$

Adding together the three components of m_e gives:

$$m_e = 0.8697 + 0.0004 + 0.1910 = 1.0611\,\text{kg/m}$$

From equation [11.1], the natural frequency of the tubes is:

$$f_n = 1.6 \sqrt{\frac{2 \times 10^{11} \times 4.637 \times 10^{-9}}{1.0611 \times 1.0^4}} = 47.3 \text{ Hz}$$

2. The Chen method will be used to determine the velocity required to initiate vibrations of the tubes. For a tube layout angle of 30° and a pitch-to-diameter ratio (P/d_o) of $25/20 = 1.25$, Table 11.1 gives the Strouhal number to be 0.18. From equation [11.2] the frequency of the pressure pulsations with flow through this arrangement is:

$$f_v = 0.18 \frac{\bar{u}}{0.020} = 9.0\bar{u}$$

Vibrations occur when $f_v > f_n$, i.e. when $\bar{u} > 47.3/9.0 = 5.25$ m/s.
Thus the velocity required to induce vibration of the tubes is 5.25 m/s.

3. To determine the velocity likely to cause baffle damage, Fig. 11.3 gives $\bar{u}_{dw}$, the critical velocity for damage by water. In this example $L_s/d_o = 1000/20 = 50$, and $d/d_o = 16/20 = 0.80$. Reading from the graph gives $\bar{u}_{dw} = 1.82$ m/s. This must be corrected for fluid properties and baffle thickness. The former is obtained from equation [11.3] as

$$\frac{\bar{u}_d}{\bar{u}_{dw}} = (1.119/1000)^{-0.265}(1000 \times 12.9 \times 10^{-6})^{-0.118} = 10.12$$

Figure 11.3 applies to baffles whose thickness is half the external diameter of the tube. In this example the baffle thickness for which the figure is applicable is $\frac{1}{2} \times 20 = 10$ mm. The actual thickness is given as 12 mm, and the critical velocity is proportional to the square root of the thickness, so the true critical velocity is:

$$\bar{u}_d = 1.82 \times 10.12 \times \sqrt{12/10} = 20.2 \text{ m/s}$$

This is the velocity likely to cause baffle damage, calculated on the assumption that there is no bypassing of the shellside fluid. The true velocity will probably be less than this, but considerably more than the 5.25 m/s required to induce vibration of the tubes.

11.3.5 Note on tube vibration induced by two-phase flow

All the work on tube vibration reported above is for single-phase flow. To apply this to condensers and boilers, it is suggested that the density and velocity be calculated on the assumption of homogeneous flow (equation [7.22]) and the viscosity to be substituted in equation [11.3] be that of the liquid.

CHAPTER 12

Materials of construction for vaporisers

by O. J. Dunmore [1]

12.1   Introduction

There are two fundamental requirements for any material of construction in
an engineering application such as a vaporiser. Firstly, it must be capable
of resisting the stresses arising from internal pressure and structural forces.
Secondly, it must be able to withstand chemical corrosion by the working
fluid being vaporised and also by the heating medium.

The most common working fluid is water, as steam is widely used for
power generation and waste heat boilers are the most common means of
heat recovery in the process industries. For these boilers the choice of
material would appear at first sight to be simple, as carbon steel has
excellent mechanical properties in the operating temperature range involved
and is also very corrosion resistant to oxygen-free water. At the same time
it is by far the cheapest of all the usable engineering materials.
Nevertheless, for some parts of high pressure boilers it may be necessary to
use a more highly alloyed material than carbon steel. There are two steels
of differing alloy content which are likely to be encountered and the
reasons for the choice of these steels will be discussed. These are the 1%
chromium–molybdenum steels and the 9% chromium–molybdenum steels,
and it is salutary to remember that they are approximately double and
treble the cost of carbon steel respectively. On nuclear boilers high nickel
alloys may be used but these are outside the scope of this book as also are
austenitic steels which are used for their superior high temperature creep
strength in the superheaters of conventional boilers.

Other evaporators which are in use in the various process industries may
be made from stainless steel or non-ferrous alloys depending on the

[1] Owen J. Dunmore, PhD, ARSM, CEng, FIM, formerly of ICI Materials
Group, and now a consultant with MARIT Limited, Mill Wynd, Yarm, Cleveland,
TS15 9AF.

corrosivity of the working fluid or the heating medium. Examples of these include ammonia vaporisers in metallurgical heat treatment plant, falling-film evaporators on fertiliser plants, distillation column reboilers on methanol plants and crude oil vaporisers in a refinery. The range is very wide but a few guiding principles and pitfalls in the choice of materials will be highlighted.

Stainless steels are the most widely used materials after carbon and low alloy steels and themselves comprise a wide range of different alloys. The advantages and disadvantages of each of the basic types of stainless steel are covered in section 12.3.

Beyond stainless steel when a higher level of corrosion resistance is required the choice will probably lie between titanium, copper alloys and nickel-base alloys. Of these three groups, titanium is selected for more detailed coverage (section 12.4) because it is rapidly expanding into the areas where copper or nickel alloys have been used in the past. Copper alloys are much less often selected now than they once were because of the relatively high price of the metal. There are so many different nickel alloys that they cannot be adequately dealt with in a short general review of materials of construction. However, some of the nickel alloys such as the 800 alloy (30/20 Ni/Cr) are often regarded as superior stainless steels and are therefore included in section 12.3.

12.2   Carbon and low-alloy steel steam generators

Pure water in the absence of oxygen is not corrosive to carbon steel, because the magnetite film which is formed remains protective from ambient temperature to the saturation temperature. This is true even in the highest pressure boilers. Nevertheless, if breakdown of this passive film occurs the corrosion rate can be quite rapid. Normally the concentration of impurities which can be carried in with the boiler feed water or added to the circulating water as part of the conditioning treatment is insufficient to move the pH into the range where breakdown of the passive film will occur. This is illustrated by Fig. 12.1 from Mann (1978) where it can be seen that quite a large shift of pH is needed before corrosion rates become significant. For example, between pH 5 to pH 10 the rate of corrosion is so low that the loss of wall thickness would not cause problems within the lifetime of the boiler. Because the boiler water is almost always within the above pH range, even when deviations from the specified water composition occurs, it is often difficult for engineers to understand why boiler corrosion should be so common.

The reason is that in certain circumstances the local concentration of impurities in the water adjacent to the steel surface is not the same as in the bulk water. Concentration factors as high as 10^4 can occur which

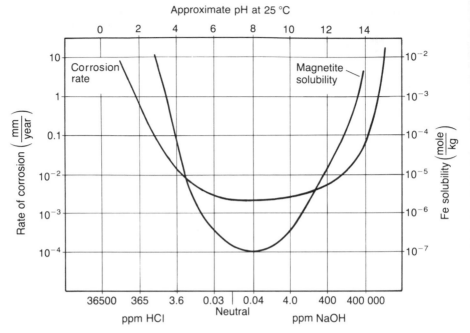

12.1 Rate of corrosion of mild steel at 300 °C in deoxygenated water (from Mann (1978))

means that 10 ppm in the bulk liquid becomes 10%. When this happens, corrosion is not unexpected.

There are a number of mechanisms which can give high concentration factors and in boilers the four which occur most frequently are the 'under deposit' effect, excessively high local heat fluxes, crevice effects and vapour/liquid interface effects. They are discussed separately in sections 12.2.1 to 12.2.4.

These types of corrosion are referred to as 'on load corrosion' to distinguish them from corrosion which can occur by ingress of oxygen when the boiler is shut down and they can be avoided by the correct design of the boiler. If the design is not correct and corrosion occurs, the factor which determines the actual rate of corrosion and the eventual life to failure of the boiler is the water quality. When the anions and cations in the water are well balanced, a high concentration factor can lead to a locally concentrated solution which is non-corrosive. This is because the pH will be near neutral. However, it is impossible to maintain a carefully balanced water quality over a period of years without occasional deviations from specification. The fact that a deviation is corrected within a day or two or even a few hours may not be relevant. The corrosion rate may be extremely rapid and once a corrosion pit has been initiated the damage is usually done. It will often remain an active site even though the bulk water reverts to its non-corrosive composition and will eventually penetrate the wall of the tube. If the water composition has been poorly controlled and in spite of this the time to failure has still been prolonged it may be possible to slow down rates of corrosion by an improvement in the water, especially if chemical cleaning is done. However, the only way to prevent failure completely is to avoid the problem at the design stage.

The higher the pressure of the boiler, the more important it is to achieve a correct design because the temperature of the metal in contact with the water rises with the saturation temperature. Corrosion reaction occurs much faster as the temperature is raised. Therefore, although a specific design may be acceptable on a low-pressure boiler, on a high-pressure boiler the same design could lead to rapid failure. This applies largely to carbon steel and 1% Cr–Mo steel as they both react similarly to aqueous corrosion.

For waste heat boiler tubes, 1% Cr–Mo is often selected where resistance to hydrogen attack by the process gas is needed. Occasionally, this alloy steel is used in high-pressure utility boilers in regions of very high heat flux, where its greater resistance to the hydrogen generated by corrosion and its superior high-temperature creep strength can be beneficial.

Another reason for the choice of low alloy steel is to improve the resistance of the very thin protective magnetite film at temperatures below about 180 °C where the rate of formation is very slow. In regions of high flow rates, especially where there is local turbulence, the passive film is eroded away faster than it can reform. This can lead to high metal wastage. Water chemistry and suspended solids in the water play a very important part, hence it is not possible to define a safe limiting velocity. Useful information is given by Bignold, Garbett and Woolsey (1983).

12.2.1 Under-deposit corrosion

This occurs in recirculating boilers at heat transfer surfaces. In areas where the water velocity is low or where the heat flux is fairly high, insoluble suspended solids in the boiler water can be deposited. As these porous deposits build up, the soluble salts in the water become concentrated at the bottom of the deposits. This is illustrated by Fig. 12.2, and is quite often referred to as 'wick boiling'. By maintaining a feed water quality with low levels of suspended solids and using sufficiently high blowdown rates to prevent the build-up of solids in the circulating water, solids deposition can be controlled. However, there are certain areas which are particularly susceptible to the build-up of solids such as downstream of excessively protruding weld roots, nose tubes where circulation rates are low and just behind the tube sheets of shell-and-tube boilers. Good design and construction techniques can eliminate most of these areas. Even if corrosion

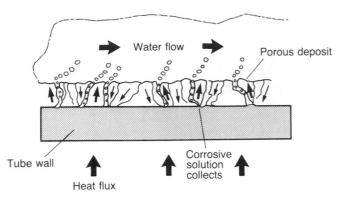

12.2 Under-deposit corrosion of boiler tube

in service does occur it is sometimes possible to improve things by a simple design modification such as the addition of a blowdown nozzle or the repositioning of a weld.

Nevertheless, there are occasions when design modifications cannot be made and then it is necessary to modify operational procedures if further failures are to be avoided. The most effective approach is to remove the deposits plus any *in situ* corrosion products by chemical cleaning. If the chemical cleaning is repeated periodically the deposits can be prevented from building up to the critical thickness at which the concentration mechanism is sufficient to trigger corrosion. A typical frequency here could be every two years.

When boiler failures occur due to corrosion it is important to diagnose the cause correctly so that the appropriate action can be taken. For instance any deposits which are present may have been directly formed *in situ* as a result of the corrosion or may have been deposited from elsewhere in the boiler and thus have caused the corrosion. For example, oxygen ingress into the economiser section of a boiler can lead to corrosion and the corrosion products can be carried forward to the evaporator section and then be deposited, leading to further corrosion. The correct action is therefore the prevention of oxygen ingress as well as chemical cleaning of the economiser and evaporator sections.

It is not possible to give more than a very general guideline for the thickness of deposit which can be unsafe as it is so dependent on the heat flux, and the design pressure of the boiler. Low-pressure boilers may not suffer corrosion with deposits up to 1 mm thick, while in some high-pressure boilers operating at about 150 bar (15 MPa) it is common practice to acid clean when the magnetite layer reaches 70 microns thickness. During the design of a boiler it is therefore very important to give careful consideration to factors such as the quality of the feed water, the way the boiler will be operated, and access for inspection of heat transfer surfaces so that deposits and corrosion can be detected at an early stage. Also important are a uniform circulation rate with no dead areas, and adequate blowdown rate, the availability of dosing chemicals and ease of chemical cleaning.

12.2.2   High heat flux corrosion

This can occur on clean tube surfaces and is often referred to as 'dryout corrosion' or 'steam blanketing' corrosion. If porous deposits are present the under-deposit wick effect will aggravate the dryout corrosion. However, even without the presence of deposits the heat flux alone may be sufficient to cause concentration of soluble salts to a level which initiates corrosion. With nucleate boiling, the tube surface is continually wetted and no concentration of soluble salts can occur. However, at a sufficiently high heat flux, intermittent dryout occurs and eventually a steam blanket forms, at the critical heat flux, which is defined in section 2.1 and which may be predicted as described in Chapter 8.

In a horizontal water-tube vaporiser, the heat flux at which corrosion may begin is much lower than in a vertical tube, due to the effect of gravity. In annular flow in a horizontal tube, the film of water may separate from the crown of the tube, as illustrated in Fig. 12.3; this may lead to the type of corrosion known as 'crown' corrosion, shown in Fig. 12.4. With a lower velocity, with stratified flow (see section 9.2.1), the corrosion may form distinct lines along the tube, known as 'tramline' corrosion and illustrated in Fig. 12.5. Horizontal and gently sloping tubes in high-temperature water-tube boilers are particularly vulnerable to these forms of corrosion.

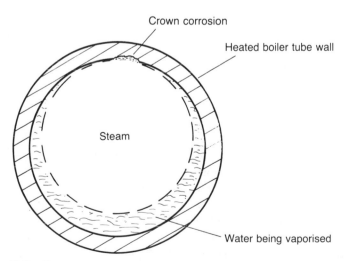

12.3   Crown corrosion in horizontal water-tube boiler (diagrammatic)

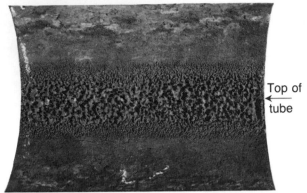

12.4 Photograph of crown corrosion in horizontal water-tube boiler

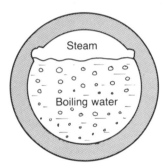

12.5 Tramline corrosion at water line

In fire-tube boilers, where turbulence of the heating medium at the end of the inlet ferrule (see Fig. 3.16) leads to an enhanced heat flux, there is also a danger of water-side corrosion. Near to the end of the ferrule the heat flux may be more than twice the mean heat flux in the tube and the heat flux to initiate corrosion can easily be exceeded. Where this type of boiler is used on a chemical processing plant, there may be little return condensate and fresh make-up water can therefore constitute a high fraction of the boiler feed water. Under these circumstances, it is particularly difficult to maintain a tightly consistent quality and corrosive conditions can occur very readily.

There are design methods for checking whether any part of the evaporation stage of a boiler is likely to be in the partial dryout stage. If partial dryout appears likely under any mode of operation of the boiler and cannot be avoided by increase in circulation rate or a reduction of heat flux there are several ways by which corrosion can be avoided. These are listed below. Needless to say, if severe steam blanketing is expected and the gas temperature is of the order of 1000 °C or more, there would be overheating of the boiler tube and rapid mechanical failure by a creep rupture process.

1. Use a more highly alloyed tube material which has better corrosion resistance. For example, 9% Cr–Mo steel has been used when the partial dryout has led to alkaline corrosion. High nickel alloys such as Incoloy 800 have also been used but these are very expensive and are only used as a last resort.

2. Twisted tapes can be inserted into the tubes of horizontal water tube boilers. These have the effect of ensuring that the crown of the tube remains fully wetted at all times and prevent the dryout which can occur at this position. These tapes are made from stainless steel strip.

3. Use rifled tubing. The rifling is in the bore and has the effect of swirling the water phase to the top of the tube in the same way as the twisted tape.

Before adopting any of these solutions, it is necessary to give very careful consideration to the water quality in the boiler.

For example, with a water treatment which can allow acidic excursions from neutral there will be no benefit in using 9% chromium–molybdenum steel tubing if dryout occurs. This is because the acid resistance of 9% Cr–Mo is little better than carbon steel even though there is a considerable improvement in alkaline corrosion resistance. Similarly, if there is a high level of suspended solids in the water these can collect in the grooves of rifled tubing and permit under-deposit corrosion.

If dryout is suspected in an existing boiler, the consequent concentration can often be detected with the boiler on line by the injection of a radioisotope such as Na^{24} into the drum water.

12.2.3 Crevice corrosion

This form of corrosion is most frequently found with materials such as stainless steel and titanium which form particularly stable oxides and are very reliant on a passive oxide film for their high corrosion resistance. However, carbon and low alloy steel can also suffer from crevice corrosion in boilers where a heat flux is present in combination with a crevice. For example, in water-tube boilers it is essential to avoid backing rings at the weld junctions in steam generating tubes. A less easily avoided crevice is between the tube and tubesheet in fire-tube boilers. It is technically possible to avoid a crevice by welding the tube to the back of the tubesheet (see Saunders (1987)). However, this is considerably more expensive than the conventional tube-tubesheet weld where the tube passes through the tubesheet, and it is only employed on very high pressure boilers. Again, this is because the higher saturation temperature involved greatly increases the danger of corrosion when boiler water becomes concentrated down the crevice. Below about 100 bar design pressure it is not necessary to consider back welding of the tube to the tubesheet merely in order to avoid crevice corrosion.

In a fire-tube boiler where the tube is welded to the front of the tubesheet, it is important to minimise the effect of the crevice whatever the design pressure of the boiler. For this reason after the tubes have been welded in, the soundness of the welds proved by a low-pressure air or helium leak test and after any post weld heat treatment, the tubes should be roller expanded into the tubesheet (see Saunders (1987)). This should ensure that the tube is a tight fit into the tubesheet, and the tube should be rolled from the back of the tubesheet to within 0.5 in of the root of the weld at the front. On very thick tubesheets, it is sufficient to expand about 2 in back from the edge of the tubesheet. No expansion grooves are required in the tubesheet and it is not necessary to aim for the same degree of tightness as when the expansion is the main method of fixing the tube to the tubesheet and no welding is being carried out. It is most important to carry out the expansion after the welding operation and not before, because this can lead to unsound welds due to gas trapped by the expansion operation leading to weld porosity. There will always be a small crevice at the back of the tubesheet due to the spring back effect. Therefore this method may not be fully effective on high-pressure boilers. However, experience has shown that on most boilers the expansion of the tube into the tubesheet after welding has been sufficient to avoid crevice problems.

When a large crevice is left, the danger to carbon and low-alloy steel boilers is caustic stress corrosion cracking down the crevice. As there will be restricted flow down the crevice the heat flux will cause the water to boil and concentrate salts. With an acidic water the corrosion product will fill the crevice and stifle further corrosion. Significant deterioration is therefore unlikely. However, with a slightly alkaline water condition there can be free caustic generated which will lead to stress corrosion cracking of the tube within the tubesheet. This will almost certainly lead to rapid failure of the boiler.

12.2.4 Vapour/liquid interface effects

When steam/water separation occurs, corrosion often takes place at the vapour/liquid interface. For example, it can occur in kettle type reboilers where incorrect level control can expose the upper rows of tubes to attack by droplet or spray impingement. It may occur in the steam pockets which can form below the upper tubesheets of vertical fire-tube boilers. In addition to corrosion, cracking of the tubesheet by thermal fatigue can occur as a result of the thermal cycling caused by the alternate wetting and drying effect.

The metal temperature approaches the gas temperature in these cases and because of splashing there is a high concentration factor. Failure may be mechanical (by creep or thermal fatigue) as a result of the overheating or corrosion may occur due to the concentration factor. Often it is a combination of the two.

12.3 Stainless steel evaporators

Apart from the carbon and low-alloy steels which are widely used for steam generators, the next most frequently used materials are the stainless steels. If the fluid being evaporated causes corrosion of carbon steel it will be necessary to choose a material which will resist this corrosion. The exact choice of material can be quite difficult in practice as the rate of corrosion will be dependent on both the heat flux and the metal temperature, as well as on the chemical composition of the corrodent. Generally stainless steel is chosen because it is the most economic choice after carbon and low-alloy steel. It may also be necessary to use stainless steel because the heating medium is corrosive, for example in a sodium carbonate reboiler on a CO_2 removal plant.

The intention here is to give some guidance in the choice of the correct grade of stainless steel and to illustrate the unexpected problems which can be met even when standard corrosion sources have been consulted. Correct engineering design can mitigate many of these problems, hence it is important for both users and designers of evaporators to have an understanding of the corrosion mechanism underlying these problems.

Steel is regarded as stainless when it contains more than about 12% of chromium, which is the minimum amount to form a passive film of chromium oxide on the surface of the steel. This leads to an order of magnitude improvement in corrosion resistance in many environments. The degree of corrosion resistance increases progressively with the further increases of chromium levels beyond 12%. Other alloying elements such as molybdenum and copper may also be added to stabilise the passive film and give greater resistance to corrosion in environments such as reducing acids.

A very wide ranging corrosion resistance is reached at 18% chromium, hence this is the alloy level most often used. However, the metallurgical problems associated with the manufacture and fabrication of the steel increase with increasing chromium content. The addition of 8–10% nickel to an 18% chromium steel allows it to be more readily processed and fabricated because the structure becomes austenitic at ambient temperatures. Therefore the workhorse of the stainless steels is the 18/8 type, which is generally referred to as 304 after its American code designation. Although widely and successfully used there are a number of problems with the steel when it is used for heat transfer equipment. Where these problems are recognised, other more recently developed grades of stainless steel of equivalent general corrosion resistance are available to solve them.

The biggest problem with type 304 stainless steel is chloride stress corrosion cracking and the results of a survey recently carried out in Japan illustrate the scale of the problem (Haruyama 1982). Out of a total of 700 stainless steel heat exchangers in service nearly 100 had failed due to chloride stress corrosion. As a result of extensive research it is now possible to predict fairly accurately the conditions which will lead to stress corrosion. A large number of factors govern the initiation and rate of propagation of cracking by stress corrosion. The most important are level of chloride, presence of oxygen, temperature and stress levels.

Below a metal temperature of about 70 °C chloride stress corrosion cracking is not a problem. At higher temperatures, when the fluid chemistry is being considered, it is much more difficult to define a minimum chloride level at which stress corrosion cracking will not occur. This is because the critical chemistry is that of the liquor actually in contact with the steel surface and not the bulk liquid composition. The design rules which have to be followed therefore are the same as for avoiding on-load corrosion in carbon steel steam boilers, i.e. avoid crevices and porous surface deposits which can result in very high concentration factors. Chloride stress corrosion is known to have occurred in equipment handling water with less than 1 ppm chloride. This is because a concentration factor of about 10^4 will give percentage levels of chloride at the metal surface which will be more than enough to crack the steel. Therefore, when water is being evaporated in a shell-and-tube heat exchanger of 304 type stainless steel, the water should be on the tubeside rather than the shellside because this will reduce the danger of concentrating chloride within the tube to tubesheet crevice. If the orientation of the heat exchanger is such that the tubes are horizontal it is particularly important to ensure that the feeding and header arrangements are such that all tubes see an equal flow because if the top tubes are starved of flow any resultant dryout will cause chloride concentration and lead to cracking.

It may seem that post-weld heat treatment of the equipment will reduce the risk of stress corrosion because it lowers the residual weld stresses in the steel to below the threshold stress required for cracking. Experience has shown that this is not correct because the very low yield stress of austenitic steel permits residual stresses to be reintroduced quite readily and although heat treatment is successful with other materials in stress corrosion situations it is of little benefit with 304 stainless steel.

The only real remedy in situations where stress corrosion cracking cannot be completely avoided by design approaches such as putting the aqueous chloride bearing fluid on the tube side or working at temperatures below the threshold for stress corrosion, is to use a more resistant grade of stainless steel.

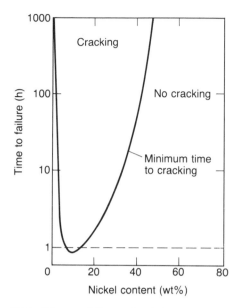

12.6 Time to failure of stainless steels due to chloride stress corrosion cracking

Figure 12.6 shows the effect of nickel content on the susceptibility of 18 % chromium stainless steel to chloride stress corrosion cracking. From this figure it is quite clear that 18/8 (304) is the worst composition and increasing or reducing the nickel content is beneficial. In many cases the most economic choice is a nickel-free stainless steel such as 18Cr/2Mo. This is a ferritic grade of stainless steel and because of the fully ferritic structure it can have poor mechanical properties adjacent to welds. The modern extra-low carbon ferritic grades are a great improvement over the earlier grades as the reduced carbon content minimises the loss of toughness at welds. Hence in the thin sections which are found in heat exchanger tubes there is generally adequate toughness at ambient temperature. This steel is completely resistant to stress corrosion cracking at all chloride levels but it must be noted that its resistance to corrosion pitting by chlorides is not better that the austenitic 304.

Where it is important to have optimum mechanical properties as well as stress corrosion resistance the high nickel grades of stainless steel are often used. Even with 30 % nickel present, the steel is not fully resistant to chloride stress corrosion cracking. However, in practice, where stress corrosion failure of a 304 heat exchanger has occurred in a few weeks a 30/20 Ni–Cr alloy (800) will not fail for many years. This type of alloy is appreciably more expensive than a 304 or an 18/2 ferritic steel but there will be occasions when the extra expense is justified. For example, 18/2 cannot be used at metal temperatures above 300 °C because it will embrittle. This reduction in ductility is due to a thermally induced precipitation effect. It is additive to the loss of notch toughness in the weld region which has already been mentioned.

One advantage of 18/2 to offset the above is the fact that it has the same thermal expansion coefficient as carbon steel and with equipment designs which require a carbon steel shell and stainless steel tubes it will often permit a cheaper box-type design instead of requiring a bellows mounted 'floating head' to compensate for the thermal expansion effects.

As a compromise a further type of stainless steel can be considered. This is a steel containing 5% nickel and it is alloyed to achieve a 50/50 austenitic/ferritic structure. For this reason, it is generally referred to as a duplex stainless steel. The duplex structure considerably improves the mechanical properties in the weld region and the thermal expansion is quite close to that of carbon steel. However, it has the same embrittlement problem as the fully ferritic steel above 300 °C and a further disadvantage in that it is not completely immune to chloride stress corrosion cracking. Nevertheless it is as good in most practical situations as the 30/20 austenitic grades and is generally regarded as a chloride stress corrosion cracking resistant grade. Also it is the strongest grade of stainless steel, with more than double the yield strength of 18/8, and for high pressure applications the saving on wall thickness can make it the most economic choice.

Another potential corrosion problem with all grades of stainless steel is local intergranular corrosion at welds. This is due to precipitation of carbides by the thermal cycle during welding, often referred to as sensitisation. It is cured by adding a stabilising element such as titanium to the steel or by reducing the carbon content.

Because of the thin sections used in many vaporisers, sensitisation is a problem which does not have to be considered very often. However, when the thickness at a weld exceeds about 10 mm, it is necessary to use a stabilised or extra-low carbon (0.03 %C) grade of stainless steel. Thick sections of stainless steel are more prone to sensitisation than thin sections because they lose heat from the welding process less rapidly and therefore remain longer in the temperature range in which sensitisation occurs. Any other operation in addition to welding which can raise the temperature of the stainless steel above about 450 °C for more than about 15 min can also cause sensitisation. Quite mild aqueous corrodents can lead to intergranular corrosion of the sensitised steel and when it has occurred there is little visible change until applied stresses lead to cracking and sudden failure. Therefore, it is important to consider this aspect when the specific grade of stainless steel is being selected.

The four basic types of stainless steel are summarised in Table 12.1. From what has been said about them it can be seen that the final choice in any particular instance is far from straightforward. Help in the choice can always be obtained from the major steel suppliers and experienced fabricators.

12.1 The four basic types of stainless steel

Type of stainless steel	Composition	Typical grades
Austenitic	18% Cr, 8% Ni, sometimes with Mo, Cu, Ti or Ni	304, 304L, 321, 316
Duplex	18% Cr, 5% Ni, sometimes with Mo, Cu, Ti or Ni	3RE60*, Ferralium*
Ferritic	18% Cr, sometimes with Mo	ELIT* 18/2
High nickel austenitic	25% Cr, 20% Ni, sometimes with Mo, Cu, +Ti and Co	904L*, Incoloy 800*

* = Trade name

12.4 Titanium evaporators

Over the last decade the price of titanium equipment has decreased considerably in real terms and it is now often competitive with the special grades of stainless steel and generally much cheaper than the nickel-base alloys. Once it was regarded as an exotic material which was very difficult to weld and fabricate. However, the time has now arrived when it should be regarded as an established material of construction which needs to be considered on its merits at the process design stage where its high corrosion resistance can be exploited. A large number of titanium alloys are available but most of these were developed for applications which demanded a high strength/weight ratio. Generally the corrosion resistance of these alloys is considerably inferior to pure titanium. However, there are a few alloys containing small amounts of palladium or molybdenum and nickel where the crevice corrosion and general corrosion resistance is enhanced.

Examples of where titanium is used include reboilers on a terephthalic acid plant and nitric acid evaporators on a nylon plant. The reason a chemically reactive material like titanium is so corrosion resistant is that a very stable passive film of titanium oxide is formed on its surface. Therefore it is very resistant to strongly oxidising environments such as nitric acid. Conversely in a strongly reducing environment such as hydrochloric acid or in crevices where oxygen access to maintain the passive film is limited there can be rapid corrosion. Palladium, nickel and molybdenum increase the resistance of the passive film to breakdown under marginally reducing conditions, hence the benefit to crevice corrosion resistance. However, iron has the opposite effect and markedly reduces the resistance of the passive film to breakdown. For this reason, it is common practice to specify a low iron (0.5% max) grade of titanium for increased corrosion resistance even though many commercial specifications for pure titanium permit considerably more than this. In addition to the requirement of a low iron content in the parent metal, it is even more important to ensure that there is no surface contamination by metallic iron. For example, a steel tool lightly striking the surface of a titanium component will leave a tiny smear of iron on the surface of the titanium. This will then be sufficient to cause local breakdown of the titanium oxide passive film under relatively mild corrosive conditions.

As a precaution against surface iron contamination it is good practice to anodise the titanium equipment. One way of doing this is to fill the equipment with an electrolyte such as 5% ammonium sulphate and pass an electric current through the solution to a submerged cathode. The polarity is such that the titanium is the anode and this increases the thickness of the oxide film and at the same time dissolves off any iron which may be contaminating the surface.

If corrosion occurs due to breakdown of the passive film the hydrogen generated at the corroding surface may diffuse into the titanium and form titanium hydride. This phase is very brittle and at temperatures above about 150 °C the diffusion rate of hydrogen is so rapid that the metal will become severely embrittled and mechanical failure may take place before the loss of metal caused by the corrosion has become noticeable. Therefore if the operating temperature is near or above 150 °C it is even more important to avoid passive film breakdown by using pre-anodised tubes or anodising after assembly.

The welding of titanium is straightforward but careful attention to inert gas shielding and cleanliness is very important. An experienced fabricator will have no difficulty in making sound welds because he will have sufficient knowledge to avoid oxygen and nitrogen from the atmosphere diffusing into the reactive weld metal and causing local embrittlement. It should not be assumed that a fabricator who is quite capable of making good welds in stainless steel by a gas shielded process can successfully make good welds in titanium by the same process. The welds will look good but metallurgical examination and destructive mechanical testing will be needed to demonstrate that dangerous embrittlement has not occurred.

Once again, it should be stated that there is no difficulty in designing and obtaining very reliable vaporising equipment in titanium or titanium alloys provided the basic properties of the metal are properly understood and no attempt is made to take shortcuts with the design standard and quality control.

CHAPTER 13

Operation, maintenance and commissioning of vaporisers

by P. J. Nicholson [1]

In this chapter some of the special problems which arise in the operation, maintenance and commissioning of vaporisers are discussed. In operation many of the problems encountered arise under conditions other than steady running at design rates, such as when starting up, when plant upsets occur and when shutting down, particularly in emergencies. Other problems arise when fluid compositions depart from the design range.

13.1 Fouling

The most difficult item to assess in the thermal design of a vaporiser is often the thermal resistance of any dirt that may be deposited on the surfaces. As mentioned in section 6.4, no rational methods are available for estimating fouling resistances. Figures based on experience are selected for each side of the heat transfer surface and used in the establishment of the overall heat transfer coefficient, from which the required surface area of the vaporiser and the tube metal temperatures can be calculated.

In actual operation the fouling resistances will be near to zero when the vaporiser is new and will gradually increase with time until either the vaporiser is cleaned or until the deposit of dirt reaches an equilibrium thickness, when the rate of deposition equals the rate at which dirt is removed by the shear stress exerted by the flowing fluid. In some instances there is no removal, so an equilibrium is never established. In other instances the rate of increase of thermal resistance may be reduced but never completely eliminated. Changes in operating conditions may alter the rate of deposition of dirt.

[1] P. J. Nicholson, BSc, PhD, MIChE, CEng is the Manager of the Vessels, Furnace & Boiler Section of Engineering Department, North East Group, ICI PLC.

The plant operator must therefore be able to cope with overall heat transfer coefficients both above and below the value used in design. Ideally where this has serious consequences the process designers of the plant will have provided him with some means of control over the situation. A further aspect, which is seldom easy to assess with the plant on-line, is the split of fouling resistance. Thus in fired vaporisers or high temperature waste heat vaporisers it is far more serious if excessive fouling occurs on the side of the heat transfer surface on which vaporisation takes place since this can lead to dangerously high metal temperatures and corrosion.

It is desirable that accurate performance tests should be carried out on vaporisers when the plant is new and after a period of operation; the former provides a check on the design calculations and the latter provides information on the rate of fouling. Unfortunately such tests can be very difficult to arrange, particularly when the plant is new and difficulties are being experienced in establishing the process under time constraints. Periodic tests with normal plant instruments are unlikely to be sufficiently accurate to provide the desired information but they are useful in providing approximate values of overall heat transfer coefficient, which can be plotted against time to assist in the prediction of when vaporiser performance will limit plant output and thus help in deciding on the need to clean the vaporiser at the next plant shutdown.

Regardless of the degree of fouling indicated by on-line measurements it is wise to inspect the boiling side of vaporisers during shutdown since the concentration of impurities inherent in the boiling process can cause corrosion or local deposits of dirt under which corrosion may start, knowledge of which can give timely warning that some aspect of design or operation is incorrect.

At the design stage, thought should be given to the means of inspecting surfaces on which vaporisation takes place. In fire-tube boilers it is desirable to provide an inspection port in the shell adjacent to the zone of peak heat flux. In water-tube equipment the problem is more difficult but it may be possible to install inspection ports in headers through which remote viewing equipment such as miniature TV cameras can be inserted into tubes. Similarly, methods of cleaning should be considered at the design stage so that suitable valves and branches can be incorporated into the vaporiser during manufacture. The design of modern vaporisers, for economic reasons, will seldom allow direct mechanical cleaning of surfaces on which vaporisation takes place. In such cases discussion with process or water chemists should be held during design to decide on effective chemical cleaning methods which will not damage the tubes. The cleaning of heat exchange equipment is discussed in section 13.9.

Because of the concentrating nature of the boiling process, a blow-down is almost always needed to remove non-volatile matter and control the overall concentration factor. It is usually economic to provide a blow-down system which the operator can regulate quantitatively to suit the prevailing level of impurities in the vaporiser feed. Blow-down of steam generating equipment is dealt with in more detail in Chapter 14.

All water fed to steam generating plant must be treated to prevent the deposition of scale on heated surfaces and to avoid corrosion by the water as mentioned in section 12.2. Boiler water treatment is discussed in Chapter 14.

13.2   Ensuring an adequate flow of vaporising liquid

In addition to loss of performance, an inadequate flow of the liquid to be vaporised may lead to overheating of metal surfaces, fouling and corrosion.

Natural circulation is often preferred to forced circulation for economic reasons and because gravity never fails. However, there are certain configurations of waste heat boiler plant where forced circulation is advantageous and there are a number of compact boiler designs on the market (e.g. the La Mont Boiler) which require forced circulation. These compact boilers often have fairly low initial cost but high running costs. Circulating pumps may fail due to power failure, motor failure, and choking or mechanical failure of the pump. In steam generating plant where the loss of circulation could lead to dryout or even burnout it is normal practice to have installed spare pumping capacity. It is important to check periodically that the spare pump will automatically come into operation immediately the normal pump shuts down.

When natural circulation is used, the operator must be aware of the conditions that might lead to a serious loss in circulation rate. Thus during start-up a significant period may be needed to overcome the inertia of the circulating liquid and establish the normal circulating velocity. It may therefore be necessary to restrict the initial heat transfer rate to cope with the lower circulation. Similarly, when the plant is operating at reduced output the circulating velocity is reduced. Fouling may lead to a considerable reduction in output, which in turn may result in a loss of circulation velocity and considerably more fouling.

In addition to ensuring that the total flow of liquid to a vaporiser is adequate, it is important that all heat transfer surfaces within the vaporiser experience an adequate flow of liquid. In boilers which have a liquid surface above a tube bundle, such as kettle types or fired shell boilers, incorrect level control can expose the upper tubes. The effects of the vapour/liquid interface are discussed in section 12.2.4. Water-tube boilers, which consist of multiple parallel paths, e.g. La Mont, can fail when a single path becomes choked with debris or a foreign body.

In steam generating systems where several steam generators are connected to a common steam drum there exists the possibility of one generator inducing reverse circulation in another, especially if the sources of heat input to the individual generators are independent. Reverse circulation can also arise within the circuits of a single fired water-tube boiler. The danger of failure may not be due so much to the reverse circulation itself as to the cessation of flow which can arise if at some point of loading the circulation reverts to the normal direction. The likelihood of reverse circulation should be studied at the design stage and steps taken to prevent it. There are several techniques which can be used. Where two or more steam generators are served by a single steam drum, the risers from each generator can be connected to physically separated groups of cyclones within the drum. Alternatively, steam can be injected into the risers during start-up to overcome reverse circulation driving forces. To prevent reverse circulation from being induced in the cold-end risers of a single steam generator by the hot-end risers, which are the first to carry steam, the cold-end risers can be connected to the steam drum above the start-up water level within the cyclone boxes. This approach can also be used to prevent reverse circulation between the circuits of a single water-tube boiler.

13.3   Foaming

The operation of a vaporiser may be seriously impaired by foaming, and downstream equipment may be fouled or damaged by the attendant carry-over of liquid.

13.3.1   Mechanism of foaming

The addition of a few drops of detergent liquid to a container of boiling water leads to the formation of froth on the surface at low boiling rates. As the heat input is increased, there is a sudden increase in the level due to foaming of the liquid. The reduced surface tension has made possible the formation of stable bubbles that are much smaller than those formed when boiling pure water. The addition of a small amount of very fine powder has a similar effect, the small particles forming nuclei for the formation of small bubbles. Small particle nucleation of bubbles is also illustrated in the foaming produced when colloidal suspensions (particles < 0.0001 mm diameter) such as milk are boiled.

Foaming is thus the continuous generation of relatively stable bubbles which in serious cases retain their form until carried out of the vaporiser. The liquid in the envelopes of the bubbles then constitutes carry-over. The liquid surface in shellside vaporisers may disappear completely and the contents become a graduation of foam (Association of Shell Boilermakers 1976).

When designing a vaporiser for fluids other than water, it may be difficult to predict whether the fluid will foam. Small changes in composition may cause an abrupt change in foaming tendency.

With a foaming liquid the minimum interfacial area between liquid and vapour to avoid excessive carry-over is considerably greater than given by section 9.8, equation [9.64]. In practical design it is seldom possible or economic to provide sufficient interfacial area to prevent foaming of a liquid with marked foaming tendencies.

13.3.2 Foaming in steam generators

The main impurities in water which can cause foaming in steam generators are listed below (Association of Shell Boilermakers 1976).

1. Detergents, oils and greases.
2. Suspended solids.
3. Dissolved solids.
4. Alkaline salts.

Detergents may be from decaying organic matter or domestic and industrial waste in raw waters. As little as 1 ppm can cause severe foaming. Oils and greases from materials of construction can be removed by chemical cleaning prior to service or during initial low rate operation of the boiler.

As a very rough guide, fine suspended solids should not exceed 200 ppm and total dissolved solids 3500 ppm if foaming is to be avoided, but these figures are dependent on the nature of the impurities.

If foaming takes place the steps which the operator of a boiler can take to minimise the problem are to reduce vaporisation rate, maximise pressure of operation, increase blowdown to reduce concentration factor and add antifoam chemicals to the water. The only certain cure, however, is to rectify the water composition and if an alternative source of raw water or feed water is available this should be tried.

Antifoam chemicals can be very effective in breaking down foam in some cases but ineffective in others, particularly where there are suspended solids on which the antifoam may be adsorbed.

Foaming is unlikely to be a problem in steam generators where a large proportion of the feed water is returned condensate. For a given water composition it is likely to be most severe in boilers operating at 7 bar (0.7 MPa) or below where the specific volume of steam is high.

In addition to producing carryover, foaming reduces the density of the liquid; this may cause low-level float switches to sink and shut down the boiler. Foam adjacent to heated surfaces will provide less effective cooling than water and may cause overheating of the metal.

The characteristic appearance of the water in a level gauge during foaming is that the level fluctuates and there is a large water flow down from the steam connection.

Goodall (1980) in his Figs 4.3 and 4.4 shows the effects of increasing the heat input on water level in a shell boiler (and in its level gauge) with and without foam formation. At low heat input there was little difference but at high heat input the foaming liquid filled the steam space and caused heavy carry-over.

13.3.3   Foaming in concentrating and crystallising evaporators

In a concentrating or crystallising evaporator the prevention of foaming can be very difficult and unpredictable. If possible a climbing-film evaporator should be used for concentrating duties as recommended in section 5.4.3. Foaming upsets the performance of crystallising evaporators.

There is little published information on foaming of fluids other than water. If foaming tendencies are suspected it may be necessary to do experimental work with the particular fluid to be vaporised.

13.4   Control of vaporisers

In the operation of a vaporiser, it is often necessary to control the rate of heat transfer to give the required vapour generation rate, and to be able to control the flow of liquid into the vaporiser to give the required liquid level inside the vaporiser. Thus instruments are required for measuring flow rate, temperature and level. If the heat source is condensing steam, the heat transfer rate is controlled by a throttle valve in the steam supply line to control the steam pressure in the heating space. If the heat source is a hot process fluid, the heat transfer rate is controlled by a bypass valve which controls the amount of process fluid passing through the vaporiser. Instrumentation and automatic control are outside the scope of this book.

13.5   Problems in start-up, change in operating conditions and shutdown

The first problem in start-up is to ensure the cleanliness of the plant. Even small amounts of debris left in a pipe or vessel can cause immense troubles and delays in commissioning. The commissioning and cleaning of steam generating equipment are discussed in sections 13.8 and 13.9.

The start-up of a large chemical plant may take several days, each part being commissioned separately, at part load, before the next part can be started. This introduces complexities in heat exchangers where one fluid must be heated or cooled before the other fluid is available. Plants relying on waste heat boilers to produce all or most of their steam requirements will need to import large quantities of steam before these waste heat boilers become operative.

The unusual operating conditions of start-up may lead to the establishment of exceptionally high or low temperatures, which might enhance fouling or induce corrosion.

Very sudden changes in operating conditions may lead to overstressing as a result of thermal shock. For example, a sudden increase in the inlet temperature of the hot fluid passing through the shell of a tubeside vaporiser with fixed tubesheets could overstress the tubesheet because the tubes would respond more rapidly than the shell to the rise in temperature.

On shut-down, plant must be purged of toxic or flammable gases, and any water should be drained from the system, to avoid freezing in cold weather. In a fired boiler, precautions must be taken to ensure that the flue ducting is fully purged, to avoid an explosion when relighting the burner.

13.6   Operation of vacuum equipment

When condensers are taking steam from an evaporator operating under vacuum, difficulty may be experienced in maintaining the required depth of vacuum. The equipment installed to provide the vacuum should be of a sufficient capacity to cope with start-up conditions, when the condenser must be evacuated sufficiently promptly for the pressure to fall almost to the required depth of vacuum in a period of not more than several minutes. If this has not been achieved, or if there is a change in process that results in an increase in the amount of gas coming out of solution, then extra vacuum-producing equipment must be installed.

13.7   Problems in the operation of a crystallising evaporator

The problems here are to produce crystals of the required size, to avoid scaling of the heated surface, and to avoid foaming. A choice is offered in section 5.4.4 between five types of heat exchanger for a crystallising evaporator, two with forced circulation, two with natural circulation and a forced-flow flash evaporator. The problems differ according to the type chosen and according to the nature of the crystals to be produced. On account of the latter, it is not possible to give more than a few guidelines about the different types.

With natural circulation, the main problem is to maintain an adequate, steady, velocity up the tubes. Operation at reduced output may be difficult, because reducing the vapour generation rate reduces the circulation velocity, but the fractional reduction in velocity is appreciably less than the fractional reduction in vapour generation rate. Experience has shown that an evaporator operating satisfactorily at constant output for several weeks would suddenly, without any apparent change in conditions, produce crystals of considerably reduced size. Altering the rate had no effect, but suddenly satisfactory operation would be resumed. No explanation of the phenomenon could be found; possibly the plant was operating near to the region of unstable flow.

With forced circulation, frequent checks should be made to ensure that the crystals are not eroding the impeller of the pump or that the impeller is not breaking the crystals.

Figure 5.6 shows a flash crystallising evaporator, in which saturated liquid is pumped through a heater and into a flash vessel, where crystallisation takes place. The design of the crystallising vessel is specific to the material being crystallised, and information on how to operate such systems has not been published.

13.8 Commissioning of steam generating plant

A boiler system may comprise feed water pumps, economisers, steam drum, water-tube boilers and/or fire-tube boilers. The preparation of the system for service can critically affect its reliability, especially if it is designed for high heat flux and high pressure operation (see section 13.9.5). In this section, the various steps in the commissioning of the water side of a high pressure boiler system are described. The exact procedure for a particular plant may vary a little depending on the type and operating pressure of the system. However, a key point to be recognised in all cases is that the commissioning of a new or extensively repaired or modified boiler system requires careful planning and allocation of sufficient time in the work programme. Short cuts during commissioning often result in very expensive down-time at a later stage.

13.8.1 Cleanliness

With respect to cleanliness, preparation for service begins at the fabrication stage. All plate material, piping and fittings used in construction should be free from millscale, excessive rust and protective coatings. During erection, care should be taken against the ingress of foreign matter, in order to avoid the problems outlined in section 13.9.5. Chemical cleaning at a later stage must not be relied upon as a cure-all for cleanliness but rather should be regarded as a final cleaning and conditioning process. Rigorous inspection must be maintained throughout construction.

13.8.2   Hydraulic testing

On mechanical completion, it is necessary to carry out a system pressure test. The design of all plant items should be examined to determine the correct test pressure and temperature of the test water. It is not uncommon for boilermakers to require the test to be carried out at temperatures considerably above ambient, in which case there must be a means on site of heating the water. The quality of the test water may also be important. Deaerated, demineralised water is generally preferable.

In some cases, where a large number of joints and welds need to be made after chemical cleaning, a second hydraulic test is necessary. In this case specialist advice is required on test water composition and procedure to ensure that the protective film produced by the final stage of chemical cleaning (see sections 13.8.3 and 13.9.5) is not destroyed.

13.8.3   Chemical cleaning

This general term covers a series of processes described in detail in section 13.9.5. Both the feed water and steam generating parts of the plant may be flushed, degreased, acid washed to remove oxides of iron and passivated. The last process forms a protective film on the bare metal produced by the acid wash, which, provided it is kept intact, will provide a good basis for the formation of a fully protective film in service.

13.8.4   Storage prior to service

It will normally be necessary to open up the water side of the system after chemical cleaning for inspection and final fitting of steam drum internals. Additionally the system may have to stand for some time while other parts of the plant are completed. Conditions within the system require careful control at this stage since the protective film formed during passivation will last only 1–2 days in a damp oxygenated atmosphere.

An effective method of storage which also allows entry into the system is to maintain a through flow of dehumidified air. Equipment to provide this is readily available. If the dry storage technique is used, initial drying can be aided by draining the system hot at the end of chemical cleaning and opening it at top and bottom to allow a strong natural draught to be established. The heat capacity of the metal and insulation will normally be sufficient to evaporate most of the residual moisture and can be augmented by warm air blowers to complete drying. The flow of dehumidified air should be started as the temperature of the system falls towards ambient.

13.8.5 Initial pressurisation of the system

In high-pressure systems with thick-walled steam drums it will usually be necessary to control the rate of pressurisation and depressurisation in order to limit the thermal gradients within the metal of the drum and hence limit thermal stressing. The procedure followed is to attach thermocouples to the outside of the drum and record temperatures as a function of pressure. Provided temperature differences remain within limits shown by stress analysis to be acceptable then the achieved rates of pressurisation and depressurisation are deemed safe. Rates established at this stage are then taken as maxima for future operation.

13.8.6 Cleaning and commissioning of steam distribution pipework

Particulate matter carried by steam can seriously damage turbine blades. It is therefore necessary to clean the pipework carrying steam to turbines before service. The most frequently used technique is steam blowing. The pipework is carefully warmed with a low flow of steam until free from condensate and then repeatedly blown at high velocity to atmosphere. Blowing is continued until a target plate of soft metal at the open end of the pipework ceases to be marked by particles. Ten to twenty blows may be needed. The thermal cycling of the pipework during blowing loosens scale and this and other material are removed by the high velocity flow. For each part of the system the flow of steam is calculated to produce a drag force on particles of at least 1.6 times the maximum which would occur in normal operation. Provided a sufficiently high flushing velocity can be achieved chemical cleaning can sometimes be used as an alternative to steam blowing (see section 13.9).

A problem which can occur in steam distribution systems whenever they are put into service is water hammer. If the pipework is pressurised with significant quantities of condensate present a potentially unstable condition arises where steam is in contact with water below the prevailing saturation temperature. A disturbance can cause mixing of the steam and water and hence rapid condensation of the steam. The following in-rush of further steam picks up the water which then impinges on pipe bends at velocities sufficiently high to cause mechanical damage. To avoid water hammer steam pipework should be thoroughly warmed at atmospheric pressure waiting until all drains blow steam rather than condensate before pressurisation begins. Pressure should then be increased in small steps especially at first, waiting at each stage for drains to blow to steam before the next increase is made. Safe pressurisation of high-pressure steam systems may take about 12 hours. Prior to initial commissioning steam pipework should be inspected to ensure that number and location of drains are adequate and that none are blocked.

13.8.7 Floating of safety valves

The final stage of commissioning is to prove the setting of safety valves by pressurising the system in service under careful control until the valves just lift off their seats.

13.9 Cleaning of heat exchange equipment

With the continuing drive to maximise the performance of plant and reduce maintenance costs, the cleaning of heat exchange equipment has assumed increased importance in recent years. In sections 13.9.1 to 13.9.4 below physical and chemical cleaning techniques applicable to a wide range of heat exchangers and vaporisers are outlined while in section 13.9.5 the special case of the chemical cleaning of the water side of steam generating equipment is discussed.

Increasingly ingenuity is being called for in dealing with fouling. There is the wish to minimise its effects on plant performance but at the same time a reluctance to increase capital costs to accommodate it or to simplify cleaning. Thus there is likely to be reluctance to increase the size of equipment by providing large tube pitches, to make tube bundles removable or to provide large numbers of flanged connections for cleaning purposes. Particularly for the shellside of exchangers this leads in the direction of chemical cleaning.

A further economic consideration is to reduce maintenance costs and avoid extensions to the duration of shutdowns arising from cleaning operations. For some equipment cleaning during shutdowns can be eliminated by the use of on-line cleaning, a good example being cooling water circuits. A benefit of on-line cleaning is that it can prevent extreme build-up of foulants which may be difficult to remove by any technique.

A comprehensive review of cleaning methods and their applicability to commonly encountered foulants is given by Lester and Walton (1982). Factors which can influence the choice of cleaning method for a particular exchanger are access to the fouled surface, the nature of the foulant, the time available, the materials of construction of the exchanger and the compatibility of any materials used for cleaning with the process. Cleaning may be by mechanical or chemical means and may be done on-line or off-line.

13.9.1 Off-line mechanical cleaning

Washing with high-pressure water jets is the most frequently used technique. The power of the jet can be varied over a wide range to suit soft deposits or very hard scale. There are many forms of the equipment ranging from a hand-held lance to automatic equipment which can clean both sides of a removable tube bundle.

Hydrosteam cleaning uses a steam–water mixture or steam-cleaning agent mixture to remove deposits. The mixture is applied in jet form and is advantageous when the foulant is temperature sensitive, such as wax or grease.

Pneumatic blasting with sand or other materials is used for hard deposits such as rust and boiler scale. It is usually important to remove the abrasive before the exchanger is returned to service and pneumatic techniques are therefore sometimes combined with vacuum equipment.

There are a variety of rotary and percussive tools available, mostly suitable for the internal surfaces of tubes. Rotary heads can be brush, scraper or drill types.

13.9.2 Off-line chemical cleaning

The use of chemical cleaning techniques is increasing. Major advantages are that surfaces which can only be reached by a flow of fluid, such as the shell side of a fixed tubesheet heat exchanger or the inside of a complex tube shape, can be cleaned and that the equipment is cleaned *in situ* with minimum dismantling.

The range of chemical treatments is far too wide for discussion here but among those frequently used are inhibited acids for the removal of oxides of iron and calcium and magnesium scales, detergents and/or alkalis for the removal of oils and grease and chlorinated hydrocarbons for the removal of heavy organic deposits such as tars and polymers. In acid cleaning the function of the inhibitor is to prevent the acid from attacking the cleaned metal surfaces, on which it forms a barrier. The ability of the acid to dissolve or loosen foulants is not affected.

In certain cases emulsions may be used to remove two foulants simultaneously. An example of this is the removal of oil and rust from the shell side of a refrigerant vaporiser by an emulsion of degreasant in inhibited acid.

The equipment needed to carry out a chemical clean is illustrated in Figs 13.1 and 13.2. These show chemical cleaning circuits for steam generating plants, but are fairly typical for many types of chemical cleaning duty.

Large quantities of chemicals, many of which are potentially harmful to personnel, equipment and the environment, must be handled during

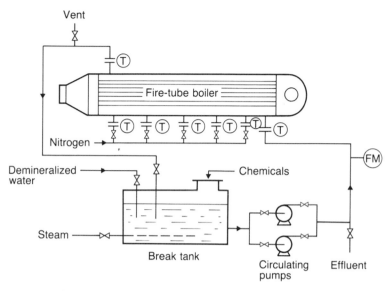

13.1 Chemical cleaning circuit for single boiler

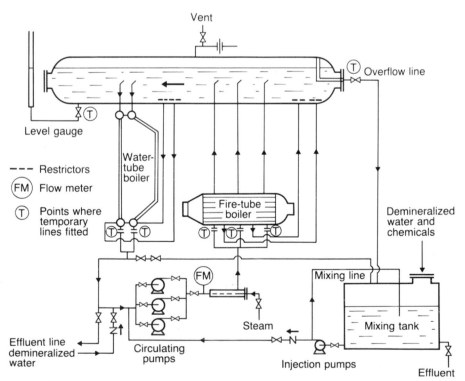

13.2 Chemical cleaning circuit for high-pressure steam generating system

chemical cleaning processes. For each cleaning operation safety aspects must be thoroughly studied and appropriate precautions taken.

13.9.3 On-line mechanical cleaning

The Taprogge system is well known for the cleaning of the tubeside of exchangers and condensers where water is the tubeside fluid. Rubber sponge balls, which may be coated with a mild abrasive, are injected into the water upstream of the exchanger, pass through the tubes and are then caught in a filter basket and recirculated. The diameter of the balls is chosen to be a close fit into the tubes so that foulant is removed as the balls pass through. It has been shown that this technique is most effective when used on an initially clean exchanger; foulants are prevented from building up. When applied to a dirty exchanger, although improvement is achieved, the exchanger is sometimes not fully cleaned.

Probably the most common on-line cleaning technique is soot blowing in boiler plant where periodic blasting of the fluegas side of tube surfaces with steam is required to remove deposits.

13.9.4 On-line chemical cleaning

Examples are the use of biocides and dispersants in cooling water systems and the addition of a chelating agent to boiler feedwater to limit scale formation. Although both of these techniques will clean fouled equipment they are most effective when the chemicals are added to clean systems where they prevent deposition of foulants.

On-line cleaning of both open and closed water circuits has been done by injecting high concentrations of chemicals on a once-through basis and for recirculation.

13.9.5 Chemical cleaning of steam generating equipment

The objective of chemically cleaning the water side of steam generating plant is to prepare the surface for service by removing all impurities and deposits and, for the more arduous duties, conditioning metal surfaces to assist in the formation of a fully protective film of magnetite by the boiler water.

Water-side impurities which may be present in new steam generating systems are millscale and oxides of iron on materials of construction, soil and silaceous matter, oils and greases, plastic and fibrous materials such as packaging, and miscellaneous scrap materials. The potentially harmful effects of these impurities are set out below.

1. The larger items of foreign matter may choke water tubes and other flow passages.

2. Particulate matter may accumulate at low points or adhere to heat transfer surfaces particularly in regions of low flow where it may cause overheating or under-deposit corrosion.

3. Oils, greases and fine solids may cause foaming (section 13.3).

4. Residual oxides of iron on heat transfer surfaces will reduce the effectiveness of the protective film formed by the boiler water during service.

5. Silaceous material can form a particularly low conductivity scale on heat transfer surfaces and at pressures above 40–50 bar it will lead to the presence of volatile $Si(OH)_4$ in the steam, which may deposit on the blades of turbines.

6. Safety valve seats can be damaged by particles of grit or scale.

Provided materials of construction are free from millscale and excessive rust and the larger items of foreign matter are removed during construction chemical cleaning can obviate all the foregoing dangers.

As part of the commissioning it is advisable to subject most boiler systems to some degree of chemical cleaning. Suggested guidelines are given in Table 13.1.

The reasons for increasing the extent of treatment with pressure are severalfold. Corrosion reactions do take place more rapidly at the higher saturation temperatures and in practice higher-pressure operation infers the use of demineralised boiler feedwater which can be rather more searching than the softened water usual at lower pressures. Almost irrespective of pressure, high heat flux renders boilers more susceptible to corrosion by increasing the concentration of impurities at concentration sites. In general

13.1 Suggested chemical cleaning criteria

Operating pressure of boiler	Cleaning processes
< 20 bar	Flushing and degreasing
20–40 bar	Flushing and degreasing; acid cleaning and passivation optional
> 40 bar	Flushing, degreasing, acid cleaning and passivation
All pressures when peak heat flux > 250 kW/m²	Flushing, degreasing, acid cleaning and passivation

the more severe the pressure/heat flux combination the greater the importance of chemical cleaning.

In order to prevent foreign matter being carried into a boiler system the feedwater system (boiler feed water pump delivery to steam drum inlet, including economisers) should also be chemically cleaned as part of commissioning especially if it is large.

Boiler systems may also require chemical cleaning after periods of service (perhaps every 2–5 years) to ensure continued reliability by removal of accumulated scale and deposits from heat transfer surfaces. In the high-pressure boilers used in the power generation and process industries, the magnetite film on tube surfaces, which is normally only a few microns thick, may grow to a thickness of several millimetres if water conditions are too acid or too alkaline, see Garnsey (1980). In doing so it loses much of its protective power and may in part become detached. Subjecting the boiler system to chemical cleaning at this stage removes all the loose magnetite and strips the surface to bare metal allowing re-establishment of a thin close-grain protective film on return to service.

Routine inspection of heat transfer surfaces can reveal the need for periodic cleaning.

The stages of chemical cleaning to be applied to a particular steam generating system will depend on the operating conditions generally as outlined in Table 13.1. However, the chemicals used for each stage will depend on the materials of construction and, for a system which has been in service, the nature of any deposits or contaminants which may be present, and should be decided in conjunction with water chemists and metallurgists.

An example of the chemical cleaning sequence for a new high-pressure steam generation system constructed in normal boiler steels is given in Table 13.2. For a system which has been in service the degreasing stage will often be omitted. (The process outlined in Table 13.2 is for a system which must be cleaned at atmospheric pressure).

The chemical treatments given in Table 13.2 are well proven but in all cases there are alternatives. For the acid cleaning stage inhibited citric acid or inhibited hydrofluoric acid may be used instead of inhibited hydrochloric acid. For carbon steels all the above acids are suitable but when austenitic steel components are present hydrochloric acid should be avoided. In recent years the hydrofluoric acid process has been increasingly used because it is as effective as the hydrochloric acid process in the removal of oxides of iron at lower temperatures, with shorter circulation times and with some safety advantages.

To carry out a chemical clean requires considerable planning and preparatory work. Once the chemical processes have been selected, temporary equipment must be manufactured or hired to allow mixing and introduction of chemicals and to provide a circulation route through the equipment to be cleaned. Typical chemical cleaning circuits for a single

13.2 Typical chemical cleaning sequence for new high-pressure steam generating systems

	Process	Remarks
Stage 1	Fill system with demineralised water, circulate at maximum rate and drain	This stage removes loose debris
Stage 2	Fill system with demineralised water, circulate at maximum rate, heat to 95 °C. Inject 1500 ppm wt trisodium phosphate and 100 ppm wt detergent. Circulate at 95 °C for 24 h then drain	This is the degreasing stage. Flushing to <2 ppm wt trisodium phosphate before Stage 3 is necessary
Stage 3	Fill system with demineralised water, circulate, heat to 75 °C. Inject HCl and inhibitor to 5% wt and 0.25% wt respectively. Circulate at 75 °C until total iron concentration constant and then further 2 h. Drain system	This is the acid cleaning stage. The acid removes oxides of iron from surfaces but is prevented from attacking metal by a surface barrier of inhibitor
Stage 4	Fill system batchwise with a mixture of 0.2% wt citric acid ammoniated to pH3.5–4.0. Circulate for 1 h. Drain system then flush to low water conductivity	This stage removes final traces of iron oxides by forming a soluble complex
Stage 5	Fill system with demineralised water and circulate. Heat to 95 °C. Inject 300 ppm wt hydrazine and 50 ppm wt ammonia and circulate at 90–95 °C for 24 h. Drain system and prepare for storage	This stage forms a passive oxide film on the bare metal produced by Stage 3 on which a fully protective film will form in service

boiler and a steam generation system are shown in Figs 13.1 and 13.2 respectively. The circuit shown in Fig. 13.1 does not allow a particularly good distribution of cleaning fluid within the boiler which is shown fitted with nitrogen injection points to increase velocity and turbulence within the tube bundle. The circuit for the three component steam generation system shown in Fig. 13.2 requires restriction of the downcomer pipes of both boilers to force the bulk of the circulating fluid over the heat transfer surfaces.

13.10 Protection of boilers during shutdown periods

Unless suitable precautions are taken, both the water side and the heating medium side of boilers can suffer corrosion during shutdowns. This section is concerned with the protection of the water side, which in most cases is by far the more important. Protection of both sides of fired boilers is discussed by the Association of Shell Boilermakers (1976). The protection techniques described below are applicable to all types of boilers and indeed to any equipment made of ferritic steel.

The protective magnetite or oxide film formed on the water side surfaces of a boiler during service can be destroyed in a short time if exposed to damp oxygenated conditions during a shutdown. The black colouration is lost and the surfaces become rusted. Once this has happened the effectiveness of the new film formed on return to service will be reduced and the boiler therefore more susceptible to on-load corrosion. In extreme cases actual corrosion during shutdowns can seriously pit or even penetrate tube walls.

If a shutdown is to last only a few days it will be sufficient to seal the boiler with the water at normal level, perhaps increasing the concentration of oxygen scavenging chemical (sodium sulphite or hydrazine) a little for a few hours before shutting down. Purging the steam space with nitrogen immediately steaming ceases and maintaining a purge or a positive pressure to exclude air will be beneficial. With a very carefully arranged nitrogen blanket and periodic sampling to confirm exclusion of oxygen, safe storage for 4–6 weeks can be achieved.

Where plant must be drained, protection for several weeks can be achieved by filling with a solution of 200 ppm sodium nitrite and 200 ppm sodium borate in demineralised water and then draining, leaving a film on the metal surfaces. Before returning the plant to service it must be flushed.

For shutdown periods of several months boilers can be stored wet by completely filling with a solution of 50 ppm hydrazine ammoniated to pH 10–10.5 in demineralised water. The solution should be analysed regularly and chemicals added to maintain composition and pH. Before returning the boiler to service the storage solution must be drained.

For indefinite storage and indeed for any extended shutdown, dry storage is the preferred method. The first step is to drain all parts of the boiler system and thoroughly dry the surfaces as quickly as possible (see section 13.8.4). Having achieved initial dryness there are several ways of preventing corrosion during standing.

1. Maintain relative humidity below 30% by a throughflow or circulation of dehumidified air.
2. Seal the boiler under a positive pressure of dry nitrogen.
3. Seal the boiler with an air or nitrogen atmosphere with an adequate supply of desiccant inside.
4. Seal the boiler under an atmosphere of air or nitrogen with an adequate supply of vapour phase inhibitor inside.

In addition to general corrosion, lack of corrosion control during shutdowns can reduce the fatigue life of boiler components (Schoch and Spahn, 1971). If the magnetite layer is damaged by mechanical stressing during depressurisation, or the tip of an existing crack is not adequately

0.05 mm
⊢——⊣

13.3   Corrosion fatigue crack showing stress-induced corrosion during shutdowns

protected by magnetite, oxygenated boiler water will attack the exposed metal during the shutdown. This is termed stress-induced corrosion. A crack exhibiting stress-induced corrosion is shown in Fig. 13.3. The rounded interruptions in the crack each represent a shutdown period when exposed metal at the crack tip has corroded. As a consequence of stress-induced corrosion during shutdowns fatigue life of the component is decreased. The mode of failure is then termed corrosion fatigue.

CHAPTER 14

The treatment of water for steam generating plant

by P. J. Nicholson

A vital requirement for the trouble-free operation of steam generating plant is a supply of water of the correct purity. The equipment to produce boiler feed water from raw water often accounts for a surprisingly large fraction of total capital cost.

The intention of this chapter is to outline why it is necessary to treat raw water, the methods and to some degree the extent of treatment needed for the various types of steam generators and operating conditions.

Water treatment is a highly complex, advancing technology and early consultation of specialists is advisable during the design of new steam generating plant or when water-related problems arise on existing plant.

The objectives of water treatment can be summarised as follows:

1. To enable the boiler to transfer heat at the design rate and efficiency by preventing deposition of scale.
2. To prevent corrosion of the feed water system, boiler and steam system.
3. To enable the boiler to produce steam of the specified purity from a given raw water.

14.1 Water treatment terminology

The following fairly generally accepted definitions are taken from British Standard 2486:1978.

Raw water:	Water as received by the user irrespective of any previous treatment.
Treated water:	Water which has received any treatment on the user's premises.

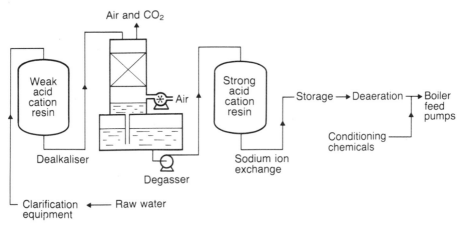

14.1 Dealkalisation/base exchange softening process

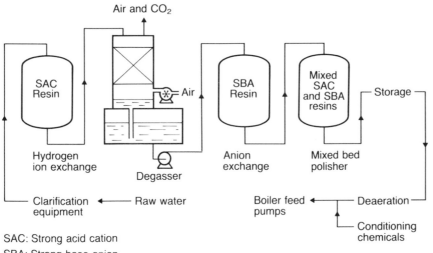

SAC: Strong acid cation
SBA: Strong base anion

14.2 Demineralisation plant for higher quality water

Softened water:	Water which has had its hardness substantially reduced.
Demineralised water:	Water from which nearly all of the ionisable dissolved solids have been removed.
Feed water:	The water passing through the boiler feed pumps (see Figs 14.1 and 14.2).
Boiler water:	The water present in a boiler when steaming is in progress or has been in progress.

Make-up water: Water which has had to be added to the system in order to make up for losses.

Feed water will often be a mixture of treated raw water and returned steam condensate. The condensate will always be contaminated to some extent and will usually need to be purified. This is especially true when the steam has been used in process plant or when a high purity feed water is required, e.g., for once-through boilers and most high-pressure recirculating boilers. Some recirculating boilers in the power generation industry do operate with untreated condensate. Contamination in condensate may be corrosion products, leakage of cooling water from a vacuum condenser, and a variety of substances where the steam has been used for process purposes. Condensate may be treated by returning it to some stage in the raw water treatment process (see Figs 14.1 and 14.2), or in separate equipment in which case it is added to the treated raw water upstream of storage or deaeration. The details of condensate treatment are beyond the scope of this book.

The operating pressure of a boiler is often described as low, medium or high. While it is not possible or necessary to define these pressure ranges precisely it may help to regard low pressure as up to about 25 bar (2.5 MPa) medium pressure as about 25–65 bar and high pressure as over 65 bar.

In the analysis of water the results will normally be obtained as mg/l concentration of ionic species (e.g. Ca^{2+}, Mg^{2+}, Na^+, HCO_3^-, CO_3^{2-}, SO_4^{2-}, NO_3^-). However, it is often convenient to refer concentrations to a common basis which is conventionally taken as equivalent calcium carbonate concentration (mg $CaCO_3$/l). Equivalent concentrations of calcium carbonate (relative molecular mass 100) are calculated as follows:

mg $CaCO_3$/l for Ca^{2+} (Ionic mass 40)

$$= \frac{100}{40} \times (\text{concentration of } Ca^{2+}, \text{mg/l})$$

mg $CaCO_3$/l for HCO_3^- (Ionic mass 61)

$$= \frac{100}{2 \times 61} \times (\text{concentration of } HCO_3^-, \text{mg/l}).$$

Equivalent concentrations can be added directly.

Grouping of ionic species by addition of the equivalent calcium carbonate concentrations is particularly useful in the expression of total hardness ($Ca^{2+} + Mg^{2+}$) and total alkalinity ($OH^- + HCO_3^- + CO_3^{2-}$). Total hardness is a measure of the potential of a water to form scale and alkalinity of its capacity to neutralise acids. Caustic alkalinity is the OH^- concentration expressed as equivalent concentration of calcium carbonate.

Electrical conductivity of water is a useful measure of total ionic content and is increasingly being used as a quality parameter for demineralised feed water and boiler water. The electrical conductivity of a sample of the water is measured either directly or after treating it by passing it through a cation resin which replaces metallic cations by hydrogen ions. This treatment increases the sensitivity of the conductivity of the sample to contaminants because the free acid has a higher conductivity than the neutral salt. It can also be used in the presence of volatile alkalising agents such as morpholine. No conductivity measurement will detect organic contaminants and cation conductivity will not detect sodium. The sodium which has not been removed by demineralisation plants is known as 'sodium slip'.

14.2 Impurities in raw water

Totally pure water is not significantly corrosive to boiler steels. Natural water, however, never exists in the pure form. It always contains dissolved gases and solids and very often will carry suspended and colloidal solids, many of which are harmful to steam generation equipment by causing corrosion, scale or foaming (see section 13.3). Almost all waters therefore need to be treated to remove impurities or render them harmless.

The impurities in raw water vary greatly in type and quantity according to the environment through which the water has passed and furthermore often vary seasonally. Water always contains dissolved oxygen and carbon dioxide from the atmosphere and possibly traces of methane, hydrogen sulphide, oxides of sulphur and ammonia. Important dissolved metallic cations are calcium, magnesium, iron, sodium and perhaps potassium, in various combinations with bicarbonate, carbonate, sulphate, chloride and nitrate ions (anions). Silica is present in colloidal and/or dissolved form. Organic matter in raw water arises from decay of vegetation and pollution. The natural organics may be the weak acids of high relative molecular mass known as fulvic and humic acids, predominantly present in soluble or colloidal form, and suspended particles of vegetation. Man-made organics may be oils and detergents. Other suspended or colloidal solids may be muds, silt and clays.

The most important dissolved gases are oxygen and carbon dioxide, both of which will cause corrosion in boiler systems. Calcium and magnesium ions are known as 'hardness' and their salts give rise to scale and sludge. Calcium and magnesium bicarbonates (temporary hardness) break down on heating with precipitation of the insoluble carbonates and liberation of carbon dioxide. Calcium sulphate (permanent hardness) forms scale, particularly on heated surfaces since its solubility decreases with increasing temperature. Oxides of iron and aluminium can form scales and sludge and certain silicates can give scales of particularly low thermal conductivity. At boiler pressures of above about 40 bars (4 MPa), silica is soluble in steam and may cause problems by being deposited on turbine blades.

At low and medium pressures the naturally occurring organic materials are unlikely to be harmful to boiler operation but in high-pressure boilers using demineralised feed water their degradation products can lead to serious corrosion. Oils and detergents can cause foaming in boilers at all pressures.

Suspended and colloidal solids can cause foaming, scales or sludges in boilers, and additionally they may foul water treatment plant, seriously affecting operation.

14.3   The treatment of raw water

The nature of the treatment of raw water depends upon the raw water composition, the fraction of make-up water in the boiler feed water, the boiler type, its operating pressure and peak heat flux, and the required purity of the steam. Economic factors may also influence selection of treatment processes. In this section the more frequently used treatments are outlined. In most cases there are variations which are not covered.

14.3.1   Internal and external treatments

In 'internal treatment' scale formation and corrosion are prevented solely by the addition of chemicals to the water fed to boilers. Hardness salts are converted to mobile sludges and corrosion minimised by chemically removing dissolved oxygen and elevating pH. Internal treatment is applicable only to low-pressure boilers and is economic only for small plants. It is not generally recommended.

The water treatment used for most boilers is 'external', in which the raw water is passed through a variety of process steps to remove impurities before reaching the boiler feed pumps.

14.3.2   Softened water

In the lime/soda precipitation softening process calcium hydroxide and either sodium carbonate or sodium hydroxide are added to raw water in a reaction vessel. Permanent hardness is precipitated as calcium carbonate and temporary hardness as calcium carbonate and magnesium hydroxide. After the softening chemicals have reacted, the precipitated hardness is removed in a clarifier following the addition of coagulants which form a floc. Silica is partially removed by adsorption onto this floc. The effectiveness of hardness/alkalinity and silica removal is increased by operating at elevated temperatures (the hot lime process) although economics often dictate ambient temperature operation. Because the process itself clarifies the water (removes suspended material) raw water is fed to it without any of the pretreatment which ion exchange processes require. It will normally be followed by a filter of some type. The use of lime/soda softening in the UK is declining because of cost, its inability to reduce hardness to zero and because of sludge disposal problems.

The lime/base exchange process is a combination of temporary hardness precipitation by lime and removal of residual temporary and permanent hardness by ion exchange (see below). A filter is installed after the precipitation stage.

The most economic and effective softening process is now base exchange. Water is passed through a strong acid cation resin on which calcium and magnesium ions are exchanged for sodium ions. The soluble sodium carbonate produced does not form boiler scales. When the resin is exhausted it is regenerated by brine and the calcium and magnesium ions flushed to drain. Where the raw water contains few suspended solids (e.g. borehole water) it may be fed directly to the ion exchange vessel but normally a pretreatment stage is required before the ion exchange vessel to prevent fouling of the resin and to protect the boiler.

The simple base exchange process has the disadvantage that there is an increase in total dissolved solids and alkalinity which for many raw waters may prevent the treated water from meeting the feed water specification (see section 14.6).

To overcome this disadvantage, dealkalisation may be used. Before reaching the base exchange vessel the water is passed through a weak acid cation resin which replaces the calcium and magnesium ions by hydrogen ions, leaving a solution of carbon dioxide in the water. This is stripped out by air in a degasser tower before the water, now containing little temporary hardness, passes to the base exchange vessel. This process, illustrated in Fig. 14.1, reduces hardness, alkalinity and total dissolved solids. As in all ion exchange processes, some pretreatment of the raw water is normally needed.

14.3.3 Demineralised water

The first method of producing the very pure water required for high-pressure boilers was evaporation, but ion exchange processes can now produce superior water, especially with respect to silica, more economically.

Demineralisation processes use both cation and anion exchange resins and can achieve almost complete removal of ionic material. A typical demineralisation plant of the type used for high-pressure high heat flux boilers (see section 14.6 for the feed water specification) is shown in Fig. 14.2. The water is first passed through a pretreatment stage to remove suspended and colloidal material and then enters a strong acid cation resin on which all salts are converted to their corresponding acids. Dissolved carbon dioxide (from carbonic acid) is then stripped out in the degasser tower. From the degasser sump water is pumped through a strong base anion resin where the anions are taken out by exchange with hydroxyl ions. The last traces of silica and carbon dioxide are also removed here. Up to this point the equipment shown in Fig. 14.2 is known as a two-stage demineralisation plant and would produce water suitable for many medium-pressure boilers. To obtain a higher quality feed water, a final stage of purification is needed. This is termed mixed bed polishing. The water is passed through an intimate mixture of strong base anion and strong acid cation resins which removes 'slip' from the upstream ion-exchange vessels.

The pretreatment stage will normally remove 60–80% of the organic material present in raw water. Most of the remainder will be adsorbed onto the strong base anion resin with the result that resin life can be reduced since the organics are not very effectively removed during the regeneration of the resin. Organic fouling can be a major problem in demineralisation plant operation.

14.3.4 Deaeration

The final stage of external treatment for both softened water and demineralised water is removal of the bulk of the dissolved oxygen, usually by stripping with steam in a deaerator.

It is possible to remove oxygen solely by chemical means but this becomes uneconomic for boilers of more than a few tonnes per hour of steam output.

14.4 Conditioning of boiler feed water

In almost all cases the water produced by external treatment will contain residual impurities which can be harmful to boiler systems. Some degree of final treatment (conditioning) is therefore needed. This is effected by the addition of small amounts of chemicals which are usually injected downstream of the deaerator.

14.4.1 Control of oxygen

Softened waters and demineralised waters require removal of the small amounts of oxygen remaining after the mechanical deaeration stage (typically 0.005 to 0.2 mg/l) if pitting of economisers, boiler tubes and pipework is to be avoided.

The operating pressure of the boiler is an important factor in the selection of oxygen scavenging chemical. Sodium sulphite is widely used at lower pressures where its rate of reaction with oxygen is faster than that of hydrazine, its competitor. As pressures rise sodium sulphite decomposes at an increasing rate, producing acidic gases. The absolute limit for its use is about 100 bar (10 MPa). Hydrazine, which offers the advantages of not adding to the dissolved solids in boiler water and raising the pH of steam condensate, tends to be used for the higher pressures. If a particularly pure steam is wanted (e.g. for turbines) hydrazine is better, even at pressures as low as 20 bar.

14.4.2   Conditioning of softened water

There are two basic conditioning treatments to prevent residual impurities from forming scale on the boiler tubes.

The first, addition of sodium phosphates, precipitates scale forming impurities in the bulk of the water as insoluble phosphates which are removed as sludge by blowdown of the boiler. Sludge conditioning agents are sometimes added to facilitate its removal. The sodium phosphates add alkalinity to maintain bulk water pH in the non-corrosive range.

The second treatment prevents the laydown of scale by keeping scale forming impurities in solution. A chelating agent is added to the feedwater to form a soluble complex with iron, calcium and magnesium ions. Sodium hydroxide is added for alkalinity control.

If the raw water has not been through a dealkalisation/degassing stage the steam produced by the boiler will contain sufficient carbon dioxide to form a highly corrosive condensate. Steam system corrosion can be prevented by addition to the feedwater of either a volatile neutralising amine such as morpholine, which reduces the acidity of the condensate, or a filming amine which prevents corrosion by forming a coating on metal surfaces.

14.4.3   Conditioning of demineralised water

There are three basic treatments:

1. The volatile treatment is applied to nearly all once-through boilers and some recirculating boilers; it aims to approach the zero dissolved solids situation and requires a very high quality feed water. Hydrazine must therefore be used for oxygen scavenging and although thermal decomposition of excess hydrazine produces ammonia the addition of a volatile alkalising agent such as morpholine or AMP is necessary to maintain bulk water pH in the non-corrosive range. The volatile

treatment does not give protection at concentration sites against non-volatile acidic or alkaline ions because the alkalising agent vaporises.

2. A solid alkalising agent such as sodium hydroxide, some form of sodium phosphate or a mixture, is used for lower quality feedwaters than required by the volatile treatment or indeed any feed water when the boiler may have concentration sites. Sodium hydroxide has the disadvantage that unless very accurately controlled it can itself cause corrosion at concentration sites. A congruent sodium phosphate treatment for alkalinity and pH control can probably offer the best chance of avoiding corrosion at concentration sites by maintaining neutrality during the concentration process. The use of sodium phosphates is very complex. Hide-out (the deposition of sodium phosphate on heat transfer surfaces) can occur, leading to metal wastage and at high temperatures a second phosphate-rich phase may separate out. The choice of phosphate system is a matter for specialists.

3. Neutral treatment (see section 14.6).

The effectiveness of the various methods of alkalinity and pH control is discussed in more detail by Lester and Hooper (1982).

14.5 Control of boiler water concentrations by blowdown

Blowdown, the purging of water from the boiler, is the operational link between concentration of impurities in the feedwater and their concentration in the boiler water. It is the latter which largely determines the success or otherwise of boiler systems as discussed in section 14.6.

It can be shown by a simple mass balance that the blowdown rate (B) expressed as a percentage of the evaporation rate of the boiler is:

$$B = 100 \frac{s}{c - s} \qquad [14.1]$$

where s is the concentration of an impurity in the feedwater and c is its concentration in the boiler water.

The same boiler water concentration can be maintained, within limits, for a variety of feed water concentrations by varying blowdown percent. Blowdown is however wasteful in feedwater, heat and conditioning chemicals and particularly with large boilers is not generally an economic substitute for better quality feed water.

Blowdown of low-pressure boilers, particularly if small and operating on softened water, is often as high as 10–15%. For large industrial fired boilers and waste heat boilers 2–5% is more usual while for high-pressure boilers operating with a very pure feed water blowdown is often less than 2%. In some once-through boilers there is no blowdown, whilst in others the feed water rate is controlled to allow 1–2 % purge between evaporation and superheat sections.

Examples of impurities which may govern blowdown percentage are total dissolved solids, suspended solids, silica and, in very high pressure boilers, chloride.

Continuous blowdown, ideally varied with evaporation rate, allows recovery of flash steam and heat. It is more efficient in water than intermittent blowdown in maintaining a maximum concentration level. It is usual to supplement continuous blowdown with some intermittent blowdown, formulae for which are given in BS 2486: 1978.

While the approximate blowdown fraction can be calculated at the design stage, the operating figure should be established by trial during the early running of the plant and thereafter adhered to.

14.6 Feed water and boiler water quality for industrial boilers

Feed water and boiler water quality specifications and guidelines are critically reviewed by Lester and Hooper (1982).

The most widely used specifications for industrial boiler water quality are those given in the publications of the various national bodies including:

(a) British Standards Institution: BS 2486 (1978)
(b) American Society of Mechanical Engineers: ASME (1979) and Simon and Fijnsk (1980)
(c) Verein der Grosskesselbesitzer: VGB (1980)
     Vereinigter Deutscher Technische Überwachungs Verein: VdTÜV (1972).

Other information is available from manufacturers of water treatment chemicals, the electrical power industry (CEGB (1983)) the technical literature and most importantly from boiler manufacturers who will usually provide feed water and boiler water specifications for their boilers and make their guarantees conditional on achievement of those specifications.

The object of all standards is to specify feed water and boiler water conditions which will prevent corrosion by production of a thin protective film of iron oxide on the internal surfaces of boiler systems and avoid formation of harmful scales and debris.

There are two approaches to water quality specifications based on different ways of preventing corrosion of the ferritic steels used in boiler construction. Corrosion rate is very slow in either:

(a) oxygenated very pure water

or

(b) deoxygenated neutral or slightly alkaline solutions (see Fig. 12.1).

The first (neutral) approach is a recent development which may well find increasing application. Of the national standards listed above, only the VGB give guidance. A very pure demineralised feed water with cation conductivity less than 0.2 μS/cm is required. Boiler water cation conductivity must be below 3 μS/cm. No chemicals are added other than oxygen in the form of hydrogen peroxide or gaseous oxygen. Under these conditions the protective oxide layer formed on metal surfaces is of the haematite type which can be more effective than the magnetite film produced in deoxygenated waters. This method of corrosion control is well suited to once-through boilers but can be used for recirculating boilers when volatile alkali is added to maintain boiler water pH between 7 and 8. In most cases the required purity of feedwater will only be achieved in systems which have a high proportion of pure condensate in the feed water.

The second (conventional) approach is based on deoxygenated water and relies on controlled limitation of hardness, total dissolved solids, alkalinity and corrosion products to avoid harmful effects. The concentrations of ionic species are typically related to the operating pressure of the boiler. The BSI and ASME standards are typical. The following are implied or stated:

1. Use of softened water of increasing purity as boiler pressure increases up to 40 bar (4 MPa).
2. A change to demineralised feed water in the range 40–60 bar.
3. Use of demineralised water of increasing purity with further increase in boiler pressure.
4. Fire-tube boilers are more tolerant to impurities than water-tube boilers at a given pressure.

The BSI and ASME standards identify similar feed water and boiler water parameters but the ASME specifications are more stringent for a given boiler pressure.

The VGB and VdTÜV view feed water specifications in a simpler manner. One specification is given for base exchanged softened water for boilers operating in the range 5–64 bar (corresponding to the ASME specification for 40 bar and the BSI specification for 60 bar) and a second for demineralised water for boilers operating at 64 bar and above. VGB and VdTÜV boiler water guidelines for demineralised feedwater are divided into two categories, above or below 125 bar.

The national standards for water quality should be regarded as guidelines which will give trouble-free operation for the majority of boilers. They may well need to be amended for particular applications. Such refinements may arise from the boiler manufacturer who has experience of the boiler design in question or from the user who has experience of a particular source of water.

All specifications are designed to limit the chemistry of the bulk boiler water to the range of conditions shown in Fig. 12.1 in which corrosion rate is at a minimum. However, this is not achieved if water is concentrated locally at concentration sites formed by any of the mechanisms discussed in section 12.2. Concentration factors can be in the range 10^4 to 10^6 at these sites. If on-line corrosion is to be avoided in a boiler with the potential for concentration site formation both the purity of the water and the method of alkalinity control need special attention. There is a limit to which improvement in water conditions can prevent corrosion if concentration sites are present. In the most severe cases corrosion can be prevented only by changing the design of the boiler or the operating conditions.

The need for water of higher quality than given in the published standards is particularly pronounced with waste heat boiler systems where the boiler design can be heavily influenced by the layout and process conditions of the chemical plant on which it is installed. For example, horizontal water-in-tube boilers and water-in-shell boilers operating at pressures of 100 bar are frequently used. Furthermore, heat fluxes are often very high. With respect to formation of concentration sites waste heat boilers often combine all the worst features of design.

Heat flux is a major factor in the reliability of boilers which it is difficult to relate directly to water quality. High heat flux in itself will not necessarily lead to on-line corrosion, although it can if dryout is approached and concentration sites are thereby formed. Intermittent dryout of heat transfer surfaces can occur below the critical heat flux defined in Chapter 8. The more common situation is that high heat flux renders boilers more sensitive to corrosion by increasing the severity of concentration under deposits and in other non-ideal circumstances. BSI, ASME and the VGB all draw attention to the importance of high heat flux and infer, respectively, that a 'high' level is $300 \, \mathrm{kW/m^2}$, $470 \, \mathrm{kW/m^2}$ and $250 \, \mathrm{kW/m^2}$. For peak fluxes above $470 \, \mathrm{kW/m^2}$ ASME recommend a feedwater quality which would normally be used for boilers operating at about 20 bar higher pressure. For peak fluxes above $250 \, \mathrm{kW/m^2}$, the VGB recommend a maximum boiler cation conductivity of less than $3 \, \mu\mathrm{S/cm}$ and that inorganic alkalis are not used to condition feedwater. Experience has shown that these measures recommended by the national bodies are not always sufficient to avoid on-line corrosion at concentration sites.

The traditional approach of linking feedwater and boiler water specifications to operating pressure of the boiler is currently being questioned. Heat flux and other factors can be shown to be more important almost regardless of pressure. Lester and Hooper (1982) suggest that in minimising on-line corrosion there are advantages in having only two comprehensive feedwater specifications based respectively on softened water and demineralised water. These are shown in Tables 14.1 and 14.2.

The above specifications are suitable for both water-tube and fire-tube boilers and are sufficiently stringent for all heat flux levels encountered in

14.1 Softened feed water specification

pH value at 25 °C	9–9.5	After conditioning
Total hardness mg CaCO$_3$/l	<0.2	
Total iron mg Fe/l	<0.02	
Total copper and nickel mg/l	<0.01	Before conditioning
Oil mg/l	<1	
Oxygen mg O$_2$/l	<0.01	
Total organic carbon mg C/l	<5	

14.2 Demineralised feed water specification

pH value at 25 °C	9–9.5	After conditioning
Conductivity μS/cm at 25 °C	<0.2	
Oxygen mg O$_2$/l	<0.007	
Total iron mg Fe/l	<0.010	
Total copper mg Cu/l	<0.003	
Silica mg SiO$_2$/l	<0.02	Before conditioning
Total organic carbon mg C/l	<0.2	
Chloride mg Cl/l	<0.005	
Sulphate mg SO$_4$/l	<0.005	
Sodium mg Na/l	<0.005	

well-designed boilers. They are achievable by current designs of water treatment plant without major increase in capital cost in the case of new plants.

Coupled with carefully chosen alkalinity and pH control, the feed water specifications given in Tables 14.1 and 14.2 provide almost the maximum economic contribution which water treatment can make to avoiding on-line corrosion.

References

Abadzic, E. E. (1974) 'Heat transfer on coiled tubular matrix', paper 74 – WA/HT-64 presented to ASME Winter Annual Meeting, November 1974.

Ackers, P. (1969) *Tables for the Hydraulic Design of Storm Drains, Sewers and Pipelines.* Ministry of Technology Hydraulic Research Paper No 4. London HMSO.

Afgan, N. and **Schlünder, E. U.** (1974) *Heat Exchangers: design and theory source book.* McGraw-Hill.

Akinjiola, P. O. and **Friedly, J. C.** (1979) 'Prediction of density-wave stability limits for evaporators – sensitivity to model assumptions'. Multiphase Transport (Proc. of the Multi-Phase Flow and Heat Transfer Symp. Workshop, Miami Beach, Fla. USA, 16–18 April 1979). Hemisphere Pub. Corp., pp. 1229–48.

Alfa-Laval (1969) *Thermal Handbook.* Alfa-Laval, Sweden.

American Society of Mechanical Engineers (1977) *Boiler and Pressure Vessel Handbook: Section VIII, unfired pressure vessels, D.W.1.* ASME, New York.

American Society of Mechanical Engineers (1979) *Consensus on Operating Practices for the Control of Feed Water and Boiler Water Quality in Modern Industrial Boilers.* ASME Document No. HOO156.

Andros, F. E. and **Florschuetz, L. W.** (1976) 'The two-phase closed thermosyphon: an experimental study with flow visualisation', *Proc. Symp. Workshop on Two-phase Flow and Heat Transfer*, Fort Lauderdale, Fla. USA, pp. 247–50.

Association of Shell Boilermakers (1976) *The Treatment of Water for Shell Boilers.* Publication of ASBM.

Austin, D. G. (1979) *Chemical Engineering Drawing Symbols.* George Godwin.

Baker, O. (1958) 'Multiphase flow in pipelines', *Oil & Gas J.* 10 Nov. 1958, 156–7.

Barbe, C., Grange, A. and **Roger, D.** (1971) 'Echanges de chaleur et pertes de charges en écoulement diphasique dans la calandre des échangeurs bobinés', *Proc. XIII Int. Congress on Refrig.* 2, pp. 223–34.

Beatty, K. O. and **Katz, D. L.** (1948) 'Condensation of vapours on outside of finned tubes', *Chem. Engng Prog.* **44**(1), 55–70.

Becker, K. M. and **Hernborg, G.** (1963) 'Measurements of burnout conditions for flow of boiling water in a vertical annulus', presented at 6th ASME/AIChE Nat. Heat Transfer Conf., Boston, Aug. 1963.

Becker, K. M. and **Letzter, A.** (1975) 'Burnout measurements for flow of water in an annulus with two-sided heating', Report KTH-NEL 23, Royal Inst. of Technology, Stockholm.

Bell, K. J. and **Ghaly, M. A.** (1972) 'An approximate generalised design method for multicomponent/partial condensers', *AIChE Symp.* **69** No. 131, pp. 72–9.

Bennett, L. D. (1976) 'A study of internal forced convective boiling heat transfer for binary mixtures'. PhD thesis, Lehigh Univ. USA.

Bergles, A. E., Collier, J. G., Delhaye, H. M., Hewitt, G. F. and **Mayinger, F.** (1981) *Two-Phase Flow and Heat Transfer in the Power and Process Industries*. Hemisphere Publishing Corp.

Bergles, A. E. and **Chyu M-C** (1981) 'Characteristics of nucleate pool boiling from porous metallic coatings', *ASME H.T.D.* **18**, 61–71, 20th ASME/AIChE Heat Transfer Conf., Milwaukee, Wisconsin.

Bignold, G. J., Garbett, K. and **Woolsey, I. S.** (1983) 'Erosion in boiler feed water', *Proc. UK Corrosion Conf.*, Birmingham, p. 127.

Bondurant, D.L. and **Westwater, J. W.** (1971) 'Performance of transverse fins for boiling heat transfer', *Chem. Engng Prog. Symp.* No. 113, **67**, 30–7.

Bowring, R. W. (1972) 'A simple but accurate round tube, uniform heat flux dryout correlation over the pressure range 0.7–17 MN/m^2 (100–2500 PSIA)'. AEEW – R789, AEE Winfrith.

Boyko, L. D. and **Kruzhilin, G. N.** (1967) 'Heat transfer and hydraulic resistance during condensation of steam in a horizontal tube and in a bundle of tubes', *Int. J. Heat Mass Transfer*, **10**, 361–73.

Brisbane, T. W. C, Grant, I. D. R. and **Whalley, P. B.** (1980) 'A prediction method for kettle reboiler performance'. ASME/AIChE Nat. Heat Transfer Conf. Orlando, Florida, July 1980, paper 80 – HT-42, 10 pp.

British Standards Institution (1976) *Specification for Unfired Fusion-Welded Pressure Vessels*. BS 5500, BSI, London.

British Standards Institution (1978) *Recommendations for Treatment of Water for Land Boilers*. BS 2846.

British Standards Institution (1980) *Boilers and Pressure Vessels: an international survey*. Published by Technical Help to Exporters, BSI Maylands ADE, Hemel Hempstead, Hertfordshire, England.

Butterworth, D. (1971) 'A model for predicting dryout in a tube with a circumferential variation in heat flux', Report AERE–M2436, UKAEA, Harwell.

Butterworth, D. and **Hewitt, G. F.** (eds) (1977) *Two-phase Flow and Heat Transfer*. Harwell Series, Oxford Univ. Press.

Butterworth, D. and **Shock, R. A. W.** (1982) 'Flow boiling', 7th Int. Heat Transfer Conference, Munich, Sept. 1982.

Carnavos, T. C. (1981) 'An experimental study: pool boiling R–11 with augmented tubes', *ASME H.T.D.* **18**, 103–8, 20th ASME/AIChE Heat Transfer Conf., Milwaukee, Wisconsin.

Central Electricity Generating Board (1983) 'Chemical control of the steam-water circuit of drum-type and once-through boilers'. Generations Operations Memorandum No. 72.

Chawla, J. M. (1967) 'Wärmeübergang und Druckabfall in Waagrechten Rohren bei der Strömung von verdamfenden Kältemitteln', *V.D.I.*, *Forschungsheft*, 523.

Chen, J. C. (1966) 'A correlation for boiling heat transfer to saturated fluids in convective flow', *Ind. and Engng Chem. – Process Design Dev.* **5**(3), 322–33.

Chen, Y. N. (1968) 'Flow-induced vibration and noise in tube-bank heat exchangers due to Von Karman Streets'. Trans. ASME Series B, *J. of Engng for Industry*, **30** (Feb.), 134–46.

Chenoweth, J. M. and **Impagliazzo, M.** (1981) 'Fouling in heat exchange equipment', *ASME, H.T.D.* **17**, 105 pp, 20th ASME/AIChE Heat Transfer Conf., Milwaukee, Wisconsin.

Chiesa, C., Ciminale, A., Elias, G. and **Silvestri, M.** (1974) 'Misure sul coeficiente di scambio termico e sulle cadute di pressione di un lungo tubo verticale', *La Termotecnica* **28**(3), 140–5.

Chilton, C. H. and **Perry, R. H.** (1973) *Chemical Engineer's Handbook* (5th edn). McGraw-Hill.

Chisholm, D. (1983) *Two-phase Flow in Pipelines and Heat Exchangers.* George Godwin.

Chojnowski, B. and **Wilson, P. W.** (1974) 'Critical heat flux for large-diameter steam generating tubes with circumferentially variable and uniform heating, *Proc. 5th Int. Heat Transfer Conf.* Tokyo **4**, pp. 260–4.

Chojnowski, B., Wilson, P. W. and **Whitcutt, R. D. B.** (1974) 'Critical Heat Flux for Inclined Steam Generating Tubes', *Symp. Multi-Phase Flow Systems*, Univ. Strathclyde, 2–4 Apr. 1974, paper E3.

Chun, K. R. and **Seban, R. A.** (1971) 'Heat transfer to evaporating liquid films', *Trans. ASME J. Heat Transfer* **93**, 391–6.

Chun, K. R. and **Seban, R. A.** (1972) 'Performance prediction of falling film evaporators', *Trans. ASME J. Heat Transfer* **94**, 432–6.

Churchill, S. W. and **Chu, H. H. S.** (1975) 'Correlating equations for laminar and turbulent free convection from a horizontal cylinder', *Int. J. Heat Mass Transfer* **18**, 1049–53.

Colburn, A. P. and **Hougen, O. A.** (1934) 'Design of cooler condensers for mixtures of vapours with non-condensing gases', *Ind. and Engng Chem.* **26**, 1178–82.

Colburn, A. P. and **Drew, T. B.** (1937) 'The condensation of mixed vapours', *Trans. AIChE* **33**, 197–211.

Colebrook, C. F. and **White, C. M.** (1937) 'Experiments with fluid friction in roughened pipes', *Proc. Roy. Soc.* Series A, **161**, pp. 367–81.

Collier, J. G. and **Wallis, G. B.** (1967) *Lecture Notes for Course on Two-Phase Flow and Heat Transfer.* Glasgow University.

Collier, J. G. (1981) *Convective Boiling and Condensation*, 2nd edn. McGraw-Hill.

Connors, H. J. (1970) 'Fluidelastic vibration of tube arrays excited by cross flow', presented at ASME Winter Annual Meeting, New York, 1 Dec. 1970.

Cornwell, K., Duffin, N. W. and **Schüller, R. B.** (1980) 'An experimental study of the effects of fluid flow on boiling within a kettle reboiler tube bundle', *ASME/AIChE Nat. Heat Transfer Conf.*, Orlando, Florida, July 1980.

Cornwell, K., Schüller, R. B. and **Einarsson, J. G.** (1982) 'The Influence of Diameter on Nucleate Boiling outside Tubes', 7th Int. Heat Transfer Conf., Munich, Paper PB8.

Coulson, J. M. and **Richardson, J. F.** (1977) *Chemical Engineering: 1 Fluid Flow, Heat Transfer and Mass Transfer.* Pergamon Press.

Cumo, M., Farello, G. E., Ferrari, G. and **Palazzi, G.** (1974) 'The influence of twisted tapes in subcritical, once-through vapour generators in counter flow', *J. Heat Transfer*, Aug. 1974, 365–70.

Czikk, A. M., O'Neill, P. S. and **Gottzmann, C. F.** (1981) 'Nucleate boiling from porous metal films. Effect of primary variable'. *ASME H.T.D.* **18**, 109–22, 20th ASME/AIChE Heat Transfer Conf. Milwaukee, Wisconsin.

Davis, E. J. and **Anderson, G. H.** (1966) 'The incipience of nucleate boiling in forced convective flow,' *AIChE J.* **12**(4), 774–80.

Davis, R. F. (1940) 'The physical aspects of steam generation at high pressure and the problem of steam contamination', *Proc. IMechE*, London, **144**, pp. 198–216.

Dhir, V. K. and **Lienhard, J. H.** (1974) 'Peak pool boiling heat flux in viscous liquids', *J. Heat Transfer, Trans ASME*, Series C, **96**, 71–8.

Engelbrecht, A. D. and **Hunter, J. B.** (1974) 'Development of a continuous process for concentration of aluminium sulphate solutions in a climbing film evaporator', Conf. on Heat Transfer and the Design and Operation of Heat Exchangers, S. African IChE.

Erskine, J. B. and **Waddington, W.** (1973) 'A review of some tube vibration failures in shell and tube heat exchangers and failure prediction methods', presented at Int. Symp. on Vibration Problems in Industry, Keswick, England, 10–12 Apr. 1973.

ESDU (1966) 'Friction losses for fully-developed flow in straight pipes', *Engineering Sciences Data* Item No. 66027.

ESDU (1967) 'Forced convection heat transfer, in circular tubes, Part 1: correlations for fully developed turbulent flow – their scope and limitations', *Engineering Sciences Data* Item No. 67016.

ESDU (1968a) 'Forced convection heat transfer in circular tubes, Part II; data for laminar and transitional flows including free-convection effects', *Engineering Sciences Data* Item No. 68006.

ESDU (1968b) 'Forced convection heat transfer in circular tubes, Part III; further data for turbulent flow', *Engineering Sciences Data* Item No. 68007.

ESDU (1972) 'Flow through a sudden enlargement of area in a duct', *Engineering Sciences Data* Item No. 72011.

ESDU (1973a) 'Pressure losses in three-leg pipe junctions: dividing flows', *Engineering Sciences Data* Item No. 73022.

ESDU (1973b) 'Pressure losses in three-leg pipe junctions: combining flows', *Engineering Sciences Data* Item No. 73023.

ESDU (1973c) 'Convective heat transfer during crossflow of fluids over plain tube banks', *Engineering Sciences Data* Item No. 73031.

ESDU (1974) 'Pressure loss during crossflow of fluids with heat transfer over plain tube banks without baffles', *Engineering Sciences Data* Item No. 74040.

ESDU (1977a) 'Pressure losses in curved ducts: interaction factors for two bends in series', *Engineering Sciences Data* Item No. 77009.

ESDU (1977b). 'Pressure losses in flow through a sudden contraction of duct area', *Engineering Sciences Data* Item No. 78007.

ESDU (1977c) 'Thermal conductivity, viscosity, heat capacity, density and Prandtl number of sea water and its concentrates', *Engineering Sciences Data* Item No. 77024.

ESDU (1978) 'Internal forced convective heat transfer in coiled pipes', *Engineering Sciences Data* Item No. 78031.

ESDU (1979) 'Crossflow pressure loss over banks of plain tubes in square and triangular arrays including effects of flow direction', *Engineering Sciences Data* Item No. 79034.

ESDU (1980) 'Thermophysical properties of heat pipe working fluids: operating range between $-60\,°C$ and $300\,°C$', *Engineering Sciences Data* Item No. 80017.

ESDU (1981a) 'Heat pipes – performance of two-phase closed thermosyphons', *Engineering Sciences Data* Item No. 81038.

ESDU (1981b) 'Forced convective heat transfer in concentric annuli with turbulent flow', *Engineering Sciences Data* Item No. 81045.

ESDU (1983a) 'Pressure losses in curved ducts: single bends', *Engineering Sciences Data* Item No. 83037.

ESDU (1983b) 'Baffled shell-and-tube heat exchangers: flow distribution, pressure drop and heat transfer coefficient on the shellside', *Engineering Sciences Data* Item No. 83038.

ESDU (1984) Shell-and-tube exchangers: pressure drop and heat transfer in shellside downflow condensation', *Engineering Sciences Data* Item No. 84023.

Fair, J. R. (1960) *Petroleum Refiner*, **39**, No. 2, 105.

Ferrell, J. K. and McGee, J. W. (1966) 'Two-phase flow through abrupt expansions and contractions', TID-23394, **3**. Rayleigh, N.C. Dept., Chem. Engng North Carolina State Univ. 213 pp.

Fitzsimmons, D. E. (1964) 'Two-phase pressure drop in piping components', H. W. 80970, Rev. 1. General Electric Harford Laboratories, Richland, Washington.

Forster, H. K. and **Zuber, N.** (1955) 'Dynamics of vapor bubbles and boiling heat transfer', *AIChE J.* **1**(4), 531–5.

Francis, J. R. D. (1975) *Fluid Mechanics for Engineering Students*. 4th Edn, Arnold.

Friedel, L. (1979) 'Improved friction pressure drop correlation for horizontal and vertical two-phase pipe flow', Paper 2 European Two-Phase Group meeting, Ispra, Italy.

Friedly, J. C., Akinjiola, P. O. and **Robertson, J. M.** (1979) 'Flow oscillations in boiling channels', presented at ASME/AIChE 18th Nat. Heat Transfer Conf. San Diego, California, Aug. 1979.

Frost, W. and **Dzakowic, G. S.** (1967) 'An extension of the method of predicting incipient boiling on commercially finished surfaces', Publn 67-HT-61, ASME/AIChE Heat Transfer Conf., Seattle, Washington.

Garnsey, R. (1980) 'The chemistry of steam-generator corrosion', *Combustion*, Aug. 1980.

Geiger, G. E. and **Rohrer, W. M.** (1966) 'Sudden contraction losses in two-phase flow', *J. Heat Transfer* **88**(1), 1–9.

Goodall, P. M. (1980) *The Efficient Use of Steam*. IPC Science and Technology Press.

Hands, B. A. (1979) 'The flow stability of a liquid-nitrogen thermosiphon with 8 mm bore riser', 18th AIChE/ASME Nat. Heat Transfer Conf., San Diego, California, Aug. 1979.

Happell, O. and **Stephan, K.** (1974) 'Heat transfer from nucleate to the beginning of film boiling in binary mixtures', Paper B7.8 5th Int. Heat Transfer Conf., Tokyo.

Haruyama, S. (1982) 'Stress corrosion cracking of stainless steel heat exchangers', *Materials Performance*, March 1982, 14.

Hawes, R. I. and **Garton, D. I.** (1967) 'Heat Exchanger Fouling with Dust Suspensions', *Chem. Process Engng Heat Transfer Survey*, 143–5, 150.

Heat Exchange Institute (1970). *Standard for Direct Contact Barometric and Low-level Condensers*. Heat Exchange Inst., New York, 5th edn.

Hetsroni, G. (ed.) (1982) *Handbook of Multiphase Systems*. Hemisphere Publishing Corp.

Hewitt, G. F. and **Roberts, D. N.** (1969) 'Studies of two-phase flow patterns by simultaneous X-ray and flash photography', Report AERE-M 2159, UKAEA, Harwell.

Hewitt, G. F., Delhaye, J. M. and **Zuber, N.** (1982) *Multiphase Science and Technology*. Hemisphere Publishing Corp.

Hinchley, P. (1977) 'Waste Heat Boilers: problems and solutions', *Chem. Engng. Prog.*, 90–6.

Hirschburg, R. I. and **Florschuetz, L. W.** (1981) 'Laminar wave-film flow: Part II – condensation and evaporation,' Pub. 81-HT-14, 20th ASME/AIChE Heat Transfer Conf. Milwaukee, Wisconsin, Aug. 1981.

Honda, H., Nozu, S. and **Mitsumori, K.** (1983) 'Augmentation of condensation on horizontal finned tubes by attaching a porous drainage plate', presented to ASME–JSME Thermal Engineering Joint Conference, Honolulu, Hawaii, March 1983.

Hopwood, P. F. (1972) 'Pressure drop, heat transfer and flow phenomena for forced convection boiling in helical coils, a literature review', Report AEEW–R757, UKAEA, Winfrith.

Hottel, A. C. and **Sarofim, A. F.** (1967) *Radiative Transfer*. McGraw-Hill.

How, H. (1956) 'How to design barometric condensers', *Chem. Engng.* **63**(2), 174–84.

Idel'chik, I. E. (1966) *Handbook of Hydraulic Resistance: coefficients of local resistance and friction*. AEC tr. 6630.

Idel'chik, I. E. and **Shteinberg, M. E.** (1974) 'Formulae, tabular data and recommendations for the choice of gas, air and water distributing headers', *Teploenergetika*, **21**(3), 118–23.

Jensen, A. and **Mannov, G.** (1974) 'Measurement of burnout, film flow, film thickness and pressure drop in a concentric annulus, 3500 × 26 × 17 mm with heated rod and tube', European Two-Phase Flow Group Meeting, Harwell, paper A5.

Jeschke, D. (1925) *Z. Ver. deut. Ing.* **69**, 1526; *Ergänzungsheft*, **24**, 1

Kays, W. M. and **London, A. L.** (1984) *Compact Heat Exchangers*, 3rd Edn. McGraw-Hill.

Kern, D. Q. (1950). *Process Heat Transfer*. McGraw-Hill.

Kern, D. Q. and **Kraus, A. P.** (1972) *Extended Surface Heat Transfer*. McGraw-Hill.

Kissell, J. H. (1972) 'Flow induced vibrations in heat exchangers – a practical look', presented at 13th AIChE/ASME Nat. Heat Transfer Conf., Denver, Colorado, 6–9 Aug. 1972.

Kutateladze, S. S. (1961) 'Boiling heat transfer', *Int. J. Heat Mass Transfer*, **4**, 31–45.

Kutateladze, S. S. (1963) *Fundamentals of Heat Transfer*. Edward Arnold Press Inc.

Labuntzov, D. A. (1957) 'Heat transfer on film condensation of pure vapours on vertical surfaces and horizontal pipes', *Teploenergetika* **4**(7), pp. 72–9, English trans. No. CTS 454.

Larkin, B. S. (1981) 'An experimental study of the temperature profiles and heat transfer coefficients in a heat pipe for a heat exchanger'. *Advances in Heat Pipe Technology: IV Int. Heat Pipe Conf. London*, Pergamon Press, Oxford, 1982.

Ledinegg, M. (1938) 'Instability of flow during natural and forced circulation', *Die Warm* **61**, 891–8 (in German) English trans., USAEC Transl. No. AEC-tr-1861.

Lee, D. H. (1966) 'Burnout in a channel with non-uniform circumferential heat flux', Report AEEW–R477, UKAEA, Winfrith.

Leong, L. S. and **Cornwell, K.** (1979) 'Heat transfer coefficients in a reboiler tube bundle', *Chem. Eng.* **343**, 219–21.

Lester, G. D. and **Hooper, D. G.** (1982) 'The application of boiler control parameters'. Paper 3 in Conf. held by the Industrial Water Society: 'Corrosion control in steam-raising plant through the '80s', 20 Apr. 1982.

Lester, G. D. and **Walton, R. W.** (1982) 'Cleaning heat exchangers'. Paper C55/82 in Conference held at the IMechE: 'Practical applications of heat transfer', 31 Mar. 1982.

Lienhard, J. H. and **Dhir, V. K.** (1973a) 'Extended hydrodynamic theory of the peak and minimum pool boiling heat fluxes', NASA Report Cr-2270.

Lienhard, J. H. and **Dhir, V. K.** (1973b) 'Peak boiling heat fluxes from finite bodies', *J. Heat Transfer, Trans. ASME*, **95**, 152–8.

Lis, J. and **Strickland, J. A.** (1970) 'Local variations of heat transfer in a horizontal steam evaporator tube', Paper 34.6, 4th Int. Heat Transfer Conf. Paris, Sept. 1970.

Lockhart, R. W. and **Martinelli, R. C.** (1949) 'Proposed correlation of data for isothermal, two-phase two-component flow in pipes', *Chem. Eng. Prog.*, **45**, 39–48.

Love, T. J. (1968) *Radiative Heat Transfer*. Merrill Publishing Co., Ohio.

McAdams, W. H. (1954) *Heat Transmission*, 3rd edn. McGraw-Hill.

Macbeth, R. V. (1968) 'The burnout phenomena in forced-convection boiling', *Adv. Chem. Engng* **7**, 207–93.

MacDuff, J. N. and **Felgar, R. P.** (1957) 'Vibration design charts', *Trans. ASME*, **79** (Oct.), 1459–74.

McDuffie, N. G. (1977) 'Vortex free downflow in vertical drains', *AIChE J.* **23**(1), 37–40.

McNaught, J. M. (1983) 'An assessment of design methods for multi-component condensation against data from a horizontal tube bundle', paper presented to IChE Symp. Series No. 75: 'Condensers: theory and practice'. UMIST 22–3 March 1983, Pergamon Press.

McQuillan, K. W. and **Whalley, P. B.**
(1984) 'A comparison between flooding
correlations and experimental flooding
data', AERE–R11267.

Mann, G. M. W. (1978) 'History and
causes of on-load waterside corrosion
in power boilers', *Combustion*, Aug.
1978, 28.

Maron, S. H. and **Lando, J. B.** (1974)
Fundamentals of Physical Chemistry.
Macmillan.

**Martinelli, R. C., Boelter, L. M. K.,
Taylor, T. H. M., Thomson, E. C.** and
Moen, R. H. (1944) 'Isothermal pressure
drop for two-phase two-component
flow in a horizontal pipe', *Trans.
ASME*, **66**(2), pp. 139–51.

Martinelli, R. C. and **Nelson, D. B.**
(1948) 'Prediction of pressure drop
during forced circulation boiling of
water', *Trans. ASME* **70**, 695–702.

Masuda, H. and **Rose, J. W.** (1985) 'An
experimental study of condensation of
refrigerant 113 on low integral-fin
tubes', Proc. Int. Symp. on Heat
Transfer, Beijing, China, Oct. 1985, **2**,
paper 32, Tsinghua University Press.

Miller, D. S. (1978) *Internal Flow
Systems*. BHRA Fluid Engineering.

Moles, F. D. and **Shaw, J. R. C.** (1972)
'Boiling heat transfer to sub-cooled
liquids under conditions of forced
convection', *Trans. IChE* **50**, pp. 76–84.

Morris, A. W. L. (1976) 'The resolution
of a dryout-induced on-load corrosion
problem in the sloping furnace tubes of
an operational boiler', *Proc. IMechE*,
London, **190**, pp. 721–7.

Mostinski, I. L. (1963) 'Application of
the rule of corresponding states for the
calculation of heat transfer and critical
heat flux', *Teploenergetika*, 1963(4), 66–
71.

Mueller, A. C. (1977) 'An enquiry of
selected topics on heat exchanger
design', *Solar and Nuclear Heat
Transfer AIChE Symp.* Series **73** (164),
pp. 273–87.

Myers, J. E. and **Katz, D. L.** (1952)
'Boiling coefficients outside horizontal
tubes', *Heat Transfer Symposium* Series
No. 5, **49**, pp. 107–14.

Nukiyama, S. (1934) J. Soc. MechE
(Japan) **37**, 367–74, 553–4.

**Owen, R. G., Sardesai, R. G., Smith,
R. A.** and **Lee, W. C.** (1983) 'Gravity
controlled condensation on horizontal
low-fin tube', paper presented to IChE
Symposium Series No. 75: 'Condensers:
theory and practice', held at UMIST
22–23 March 1983, Pergamon Press.

Owhadi, A., Bell, K. J. and **Crain, B.**
(1968) 'Forced convection boiling inside
helically-coiled tubes', *Int. J. Heat
Mass Transfer*, **11**, 1779–93.

Palen, J. W. and **Taborek, J.** (1962)
'Proposed method for design and
optimization', *Chem. Engng Prog.*
58(7), 37–46.

Palen, J. W. and **Small, W. M.** (1964)
'A new way to design kettle and
internal reboilers', *Hydrocarbon
Processing*, **43**(11), 199–208.

Palen, J. W., Yarden, A. and **Taborek,
J.** (1972) 'Characteristics of boiling
outside large-scale horizontal multiple
bundles', *AIChE Symp.* Series **68**(118),
pp. 50–61.

Palen, J. W., Shih, C. C. and **Taborek,
J.** (1982) 'Mist flow in thermosyphon
reboilers', *Chem. Engng Prog.*, July
1982, 59–61.

Premoli, A., Francesco, D. and **Prima**
(1970) 'An empirical correlation for
evaluating two-phase mixture density
under adiabatic conditions', European
two-phase flow group meeting, Milan,
8–10 June 1970.

Reynolds, O. (1874) Proc. Manchester
Lit. Phil. Soc., **8** reprinted in *Scientific
Papers of Osborne Reynolds*. II
Cambridge, 1901.

Robertson, J. M. (1973) 'Dryout in
horizontal hairpin waste heat boiler
tubes', presented to ASME/AIChE 13th
Nat. Heat Transfer Conf., Denver,
Colorado, 6–9 Aug. 1972. *AIChE
Symp. Series* **69**, pp. 55–62.

Robertson, J. M. and **Clarke, R.** (1981)
'The onset of boiling of liquid nitrogen
in plate – fin heat exchangers',
ASME/AIChE Nat. Heat Transfer
Conf. Milwaukee, Wisconsin, USA,
Aug. 1981.

Rohsenow, W. M. (1952) 'A method of
correlating heat-transfer data for

surface boiling of liquids', *Trans. ASME* **74**, pp. 969–76.

Ross, G. (1967) 'Calculation of non-uniform radiant flux to tube banks', *The Chem. Engr.* Dec., CE272–4.

Rounthwaite, C. (1968) 'Two-phase heat transfer in horizontal tubes', *J. Inst. Fuel.* Feb. 1968, 66–76.

Rudy, T. M. and **Webb, R. L.** (1981) 'Condensate retention of horizontal integral fin tubing', presented at ASME/AIChE 20th Nat. Heat Transfer Conf. Milwaukee, Wisconsin, 2–5 Aug., 1981. *ASME, H.T.D.*, **18**, 35–41.

Ruffell, A. E. (1974) 'The application of heat transfer and pressure drop data to the design of helical coil once-through boilers', Symp. Multi-phase Flow Systems, Univ. Strathclyde, Paper I5, *IChE Symp.* Series No. 38.

Saunders, E. A. D. (1987) *Heat Exchangers – Construction & thermal design (single-phase)*. Longman.

Schoch, W. and **Spahn, H.** (1971) 'Corrosion fatigue: chemistry, mechanics and microstructure', *NACE Int. Conf.* Series, **2** University of Connecticut, 14–18 June 1971.

Schlünder, E. U. (1983) *Heat Exchanger Design Handbook*. Hemisphere Publishing Corp.

Schüller, R. B. and **Cornwell, K.** (1984) 'Dryout on the shell-side of tube bundles', 1st UK Nat. Conf. on Heat Transfer, Leeds, July 1984, **2**, pp. 795–804.

Shah, M. M. (1976) 'A new correlation for heat transfer during boiling flow through pipes', *ASHRAE Trans.*, **82**(2), pp. 66–86.

Silver, L. (1947) 'Gas cooling with aqueous condensation', *Trans. IChE*, **25**, pp. 30–42.

Simon, D. E. II and **Fijnsk, A. W.** (1980) 'Suggested control values for feed water and boiler water quality in industrial boilers', Industrial Power Conference, Houston, Texas.

Smith, E. M. (1964) 'Helical-tube heat exchangers', *Engng*, **197**, 232.

Smith, R. A. (1981) 'Economic velocity in heat exchangers', presented at ASME/AIChE 20th Nat. Heat Transfer Conf., Milwaukee, Wisconsin, 2–5 Aug. 1981.

Starczewski, J. (1965) 'Generalised design of evaporators – heat transfer to nucleate boiling liquids', *Brit. Chem. Engng*, **10**(8), 523–31.

Stearman, F. and **Williamson, G. J.** (1972) 'Spray elimination' Ch. 16 of *Gas Purification Processes for Air Pollution Control* 2nd edn, ed. G. Nonhebel, IChE, London, 564–77.

Steinmeyer, D. E. (1972) 'Fog formation in partial condensers', *Chem. Engng Prog.* **68** No. 7, 64–8.

Stephan, K. (1963) 'Influence of oil on heat transfer of boiling refrigerant–12 and refrigerant–22', Paper 11–6, Proc. 13th Int. Congress of Refrig.

Stephan, K. and **Mitrovic, J.** (1981) 'Heat transfer in natural convective boiling of refrigerants and refrigerant-oil mixtures in bundles of T-shaped finned tubes', *ASME H.T.D.* **18**, 131–146, 20th ASME/AIChE Heat Transfer Conf., Milwaukee, Wisconsin.

Sun, K. H. and **Lienhard, J. H.** (1970) 'The peak pool boiling heat flux on horizontal cylinders', *Int. J. Heat Mass Transfer*, **13**, 1425–39.

Taborek, J. (1980) 'Design of vapor generators with boiling on the shell side', Lecture CVG-9 in course on condensers and vapour generators, Houston, Texas, USA May 1980, Hemisphere Pub. Corp.

Taitel, Y. and **Dukler, A. E.** (1976). 'A model for predicting flow regime transitions in horizontal and near-horizontal gas–liquid flow', *AIChE J.* **22**(1), 47–55.

TEMA (1978) *Standards of Tubular Exchanger Manufacturers' Association.* 6th edn, TEMA, New York.

Thomas, D. G. (1968) 'Enhancement of film condensation rate on vertical tubes by longitudinal fins', *AIChE J.* **14**, No. 4, 644–9.

Thomson, A. S. T., Scott, A. W., Laird, A. McK. and **Holden, H. S.** (1951) 'Variation in heat transfer rates around tubes in cross flow', *Proc. General Discussion on Heat Transfer*, 11–13 Sept., IMechE, London, pp. 177–80.

Thorngren, J. T. (1970) 'Predict exchanger tube damage', *Hydrocarbon Processing*, April 1970, 129–31.

Tinker, T. (1951) 'Shellside characteristics of shell-and-tube heat exchangers', *Proc. Gen. discussion on Heat Transfer*, 11–13 Sept., IMechE, pp. 89–116.

VdTÜV (1972) *Richtlinien für die Speise-und Kesselwasser-beschaffenheit bei Dampferzeuger*, Ausgabe Apr. 1972.

VGB (1980) *Richtlinien für Kesselspeisewasser, Kesselwasser und Dampf von Wasserrohrkesseln der Druckstufen ab 64 bar*, Oct. 1980.

Wallis, G. B. (1969) *One Dimensional Two-phase Flow*. McGraw-Hill.

Watson, G. B., Lee, R. A. and **Weiner, M.** (1974) 'Critical heat flux in inclined and vertical smooth and ribbed tubes', *Heat Transfer*, **4**, 275–9.

Webb, R. L. (1983) 'Nucleate boiling on porous coated surfaces', *Heat Transfer Engng*, **4**, July–Dec. 1983, 71–81.

Wheatley, M. J. (1972) 'Calculations of radiative heat transfer from flame to walls of the combuster in a gas turbine combustion chamber', Univ. Sheffield Dept. Chem. Engng and Fuel Tech.

Whitt, F. R. (1966) 'Performance of falling film evaporators', *Brit. Chem. Engng*, **11**, 1523–5.

Whiteway, R. N. (1977) 'Heat transfer correlations for the nucleate, convective and dispersed regions of horizontal in-tube evaporation', PhD thesis, Dept. Mech. Engng., Duke University.

Yau, K. K., Cooper, J. R. and **Rose, J. W.** (1986) 'Horizontal low-fin condenser tubes – effect of fin spacing and drainage strips on heat transfer and condensate retention', *ASME J. Heat Transfer*, in press.

Yilmaz, S., Palen, J. W. and **Taborek, J.** (1981) 'Enhanced boiling surfaces as single tubes and tube bundles', *ASME H.T.D.* **18**, 123–9, 20th ASME/AIChE Heat Transfer Conf., Milwaukee, Wisconsin.

Yilmaz, S., Moliterno, A. and **Samuelson, B.** (1983) 'Vertical thermosyphon boiling in spiral plate heat exchangers', *AIChE Symp. Series*, **79** No. 225, pp. 47–53.

Zuber, N. (1958). 'On the stability of boiling heat transfer', *Trans. ASME*, **80**, pp. 711–20.

Glossary

Adiabatic: no heat transferred to or from a system.

Azeotropic mixture: a mixture with a boiling range of zero.

Black body: a body which at all temperatures absorbs all the radiant heat falling on it.

Blow-down: purge of boiling liquid from a vaporiser to prevent build-up of contaminants and dissolved solids to an excessive level.

Boiler: a vaporiser that generates steam (see also 'steam generator').

Boiling: the formation of bubbles of vapour in a liquid at a submerged heated surface, or at nuclei in a stream of saturated liquid.

Boiling range: the difference between the dew point and the bubble point.

Boiling temperature: the temperature at which the vapour pressure of a single-component liquid, or azeotropic mixture, equals the external pressure on the liquid.

Bubble point: the temperature at which bubbles first appear when a multi-component liquid is being heated.

Burnout: condition of boiling with a controlled heat flux when a small increase in heat flux produces a very great increase in the temperature of the heated surface.

Carry-over: removal of drops by the vapour as it rises from the liquid.

Carry-under: removal of bubbles by the recirculating liquid as it enters the downcomer.

Chelating agent: an organic compound with which metallic ions form a soluble complex.

Condensation: conversion of vapour to liquid.

Condenser: equipment for producing condensation by cooling a vapour.

Critical heat flux: maximum heat flux obtainable with nucleate boiling, as described in section 2.1.

Critical pressure: the vapour pressure at the critical temperature.

Critical temperature: the temperature above which it is not possible to condense a gas by isothermal compression.

Cyclone: a cylindrical vessel with a tangential inlet, into which a two-phase mixture is introduced, so that the denser phase is separated by centrifugation (see section 4.1.2).

Dew point: temperature at which condensation begins when a multi-component vapour or a gas containing vapour is being cooled.

Direct-contact heat exchanger: an exchanger in which heat is transferred by intimate mixing of hot and cold streams.

Dryness: see 'quality'.

Dryout: incomplete wetting of a heated surface by a boiling liquid.

Economiser: heat exchanger for preheating boiler feed water with heat from flue gas.

Elutriator: device for removing crystals from a magma (see section 4.6).

Emissivity: ratio of the amount of heat radiated by a surface to the maximum amount that could be radiated from a surface of the same area at the same temperature.

Evaporation: vaporisation at the surface of a liquid.

Evaporator: vaporiser for removing a vapour from a liquid.

Film boiling: condition of vaporisation when a stable, continuous film of vapour separates the liquid from the heated surface.

Flash evaporation: evaporation resulting from a reduction in the pressure of a previously heated liquid.

Fluid: a substance that flows whenever it is subjected to shear.

Fog: droplets of liquid formed thermally in a vapour or in a gas containing a vapour, either by reducing the pressure or by cooling a gas saturated with vapour at a high rate of sensible heat transfer relative to latent heat transfer. The droplets are mostly less than 40 μm in diameter.

Forced convective boiling: boiling at nuclei in a stream of saturated liquid.

Friction factor: term f in equation [9.1] for pressure gradient in a pipe due to fluid friction at the wall.

Gas: a fluid whose volume depends on the external pressure.

Heat flux: rate of heat transfer divided by heated surface area.

Heat transfer coefficient: heat flux divided by temperature difference.

Isothermal: occurring at a constant temperature.

Leidenfrost temperature: temperature above which a surface cannot be wetted by a liquid (see section 2.1).

Ligament: narrowest width of a tubesheet between adjacent holes.

Liquid: a fluid that has a definite volume and that forms its own level in a container. Unlike gases, liquids do not always mix with each other.

Magma: a suspension of crystals in the saturated solution from which they were formed by evaporation.

Mass flux: mass flow rate divided by cross-sectional area.

Mean bulk temperature: the temperature that would be obtained by a fluid flowing through a heat exchanger if it were mixed together adiabatically, i.e. the mean with respect to length.

Mist: a mixture of small drops of spray and agglomerates of droplets of fog, usually in the size range 40 to 100 μm.

Mixed bed polisher: an intimate mixture of strong base anion and strong acid cation resins which removes 'slip' from the upstream ion-exchange vessels.

Momentum flow rate: the mass flow rate multiplied by the mean velocity of a fluid in a stream.

Momentum flux: momentum flow rate divided by the cross-sectional area for flow.

Multiple-effect evaporation: the arrangement of several evaporators in series, as described in section 5.4.5.

Nucleate boiling: formation of bubbles at suitable cavities on a heated surface, as described in section 2.1.

Phase: a homogeneous, physically distinct, and mechanically separable portion of a system – gas, liquid or solid.

Pitch: centre-to-centre distance between adjacent tubes or baffles.

Plate heat exchanger: an exchanger in which heat is conducted through plates which separate hot from cold streams; the plates may be permanently joined together by welding or brazing, or they may be sealed by removable gaskets.

Pool boiling: boiling at a heated surface submerged in a pool of liquid, without any forced flow of liquid.

Pressure loss: irreversible pressure drop due to fluid friction at the wall of the duct and form drag in flow past obstructions.

Pressure loss coefficient: pressure loss divided by velocity pressure.

Quality: ratio of the mass flow rate of vapour (plus any gas) to the total mass flow rate.

Reboiler: vaporiser at the bottom of a distillation column, to revaporise some of the liquid collected, and to provide a source of vapour.

Saturated solution: a solution that contains as much solute as it can hold in the presence of the dissolving substance at a given temperature.

Saturation temperature: temperature at which the vapour pressure of a liquid equals the external pressure on the liquid.

Slip (water treatment): ions which have not been removed by demineralisation plants.

Slip (two-phase flow): ratio of mean gas velocity to mean liquid velocity.

Solid: a body possessing a definite volume and a definite shape.

Solute: a substance dissolved in a solvent to form a solution.

Spray: drops produced mechanically by sprays, by splash-type distributors, by vapour shear at a gas/liquid interface, or by rising bubbles bursting at a liquid surface; they are mostly in the size range 100 to 500 μm.

Spray eliminator: a separator for removing spray from a vapour.

Static instability: a condition of flow through identical parallel channels when the mass flux through some is considerably greater than through others (see section 9.2.4).

Steam generator: a vaporiser that generates steam, either in a fired boiler, or in a waste heat boiler.

Subcooled: below saturation temperature.

Superheated: above saturation temperature.

Tubesheet: perforated plate to which the ends of the tubes of a shell-and-tube heat exchanger are attached (sometimes called a 'tubeplate').

Tubular heat exchanger: an exchanger in which heat is conducted through the walls of tubes which separate hot from cold streams.

Vaporisation: conversion of liquid to vapour, either by boiling or by evaporation.

Vaporiser: any equipment for producing vapour by heating, for whatever purpose it is required.

Vapour: a gas below its critical temperature.

Vapour recompression: the use of a compressor to increase the saturation temperature of a vapour generated in an evaporator so that it can be used to contribute to the heating of the evaporator.

Velocity head: the height to which a liquid will rise if projected vertically upwards, in the absence of friction; it equals $u^2/2g$, where u is the velocity and g the acceleration due to gravity.

Velocity pressure: velocity head converted into units of pressure; it equals $\frac{1}{2}\rho u^2$ or $\dot{m}^2/\rho$, where $\dot{m}$ is the mass flux and ρ the density of the fluid.

Window: space for flow between a baffle and the shell of a shell-and-tube heat exchanger.

Conversion factors

1.	**Acceleration**	$1\,cm/s^2$	: $1.000 \times 10^{-2}\,m/s^2$
		$1\,m/h^2$	: $7.716 \times 10^{-8}\,m/s^2$
		$1\,ft/s^2$	: $0.3048\,m/s^2$
		$1\,ft/h^2$	: $2.352 \times 10^{-8}\,m/s^2$
2.	**Area**	$1\,cm^2$	: $1.000 \times 10^{-4}\,m^2$
		$1\,ft^2$	: $9.290 \times 10^{-2}\,m^2$
		$1\,in^2$	: $6.452 \times 10^{-4}\,m^2$
3.	**Density**	$1\,g/cm^3$	: $1000\,kg/m^3$
		$1\,lb/ft^3$	: $16.02\,kg/m^3$
		$1\,kg/ft^3$	: $35.31\,kg/m^3$
4.	**Energy**	$1\,cal$	: $4.187\,J$
		$1\,kcal$	: $4187\,J$
		$1\,Btu$	: $1055\,J$
		$1\,erg$	: $1.000 \times 10^{-7}\,J$
		$1\,kWh$	: $3.600 \times 10^6\,J$
		$1\,ft\,pdl$	: $4.214 \times 10^{-2}\,J$
		$1\,ft\,lbf$	: $1.356\,J$
		$1\,Chu$	: $1899\,J$
		$1\,therm$	: $1.055 \times 10^8\,J$
5.	**Force**	$1\,dyne$	: $1.000 \times 10^{-5}\,N$
		$1\,kgf$	: $9.807\,N$
		$1\,pdl$	: $0.1383\,N$
		$1\,lbf$	: $4.448\,N$
6.	**Heat flux**	$1\,cal/s\,cm^2$	: $4.187 \times 10^4\,W/m^2$
		$1\,kcal/h\,m^2$	: $1.163\,W/m^2$
		$1\,Btu/h\,ft^2$	: $3.155\,W/m^2$
		$1\,Chu/h\,ft^2$	: $5.678\,W/m^2$
		$1\,kcal/h\,ft^2$	: $12.52\,W/m^2$

7.	**Heat transfer coefficient**	1 cal/s cm^2 °C : 4.187 × 10^4 W/m^2 K
		1 kcal/h m^2 °C : 1.163 W/m^2 K
		1 Btu/h ft^2 °F : 5.678 W/m^2 K
		1 Chu/h ft^2 °C : 5.678 W/m^2 K
		1 kcal/h ft^2 °C : 12.52 W/m^2 K

8. **Latent heat** see Specific enthalpy

9. **Length**

1 cm	:	1.000 × 10^{-2} m
1 ft	:	0.3048 m
1 micron	:	1.000 × 10^{-6} m
1 in	:	2.540 × 10^{-2} m
1 yard	:	0.9144 m
1 mile	:	1609 m

10. **Mass**

1 g	:	1.000 × 10^{-3} kg
1 lb	:	0.4536 kg
1 tonne	:	1000 kg
1 grain	:	6.480 × 10^{-5} kg
1 oz	:	2.835 × 10^{-2} kg
1 ton	:	1016 kg

11. **Mass flow rate**

1 g/s	:	1.000 × 10^{-3} kg/s
1 kg/h	:	2.778 × 10^{-4} kg/s
1 lb/s	:	0.4536 kg/s
1 tonne/h	:	0.2778 kg/s
1 lb/h	:	1.260 × 10^{-4} kg/s
1 ton/h	:	0.2822 kg/s

12. **Mass flux**

1 g/s cm^2	:	10.00 kg/s m^2
1 kg/h m^2	:	2.778 × 10^{-4} kg/s m^2
1 lb/s ft^2	:	4.882 kg/s m^2
1 lb/h ft^2	:	1.356 × 10^{-3} kg/s m^2
1 kg/h ft^2	:	2.990 × 10^{-3} kg/s m^2

13. **Power**

1 cal/s	:	4.187 W
1 kcal/h	:	1.163 W
1 Btu/s	:	1055 W
1 erg/s	:	1.000 × 10^{-7} W
1 hp (metric)	:	735.5 W
1 hp (British)	:	745.7 W
1 ft pdl/s	:	4.214 × 10^{-2} W
1 ft lbf/s	:	1.356 W
1 Btu/h	:	0.2931 W
1 Chu/h	:	0.5275 W

14. **Pressure**

1 dyne/cm^2	:	0.100 Pa (N/m^2)
1 kgf/m^2	:	9.807 Pa
1 pdl/ft^2	:	1.488 Pa
1 standard atm.	:	1.0133 × 10^5 Pa
1 bar	:	1.000 × 10^5 Pa
1 kgf/cm^2 (1 at)	:	9.807 × 10^4 Pa
1 lbf/ft^2	:	47.88 Pa

14.	**Pressure** (*continued*)	$1\,\text{lbf/in}^2$	: $6895\,\text{Pa}$
		$1\,\text{mm water}$	: $9.807\,\text{Pa}$
		$1\,\text{in water}$	: $249.1\,\text{Pa}$
		$1\,\text{ft water}$	: $2989\,\text{Pa}$
		$1\,\text{mm Hg}$	: $133.3\,\text{Pa}$
		$1\,\text{in Hg}$	: $3387\,\text{Pa}$
15.	**Specific enthalpy**	$1\,\text{cal/g}$	: $4187\,\text{J/kg}$
		$1\,\text{Btu/lb}$	: $2326\,\text{J/kg}$
		$1\,\text{Chu/lb}$	: $4187\,\text{J/kg}$
16.	**Specific heat capacity**	$1\,\text{cal/g}\,°\text{C}$	: $4187\,\text{J/kg K}$
		$1\,\text{Btu/lb}\,°\text{F}$	: $4187\,\text{J/kg K}$
17.	**Specific volume**	$1\,\text{cm}^3/\text{g}$	: $1.000 \times 10^{-3}\,\text{m}^3/\text{kg}$
		$1\,\text{ft}^3/\text{lb}$	: $6.243 \times 10^{-2}\,\text{m}^3/\text{kg}$
		$1\,\text{ft}^3/\text{kg}$	: $2.832 \times 10^{-2}\,\text{m}^3/\text{kg}$
18.	**Surface tension**	$1\,\text{dyne/cm}$	: $1.000 \times 10^{-3}\,\text{N/m}$
		$1\,\text{pdl/ft}$	: $0.4536\,\text{N/m}$
		$1\,\text{lbf/ft}$	: $14.59\,\text{N/m}$
19.	**Temperature difference**	$1\,\text{deg F}$	: $5/9\,\text{K}$
20.	**Thermal conductivity**	$1\,\text{cal/s cm}\,°\text{C}$	: $4187\,\text{W/m K}$
		$1\,\text{kcal/h m}\,°\text{C}$	: $1.163\,\text{W/m K}$
		$1\,\text{Btu/h ft}\,°\text{F}$	: $1.731\,\text{W/m K}$
21.	**Time**	$1\,\text{h}$	: $3600\,\text{s}$
		$1\,\text{day}$	: $8.640 \times 10^4\,\text{s}$
		$1\,\text{year}$	: $3.156 \times 10^7\,\text{s}$
22.	**Velocity**	$1\,\text{cm/s}$	: $1.000 \times 10^{-2}\,\text{m/s}$
		$1\,\text{m/h}$	: $2.778 \times 10^{-4}\,\text{m/s}$
		$1\,\text{ft/s}$	: $0.3048\,\text{m/s}$
		$1\,\text{ft/h}$	: $8.467 \times 10^{-5}\,\text{m/s}$
		$1\,\text{mile/h}$	: $0.4470\,\text{m/s}$
23.	**Viscosity (dynamic)**	$1\,\text{g/cm s (poise)}$	: $0.1000\,\text{kg/m s (N s/m}^2)$
		$1\,\text{kg/m h}$	: $2.778 \times 10^{-4}\,\text{kg/m s}$
		$1\,\text{lb/ft s}$	: $1.488\,\text{kg/m s}$
		$1\,\text{lb/ft h}$	: $4.134 \times 10^{-4}\,\text{kg/m s}$
24.	**Volume**	$1\,\text{cm}^3$	: $1.000 \times 10^{-6}\,\text{m}^3$
		$1\,\text{ft}^3$	: $2.832 \times 10^{-2}\,\text{m}^3$
		$1\,\text{litre}$	: $1.000 \times 10^{-3}\,\text{m}^3$
		$1\,\text{in}^3$	: $1.639 \times 10^{-5}\,\text{m}^3$
		$1\,\text{UK gal}$	: $4.546 \times 10^{-3}\,\text{m}^3$
		$1\,\text{US gal}$	: $3.785 \times 10^{-3}\,\text{m}^3$

25.	**Volumetric flow**	$1\,\mathrm{cm^3/s}$	: $1.000 \times 10^{-6}\,\mathrm{m^3/s}$
		$1\,\mathrm{m^3/h}$	: $2.778 \times 10^{-4}\,\mathrm{m^3/s}$
		$1\,\mathrm{ft^3/s}$	: $2.832 \times 10^{-2}\,\mathrm{m^3/s}$
		$1\,\mathrm{ft^3/h}$	: $7.866 \times 10^{-6}\,\mathrm{m^3/s}$
		$1\,\mathrm{UK\ gal/min}$	: $7.577 \times 10^{-5}\,\mathrm{m^3/s}$
		$1\,\mathrm{US\ gal/min}$	: $6.309 \times 10^{-5}\,\mathrm{m^3/s}$

26. **Wetting rate (mass)**   see Viscosity (dynamic)

27. **Work**   see Energy

APPENDIX C

Notation

Symbol	Quantity	SI Units
a	Surface area per unit length	m
A	Heat transfer surface area	m^2
A_f	Surface area of fins	m^2
A_h	Surface area in contact with the heating fluid	m^2
A_o	External surface area	m^2
A_r	Surface area at the base of finned tubes	m^2
A_t	Tubesheet area per tube	m^2
A_v	Surface area in contact with vaporising fluid	m^2
Bo	Shah's boiling number	—
c	Velocity of propagation of a pressure disturbance (velocity of sound)	m/s
c_p	Specific heat capacity at constant pressure	J/kg K
Co	Convection number	—
Co_h	Convection number modified to allow for stratification in a horizontal pipe	—
$\left.\begin{array}{l}C_1\\C_2\\C_3\end{array}\right\}$	Parameters for estimating slip ratio (section 9.4.1).	—
C_{cr}	Constant in equation [8.1] for critical heat flux in pool boiling	—
d	Internal diameter of tube	m
d_e	Equivalent diameter of non-circular duct (see sections 6.2.3 and 9.1)	m
d_f	External diameter of circular fin	m
d_o	Outside diameter of tube	m
D	$\left\{\begin{array}{l}\text{Internal diameter of shell}\\\text{Mean diameter of a coil}\end{array}\right.$	m m
f	Friction factor, defined in section 9.1	—
f_n	Natural frequency of a tube	s^{-1}
f_v	Vortex frequency	s^{-1}
F	Enhancement factor for heat transfer coefficient in convective boiling	—
F_a	Arrangement factor for flow across tube banks	—

Symbol	Quantity	SI Units
F_c	Correction factor for boiling mixtures	—
F_p	Correction factor for radial variations in fluid properties due to temperature gradient	—
Fr	Froude number, defined in equations [7.16] and [9.18]	—
g	Acceleration due to gravity (standard value 9.81)	m/s^2
h	Specific enthalpy	J/kg
H	Height	m
	Static head	m
K	Pressure loss coefficient (section 9.1)	—
	Parameter defined by equation [6.73]	—
K_f	Pressure loss coefficient for a fitting in a pipeline	—
L	Length of tube	m
	Parameter defined by equation [6.74]	—
m	Parameter defined by equation [6.54]	m^{-1}
$\dot{m}$	Mass flux $(=\dot{M}/S)$	$kg/s\,m^2$
$\dot{M}$	Mass flow rate	kg/s
$\tilde{M}$	Relative molecular mass	kg/k mole
n_t	Total number of tubes in a bundle	—
n_v	Number of tubes vertically above each other	—
N	Molar fraction	K mole of solute per k mole of solution
Nu	Nusselt number, defined by equation [6.6]	—
p	Static pressure	Pa (i.e. N/m^2)
p_{cr}	Critical pressure	Pa
p_v	Vapour pressure	Pa
p_r	Pressure ratio $(=p/p_{cr})$	—
p_R	Ratio of saturation pressure to reference pressure of 6895×10^3 Pa (used in section 8.3.1)	—
p_s	Saturation pressure	Pa
P	Perimeter	m
	Pitch of tubes	m
Pr	Prandtl number, defined by equation [6.7]	—
$\dot{q}$	Heat flux $(=\dot{Q}/A)$	W/m^2
$\dot{q}_{cr}$	Critical heat flux	W/m^2
$\dot{Q}$	Heat transfer rate	W
r	Radius	m
R	Gas constant $(\tilde{R}/\tilde{M})$	J/kg K
Ra	Rayleigh number, defined by equation [7.4]	—
Re	Reynolds number, defined by equation [6.8]	—
S	Cross-sectional area for flow	m^2
	Chen's suppression factor	—
S_r	Area ratio, defined by equation [9.60]	—
Sr	Strouhal number, defined by equation [11.2]	—
t	Time	s
T	Temperature	K
u	Velocity	m/s
$\bar{u}$	Mean velocity	m/s

Symbol	Quantity	SI Units
U	Overall heat transfer coefficient defined by equation [6.2]	$\text{W/m}^2\,\text{K}$
v	Specific volume $(=1/\rho)$	m^3/kg
$\dot{v}$	Volume flux $(=\dot{V}/S)$	m/s
$\dot{v}^+$	Dimensionless volume flux defined by equations [9.12] and [9.68]	—
V	Volume of a space	m^3
$\dot{V}$	Volumetric flow rate $(=\dot{M}/\rho)$	m^3/s
$\dot{V}_\text{r}$	Volumetric flow ratio, defined by equation [9.25]	—
We	Weber number, defined by equation [9.19] or [9.28]	—
x_c	Fraction of circumference of finned horizontal tubes flooded by condensate	—
x_g	Quality (ratio of mass flow rate of vapour (plus any incondensable gas) to total mass flow rate)	—
x_h	The fraction of the heat transfer rate that takes place from inlet of the liquid to be vaporised to the specified location	—
X_tt	Lockhart–Martinelli parameter for turbulent flow	—
x, y, z	Coordinate lengths	m
α	Individual heat transfer coefficient	$\text{W/m}^2\,\text{K}$
α_b	Heat transfer coefficient in boiling	$\text{W/m}^2\,\text{K}$
α_bR	Black body radiation heat transfer coefficient	$\text{W/m}^2\,\text{K}$
$\bar{\alpha}_\text{c}$	Mean heat transfer coefficient in condensation	$\text{W/m}^2\,\text{K}$
α_cb	Heat transfer coefficient in convective boiling	$\text{W/m}^2\,\text{K}$
$\bar{\alpha}_\text{cl}$	Mean heat transfer coefficient in condensation with laminar flow	$\text{W/m}^2\,\text{K}$
α_dh	Heat transfer coefficient of dirt from heating fluid	$\text{W/m}^2\,\text{K}$
α_dv	Heat transfer coefficient of dirt from vaporising fluid	$\text{W/m}^2\,\text{K}$
α_g	Heat transfer coefficient for gas flowing alone, without any liquid	$\text{W/m}^2\,\text{K}$
$\bar{\alpha}_\text{gr}$	Mean heat transfer coefficient in condensation in gravity-controlled flow region	$\text{W/m}^2\,\text{K}$
α_h	Heat transfer coefficient for the heating fluid	$\text{W/m}^2\,\text{K}$
α_l	Heat transfer coefficient for liquid flowing alone, without any gas	$\text{W/m}^2\,\text{K}$
α_lo	Heat transfer coefficient for total flow as liquid	$\text{W/m}^2\,\text{K}$
α_nb	Heat transfer coefficient in nucleate boiling	$\text{W/m}^2\,\text{K}$
α_nbl	Heat transfer coefficient in nucleate pool boiling outside a single horizontal tube	$\text{W/m}^2\,\text{K}$
α_s	Heat transfer coefficient in condensation in shear-controlled flow region	$\text{W/m}^2\,\text{K}$
α_v	Heat transfer coefficient for vaporising fluid	$\text{W/m}^2\,\text{K}$
α_w	Heat transfer coefficient for wall	$\text{W/m}^2\,\text{K}$
Γ	Mass flow rate of condensate per unit length (section 6.3.1)	kg/s m
δ	Film thickness	m
Δh_v	Latent heat of vaporisation	J/kg K
$\Delta \dot{M}_\text{max}$	Maximum maldistribution, defined by equation [9.59]	—

Symbol	Quantity	SI Units
Δp_{acc}	Component of two-phase pressure drop due to acceleration	Pa
Δp_{fr}	Component of two-phase pressure drop due to friction	Pa
Δp_{gr}	Component of two-phase pressure drop due to gravity	Pa
Δp_{tp}	Total two-phase pressure drop	Pa
ΔT	Local temperature difference between fluids $(= T_h - T_v)$	K
ΔT_m	Mean temperature difference	K
ΔT_{lm}	Log mean temperature difference	K
ε	Roughness of surface	m
ε_g	$\begin{cases}\text{Void fraction (fraction of volume occupied by vapour)} \\ \text{Emissivity of gas}\end{cases}$	— —
ε_s	Emissivity of surface	—
ζ	Dimensionless group defined by equation [6.45]	—
η	Dynamic viscosity	N s/m² (kg/s m)
θ	$\begin{cases}\text{Temperature difference across a thermal resistance} \\ \text{Angle}\end{cases}$	K deg.
λ	Thermal conductivity	W/m K
ρ	Density	kg/m³
σ	$\begin{cases}\text{Surface tension} \\ \text{Ratio of smaller to larger cross-sectional area for flow at a change in section}\end{cases}$	N/m —
τ	Shear stress on a surface due to fluid flow	N/m²
ϕ_{lo}^2	Two-phase pressure drop enhancement factor	—
ϕ	General parameter for physical properties	—

Additional subscripts

a	Ambient
acc	Acceleration
b	$\begin{cases}\text{Bulk} \\ \text{Boiling}\end{cases}$
c	$\begin{cases}\text{Coil} \\ \text{Cold fluid} \\ \text{Condensate}\end{cases}$
cb	Convective boiling
cr	Critical
d	$\begin{cases}\text{Dirt} \\ \text{Downstream}\end{cases}$
dr	Driving force in a thermosyphon
f	$\begin{cases}\text{Film} \\ \text{Fin} \\ \text{Free convection}\end{cases}$
fc	Forced convection

Additional subscripts (*continued*)

fr	Friction
g	Gas or vapour
go	All flow as vapour
gr	Gravity
h	$\begin{cases} \text{Homogenous} \\ \text{Hot fluid} \end{cases}$
i	Inside of tubes
in	At inlet
ins	Insulating material
l	$\begin{cases} \text{Liquid} \\ \text{Longitudinal} \end{cases}$
lo	All flow as liquid
m	Mean
max	Maximum
min	Minimum
nb	Nucleate boiling
o	$\begin{cases} \text{Outside of tubes} \\ \text{Relating to conditions at zero quality} \end{cases}$
out	At outlet
ONB	Onset of nucleate boiling
p	Pool boiling
r	Reduced
R	Radiant
s	$\begin{cases} \text{Saturated} \\ \text{Shell} \end{cases}$
st	$\begin{cases} \text{Conditions at base of a distillation column} \\ \text{Straight pipe} \end{cases}$
t	Transverse
tp	Two-phase
T	Total
u	Upstream
v	Vaporising liquid
w	Wall
0	Where $x_h = 0$, i.e. where the fluid to be vaporised enters the heater
1	Where $x_h = 1$, i.e. where the fluid to be vaporised leaves the heater

Index